W0232297

Contributions to Statistics

Christos P. Kitsos
Werner G. Müller (Eds.)

MODA 4 –
Advances in
Model-Oriented
Data Analysis

Proceedings of the 4th International
Workshop in Spetses, Greece
June 5-9, 1995

With 37 Figures

Springer-Verlag Berlin Heidelberg GmbH

Series Editors
Werner A. Müller
Peter Schuster

Editors
Dr. Christos P. Kitsos
Department of Statistics
Athens University of Economics
and Business
76 Patission Street
GR-Athens 104-34, Greece

Dr. Werner G. Müller
Department of Statistics
University of Economics and
Business Administration
Augasse 2-6
A-1090 Vienna, Austria

ISBN 978-3-7908-0864-3

CIP-Titelaufnahme der Deutschen Bibliothek
Advances in model oriented data analysis: proceedings of the
4th international workshop in Spetses, Greece, June 5 - 9, 1995
/ MODA 4. Christos P. Kitsos; Werner G. Müller (ed.). –
(Contributions to statistics)
ISBN 978-3-7908-0864-3   ISBN 978-3-662-12516-8 (eBook)
DOI 10.1007/978-3-662-12516-8
NE: Kitsos, Christos P. [Hrsg.]; MODA <4, 1995, Spétsai>

PREFACE

This volume is the proceedings of the 4th International Workshop on Model-Oriented Data Analysis. This series of events originated in 1987 at a meeting in Eisenach, that successfully brought together scientists from numerous countries of the 'East' and 'West'. Now that this distinction is obsolete dialogue has been greatly facilitated, providing opportunities for this dialogue, however, is as vital as ever.

The present meeting at Spetses, Greece from 5th to 9th of June 1995 again assembles statisticians from all over the world as this book documents. The hospitality offered by the University of Economics of Athens and the Korgialenios School made it possible to organize this workshop. The editors are also grateful to Intracom (Greece), the Ionian Bank and the Procter & Gamble Company (USA) for their generous support. We would particularly like to mention Dr. Michael Meredith, who being our contact person at Procter & Gamble, enabled us to publish these proceedings. Further thanks go to Dr. Peter Schuster from Physica Verlag Heidelberg for his continuing support of the project.

The contributions to this volume were carefully selected from the submissions by the editors after a one stage refereeing process. We would like to thank the members of the MODA committee, A.C. Atkinson, R.D. Cook, V.V. Fedorov, P.Hackl, H. Läuter, B.Torsney, I.N. Vuchkov, H.P.Wynn,and A.A. Zhigljavsky, who not only defined the main topics of the workshop, but also served as the referees.

I. Optimal Design

The dominant topic in the submissions was the optimal design of statistical experiments a traditional 'hot' theme at the MODA meetings. The papers of this section deal with such various topics as constrained optimization, Bayesian viewpoint, algorithms, etc. and cover both branches response surface and categorical designs.

The introductory article by Fedorov and Nachtsheim presents methods for designing experiments when there is a dynamic element in the model. Among other results they succeed in illuminating the close relations to marginally constrained designs. Montepiedra also makes use of techniques originally developed for constrained designs by applying them for the purpose of model robustification. As Clyde shows in a Bayesian context constrained designs can also be applied to improve normal approximation in statistical models. The Bayesian perspective is maintained by P. Müller and Parmigiani when they provide illuminative 'real life' examples of one stage and sequential optimal designs. A different sequential design strategy builds the framework in which Galtchouk and Maljutov investigate the lower bound of its duration.

The next couple of papers cover optimal designs for various criteria and types of responses. For a generalized minimax criterion in the simple linear regression model Torsney and Fidalgo characterize optimal designs with not more than two points of support. The D-criterion is referred to by Kitsos who calculates support points of optimal designs for various nonlinear models used in chemical kinetics. That product designs are D-optimal in additive nonlinear models is demonstrated by Schwabe. For generalized linear models the properties of D-optimal designs are discussed along with a parameter dimensionality reduction technique by Sitter and Torsney.

Hilgers examines properties of generalized tic polynomials on the simplex that have applications in mixture experiments, e.g. in chemistry. Weighing designs which are treated by Ceranka and Katulska have applications in the same field. The question of whether the risk of loosing observations results in dramatic changes of the optimal design is investigated by Hackl. In contrast, the model robustness of designs is the aim of the contribution by Ibrahimy and Cook. They consider models where the form of the conditional distribution of the response on the linear predictor is unknown.

The next block of papers is related to categorical models, Koukouvinos presents a collection of new D-optimal first order saturated designs with $n \equiv 2\mod4$ observations. Block designs are considered in the three papers that follow by Bogacka, Ceranka and Kacmarek, and Mejza and Kageyama. The first gives a discussion of information matrices in generally balanced block designs, the second treats a problem from crossing experiments, the parameter estimation in factorial triallel analysis, and the third gives statistical properties of certain nested block designs. This latter provides an extension of the ideas of optimality to random effects models.

The concluding papers of the section by Donev and Jones, and W.G. Müller and Zimmerman devote themselves to algorithms for finding optimal designs. The former describes a flexible algorithm for the construction of optimum cross-over designs, whereas the latter attempts to overcome the problem of correlated observations by a suitable separation technique.

II. Estimation and Optimization

Atkinson provides an extensive discussion of diagnostic techniques for the effect of individual observations on multivariate transformations to normality and adds a collection of illuminating examples. Jureckova extends her results on the uniform asymptotic linearity of regression rank scores to a broader class of error distributions. Further results on the asymptotic distribution of regression parameters and a corresponding simulation experiment are offered by Parring. Droge also gives simulation results on the behaviour of model selection criteria under cross validation.

The following two papers are concerned with robust estimation. The first of these papers by Ch. Müller and Kitsos discusses the asymptotics of one-step M-estimators and gives applications from rhythmometry. The second by Shevlyakov and Vil'chevskiy provides asymptotic and finite sample properties of adaptive minimax M-estimators. A Bayesian perspective to heterogeneity and overdispersion via Gibbs sampling is offered by Dey, Peng and Larose. This perspective is illustrated by an example on toxoplasmosis. Gibbs sampling is also employed by Polasek and P. Müller for the estimation of volatility models in finance.

A recursive algorithm based on stochastic approximation and kernel estimation accompanied by an application from chemical engineering is introduced by G.Yin and K.Yin. A similar application field might apply to the work of Vuchkov and Boyadijeva, who use optimization techniques for quality improvement of processes which depend upon quantitative and qualitative factors and describe the effect of different target locations. Optimization is also central to the final paper by Pronzato, Wynn and Zhigljavsky who improve the Golden-section algorithm for locally symmetric functions by renormalization.

It is our hope that the present collection proves to be of interest not only for the participants of the meeting but also for a greater range of scientists from the particular subject fields.

C.P. Kitsos and W.G. Müller   Vienna, January 1995

Contents

Part I

Optimal Design

Optimal Designs for Time–Dependent Responses

VALERY FEDOROV and CHRIS NACHTSHEIM

Mathematical Sciences Section, Oak Ridge National Laboratories, PO Box 2008
TN 37831-6367, USA
and
Department of Operations and Management Science, Carlson School of Management,
University of Minnesota, Minneapolis, MN 55455, USA.

1 Introduction

The purpose of this paper is to investigate methods for the design of optimal dynamic experiments. In Section 2, we introduce notation and a determinant criterion for dynamic experiments. In Section 3, optimal dynamic designs are constructed analytically in a number of simple cases. As will be seen, the dimension of the design problem increases with the dimension of the control vectors (equivalently, the response vector). In Section 4, we discuss the use of suitable parameterizations of the control variable trajectories to reduce the dimensionality of the optimization problem. The relationship between the design of optimal dynamic experiments and the results developed for marginally restricted designs is considered in Section 5. Numerical methods are discussed in Section 6. We close, in Section 7, with an extension of the methodology to the case where one or more linear combinations of the response trajectory is observed for each control trajectory.

2 Model and Design Criterion

We assume that the process $\eta(x,t)$ can be observed at any time $t \epsilon T$ without additional expense, once the process is in operation. The control variables $x(t)$ are chosen from a set $X(t)$ which is assumed to be given at any $t \epsilon T$. Typically, $X(t)$ is a hyper-rectangle that does not depend on t. For the sake of simplicity, the set T is assumed to be discrete, i.e. $T = \{t_j\}_1^r$. Generalization to the continuous case is straightforward, although it requires more sophisticated mathematics, compare, for instance, with Spruil and Studden, 1979.

*Kitsos, C.P., and Müller, W.G., Eds., *Proceedings of MODA4*, Physica Verlag, Heidelberg, 1995

4

We shall consider models with additive error:

$$y_i(t) = \eta(x_i(t), t) + \varepsilon_i(x_i(t), t), \quad i = \overline{1, n} \tag{1}$$

where $x_i(t) \epsilon X(t)$, $t \epsilon T$ are considered in the paper. Usually $\eta(x, t)$ depends upon unknown parameters θ and this dependence is assumed to be linear:

$$\eta(x, t) = f^T(x, t)\theta, \quad \theta \epsilon R^p. \tag{2}$$

where the functions $(f(x, t), \ldots, f_p(x, t)) = f^T(x, t)$ are given a priori.

If

$$\begin{aligned}
\mathbf{y}_i^T &= (y_i(t_1), \ldots, y_i(t_r)), \\
\mathbf{x}_i^T &= (x_i(t_1), \ldots, x_i(t_r)), \\
\varepsilon_i^T &= (\varepsilon_i(x_i(t_1), t_1), \ldots, \varepsilon_i(x_i(t_r), t_r)), \\
F(\mathbf{x}_i) &= (f(x_i(t_1), t_1), \ldots, f(x_i(t_r), t_r)),
\end{aligned}$$

then (1) together with (2) can be represented in matrix form as:

$$\mathbf{y}_i = F^T(\mathbf{x}_i)\theta + \varepsilon_i. \tag{3}$$

It is generally assumed that

$$E(\varepsilon_i) = 0, \ E(\varepsilon_i \varepsilon_j^T) = \delta_{ij} \, C(\mathbf{x}_i), \tag{4}$$

where δ_{ij} is the Kronecker's symbol.

The last assumption implies that the components of vector ε_i and subsequently $\mathbf{y}_i$ can be correlated; the correlation may depend on the chosen control $x_i(t), t \epsilon T$, but the different series of trajectories y_i and y_j are not correlated.

It is well known (see for instance, Atkinson and Fedorov, 1990) that for model (3) the information matrix is given by:

$$M = \sum_{i=1}^{n} w_i \, m(\mathbf{x}_i), \tag{5}$$

where $w_i = \frac{k_i}{N}$, $N = \sum_{i=1}^{n} k_i$, k_i is the number of observations at x_i, and

$$r \cdot m(\mathbf{x}_i) = F(\mathbf{x}_i) C^{-1}(\mathbf{x}_i) F(\mathbf{x}_i). \tag{6}$$

In the context of the continuous (or approximate) design theory

$$M(\xi) = \int_X m(\mathbf{x})\xi(dx) \tag{7}$$

where $\xi(dx)$ is a probability (or design) measure with the supporting points $\mathbf{x} = (x(t_1), \ldots, x(t_r))$, $x(t) \epsilon X(t)$.

The design ξ^* is optimal if

$$\xi^* = \arg \min_{\xi} \ \Psi[M(\xi)], \tag{8}$$

where Ψ is an objective function. In this paper, we shall focus on the determinant criterion:

$$\Psi[M] = -ln|M|.$$

Hence, the D–optimal design ξ^* is defined by:

$$\xi^* = \max_{\xi} \ ln|M(\xi)|. \tag{9}$$

3 Analytic Approaches

From the mathematical point of view there is nothing new in (9) and it can be considered as a particular case of the experimental design when the observed response is a vector (see for instance, Fedorov, 1972). It is known that a necessary and sufficient condition for ξ^* to be D–optimal is the fulfillment of the following inequality:

$$tr \ m(\mathbf{x})M^{-1}(\xi^*) \le p \tag{10}$$

for all $x(t)\epsilon X(t)$, $t\epsilon T$ and equality holds almost everywhere on the supporting set of ξ^* (i.e. at points where observations have to be made). Notice that the problem under consideration differs from that of the standard situation in one important respect: the dimension of $\mathbf{x}$ can be great even for moderate r : $dim \ \mathbf{x} = r \ dim \ x(t)$. This makes (9) difficult to handle, especially numerically.

Example 1. Let

$$f^T(x,t) = (1, x, x\psi_1(t), \psi_2(t)),$$

where ψ_1 and ψ_2 are given, $X(t) \equiv [-1,1]$, $T = \{t\}_1^r$ and $C(\mathbf{x}) \equiv C$. Usually, in practice $t_l = l\tau$.

For any $\mathbf{x}$ matrix $m(\mathbf{x})$ is non–negative definite. Therefore for any design ξ and any $\mathbf{x}$

$$0 \le \ tr \ m(\mathbf{x})M^{-1}(\xi) = r^{-1} \cdot \ tr \ F(\mathbf{x})C^{-1}F^T(\mathbf{x})M^{-1}(\xi) = q(x_1,\ldots,x_r),$$

where $x_l = x(t_l)$, $l = 1,\ldots,r$. From this expression we see that $q(x_1,\ldots,x_r)$ is a non-negative, second order, polynomial with respect to $x_1,\ldots,x_r$. For an optimal design we have, from (10):

$$\max_{x_1,\ldots,x_r} \ q(x_1,\ldots,x_r) = p = 4,$$

and this maximum is attained at the supporting points of an optimal design. Thus, only points with coordinates $x_l = \pm 1, l = \overline{1,r}$ may be considered as candidates for the corresponding components of any vector $\mathbf{x}$ from the supporting set of D–optimal design.

For instance, the following two designs are D–optimal when $C = I$:

$$\xi_1^* = \left\{ \begin{array}{cc} x_1(t_l) = (-1)^{l-1} \ , & x_2(t_l) = (-1)^l \\ w_1 = \frac{1}{2}, & w_2 = \frac{1}{2} \end{array} \right\} \ \text{and}$$

$$\xi_2^* = \left\{ \begin{array}{cc} x_1(t_l) = 1 & x_2(t_l) = -1 \\ w_1 = \frac{1}{2} & w_2 = \frac{1}{2} \end{array} \right\} ,$$

Any convex mixture of them is also D–optimal.

Example 2. Condition (10) allows us to analyze some simple regression models involving correlated observations. For instance, suppose that errors are described by an autoregression model of the first order:

$$\mathbf{y}_i = \theta^T f(\mathbf{x}_i) + \varepsilon_i,$$

6

where $|x_{il}| \leq 1$, $\varepsilon_{i1} = \nu_{i1}$, $\varepsilon_{il} = \rho\varepsilon_{i(l-1)} + \nu_{il}$, $r \geq l > 1$, and $E(\nu_i) = 0$, $E(\nu_i\nu_i^T) = I$. Since:

$$L\varepsilon_i = \nu_i$$

where

$$L = \begin{pmatrix} 1 & 0 & 0 & \ldots & 0 & 0 \\ -\rho & 1 & 0 & \ldots & 0 & 0 \\ \ldots & \ldots & \ldots & \ldots & \ldots & \ldots \\ 0 & 0 & 0 & \ldots & -\rho & 1 \end{pmatrix},$$

one can easily find that

$$E(\varepsilon_i\varepsilon_i^T) = C = (L^TL)^{-1}$$

and subsequently

$$C^{-1} = L^TL .$$

Therefore:

$$r \cdot tr \; m(x)M^{-1}(\xi) = tr \; F(x)L^TLF^T(x)M^{-1}(\xi)$$

$$= tr \; LF^T(x)M^{-1}(\xi)F(x)L^T$$

$$= \sum_{i=1}^{r} d(x_i,\xi) + \rho^2 \sum_{i=2}^{r-1} d(x_i,\xi) - 2\rho \sum_{i=1}^{r-1} d(x_i, x_{i+1}, \xi)$$

$$= r \cdot q(x_1, \ldots, x_r, \xi),$$

where $d(x_l, x_k, \xi) = f^T(x(t_l), t_l)M^{-1}(\xi)f(x(t_l), t_k)$ and $d(x_l, \xi) = d(x_l, x_l, \xi)$. To apply (10), consider the behavior of the function $q(x_1, \ldots, x_r, \xi)$ for simple linear regression:

$$f^T(x,t) = (1, x).$$

In this case, $q(x_1, \ldots, x_r, \xi^*)$ is a second order polynomial with respect to $x_i, \ldots, x_r$. Therefore, for the optimal design ξ^*, all the components of any supporting vector must be ± 1. Since the design region is symmetric about zero, the information matrix of an optimal design must be diagonal. Hence:

$$d(x_{l-1}, x_l, \xi^*) = M_{11}^{-1}(\xi^*) + M_{22}^{-1}(\xi^*)x_{l-1}x_l.$$

If $\rho > 0$, then $d(x_{l-1}, x_l, \xi^*)$ makes positive contribution to $q(x_1, \ldots, x_r, \xi^*)$ only when $x_{l-1} = -x_l$. Therefore for negative ρ (to avoid any nonessential complication we assume that l is even) there exists an optimal design which consist of only the trajectory:

$$\xi^* = \{x_l^* = (-1)^l, w = 1\}_1^r.$$

Similarly, when $\rho < 0$ then one can show that there exists an optimal design with two supporting trajectories (in this case the solution is unique):

$$\xi^* = \left\{ \begin{array}{ll} x_{1l} \equiv 1, & x_{2l} \equiv -1 \\ w_1 = \frac{1}{2}, & w_2 = \frac{1}{2} \end{array} \right\}.$$

4   Trajectory Parameterization

When r is large one may try to reduce the dimensionality of optimization problem (9) by introducing some parametric approximations of the trajectories $x(t)$. Consider, for example, any reasonable approximation of $x(t)$:

$$x(t) = \varphi(u, t), \tag{11}$$

where $u \epsilon R^q$, $k = dim\ x(t)$. Clearly, (11) has to be sufficiently flexible to approximate $X(t)$. To avoid technicalities $x(t)$ is assumed to be a scalar. Equation (7) is now:

$$M(\xi) = \int_U F(u)C^{-1}(u)F(u)\xi(du), \tag{12}$$

$$U = \{u : \varphi(u, t) \in X(t),\ t \epsilon T\}.$$

Optimality condition (10) becomes:

$$tr\ m(u)M^{-1}(\xi^*) \leq p, \tag{13}$$

Note that the dimension of the resulting optimization problem is given by q, the number of parameters in $\varphi(u, t)$. The ideas developed in studies related to the frequency domain approach in optimal control theory are closest to (12) - (14), see Zarrop, 1979.

Example 3.   Let us continue the concluding part of Example 1 with use of the latter approach. Assume that

$$x(t) = u_1 \cos 2\pi u_2 t,\ \ 0 \leq t \leq 1.$$

From the definition of $X(t)$ it follows that $-1 \leq u_1 \leq 1$ and $-\infty \leq u_2 \leq \infty$. The following solution to this two–dimensional design problem corresponds to ξ_1^* of Example 1:

$$\xi_r^* = \left\{ \begin{array}{cc} (u_{11}^* = 1, u_{21}^* = \frac{r-1}{2}) & (u_{12}^* = -1, u_{22}^* = \frac{r-1}{2}) \\ w_1^* = \frac{1}{2} & w_2^* = \frac{1}{2} \end{array} \right\}.$$

Note that for large r, both supporting trajectories oscillate between points in $X(t)$, $t \epsilon T$, which are supporting points for D–optimal designs in the standard layout. When $r \to \infty$, $x^*(t)$ converges to a particular case of what are known in the control theory (see Balakrishnan, 1976) as "generalized functions or curves": for any finite time interval $\Delta t \subset T$, half of the time is spent at $x = 1$ and another half at $x = -1$.

One can also notice that

$$\lim_{r \to \infty} \frac{|M(\bar{\xi}_r)|}{|M(\xi_r^*)|} = 1, \tag{14}$$

where the design $\bar{\xi}_r$ consists of only one (in U) supporting point: $u_1^* = 1$, $u_{12}^* = \frac{r-1}{2}$.

In addition to reducing the dimension of the optimization problem, parameterization (11) permits the incorporation of various constraints on $x(t)$. For instance, suppose that the cost of changing the level of a control variable is proportional to the distance between

8

$x(t_i)$ and $x(t_{i+1})$. Then one may add to (12) the following constraint ($r \to \infty$ for the sake of simplicity):

$$\int_X \xi(d\mathbf{x}) \int_0^T |\dot{x}(t)|\, dt \leq Q, \quad \dot{x} = \frac{dx}{dt}. \tag{15}$$

If one looks for "shorter" trajectories , then the constraint

$$\int_X \xi(d\mathbf{x}) \int_0^T \sqrt{1 + x^2(t)}\, dt \leq L \tag{16}$$

looks very natural.

For either of the constraints (15) or (16), the optimal design reported in Example 2 is inadmissible. Obviously, the standard D–optimal design

$$\xi^* = \left\{ \begin{array}{cc} (u_{11}^* = 1,\ u_{21}^* = 0) & (u_{12}^* = -1,\ u_{22}^* = 0) \\ w_1 = \frac{1}{2} & w_2 = \frac{1}{2} \end{array} \right\}$$

is convenient in the sense of either (15) or (16).

Linear constraints of the form:

$$\int_X \phi(\mathbf{x})\xi(d\mathbf{x}) \leq Q \tag{17}$$

were considered by Fedorov and Gaivoronsky ,1984, see also Cook and Fedorov,1994. In neither of these papers is anything special about the nature of x assumed. Only the compactness of X is required for results reported. For this reason, constraints imposed on control $x(t)$ or on the parametric version $\varphi(u, t)$ do not lead to theoretical difficulties. Unfortunately it is not usually true for the computational methods.

5 Generalized Trajectories as Conditional Distributions

First, assume that in every series of observations the errors are independent and have equal variances i.e. $C(\mathbf{x}_i) = I_r$. Then, instead of (5) and (6), one has

$$M = \sum_{i=1}^n w_i\, m(x_i) = \sum_{i=1}^n \sum_{j=1}^r w_i\, w_{ij}\, f(x_i(t_j), t_j) f^T(x_i(t_j), t_j), \tag{18}$$

where $w_{ij} \equiv r^{-1}$ in this particular setting, but generally $w_{ij} \geq 0$ and $\sum_{j=1}^r w_{ij} = 1$ may be considered. In the continuous design framework we have:

$$M(\zeta) = \int_T \int_{X(t)} m(x(t), t)\, \zeta(dx, dt), \tag{19}$$

where

$$m(x(t), t) = f(x(t), t) f^T(x(t), t).$$

Introducing the marginal distribution:

$$\zeta_T(dt) = \int_{X(t)} \zeta(dx, dt). \tag{20}$$

and the conditional distribution (or measure)

$$\zeta(dx|t) = \frac{\zeta(dx, dt)}{\zeta_T(dt)}, \qquad (21)$$

one can change (8) for the following optimization problem

$$\zeta^* = \arg \min_\zeta \ \Psi[M(\zeta)], \qquad (22)$$

$$s.t. \quad \zeta_T(dt) = \zeta_T^0(dt),$$

where $\zeta_T^0(dt)$ is assumed to be given. For instance (compare with the comments to (17))

$$\zeta_T^0(dt) = \sum_{j=1}^r r^{-1}\delta(t - t_j)dt,$$

where $\delta(t)$ is the standardly defined δ–function.

A time–dependent measure

$$\zeta^*(dx|t) = \frac{\zeta^*(dx, dt)}{\zeta_T^0(dt)} \qquad (23)$$

can be considered as the optimal generalized trajectory and is very closely related to the generalized optimal controls considered in the optimal control theory (see, for instance, Balakrishnan, 1971).

It is evident that $\zeta^*(dx|t)$ does the same job as optimal designs defined by (8), but for the narrower class of regression models (all observations are independent, see comments to (17)). The complete additiveness of type (18) is essential for the whole approach considered in this section. Optimization problem (18) was already studied in the optimal experimental design theory by various authors (see Cook and Thibodeau, 1980, Nachtscheim, 1989, Huang and Hsu, 1993). Noting that (see (10) and (18))

$$m(\mathbf{x}) = \int_T m(x(t), t)\, \zeta_T(dt) \qquad (24)$$

one can replace (10) with the following, a necessary and sufficient condition for $\zeta^*(dx|t)$ to be optimal: almost everywhere with respect to $\zeta_T^0(dt)$ and $\zeta^*(dx|t)$

$$d(x, t, \zeta^*) \le d(x^*, t, \zeta^*), \quad \text{for any } x^* \epsilon \ supp \ \zeta^*(dx|t), \qquad (25)$$

where

$$tr \ m(x, t)M^{-1}(\zeta) = f^T(x, t)M^{-1}(\zeta)f(x, t) = d(x, t, \zeta)$$

and $supp \ \zeta(dx|t)$ is a supporting set of a design $\zeta(dx|t)$.

While condition (25) is useful (see Cook & Thibodeau, 1980, Huang & Hsu, 1993) for delineating various properties of $\zeta^*(dx|t)$, it facilitates the decomposition of the $k \times r$ dimensional optimization problem (22) into a sequence of r k–dimensional optimization problems (see section on numerical procedures).

Example 4. Let the conditions formulated in the beginning of this section hold, $X = [-1,1]$, $T = [-1,1]$, and

$$\eta(x,t) = \theta_1 f_1(t) + \theta_2 x f_2(t) + \theta_3 x^2 f_3(t) + \theta_4 f_4(t) + \ldots + \theta_p f_p(t).$$

It is evident that for every $t \epsilon T$ and $M(\zeta)$ function $d(x,t,\zeta)$ is a polynomial of the fourth order. From the symmetry of X, convexity of the objective function, and condition (25) one easily finds that $\zeta^*(dx|t)$ may oscillate between points $x_1 = -1$, $x_2 = 0$, $x_3 = 1$ with atomwise measure $\zeta^*(-1|t) = \zeta^*(1|t) = w^*(t)$, $\zeta^*(0|t) = 1 - 2w^*(t)$.

6   Algorithms

We consider only the modifications of the first order algorithms. These procedures are relatively simple to implement and yet are efficient in many applications. For the D–criterion the first order algorithms can be described by the following scheme:

i.   There is a design ξ_s. Find

$$\mathbf{x}_s = \arg \max_{\mathbf{x}} tr\ m(\mathbf{x})M^{-1}(\xi_s), \tag{26}$$

where $\mathbf{x}$ may be chosen out of all possible trajectories defined on $X(t)$ and T, see comments to (10).

ii.   Choose $0 \leq \alpha_s \leq 1$ and construct

$$\xi_{s+1} = (1 - \alpha_s)\xi_s + \alpha_s \xi(x_s). \tag{27}$$

The choice of a sequence $\{\alpha_s\}$ defines a particular version of the algorithm. The following sequences are used frequently in the standard software for the D– criterion:

1.

$$\lim_{\alpha \to \infty} \alpha_s = 0, \ \sum_{s=0}^{\infty} \alpha_s = \infty;$$

2.

$$\alpha_s = \frac{\nu_s}{(\nu_s + m - 1)m}, \ \nu_s = tr\ m(\mathbf{x}_s)M^{-1}(\xi_s).$$

All that we need for the convergence of the sequences defined by (27) for these two cases is the compactness of the set of all possible information matrices (Ermakov, Fedorov, et. al 1983, Pukelsheim, 1993). The validity of this fact is evident for any continuous $f(x,t)$ and compact $X(t)$, $t \epsilon T$.

Thus there are no difficulties in the theory, but to solve (26) we need to maximize either in the functional space $X(t), t \in T$ in the time–continuous case or in the space of $r \cdot dim\ x(t)$ dimension (see comment to (10)).

The following two approaches look promising.

<u>Trajectory Parameterization</u> Applying (11) one can replace (26) by the significantly simpler optimization problem:

$$u_s = \arg \max_u \ tr \ m(u)M^{-1}(\xi_s) \tag{28}$$

where $u_s \epsilon U$, see (12). Of course,

$$\max_{\xi(du)} |M(\xi(du))| < \max_{\xi(dx)} |M(\xi(dx))|,$$

but the repeated calculations with the different expansions $u^T\psi(t)$ give a hope for the proximity of $|M(\xi^*(du))|$ and $|M(\xi^*(dx))|$.

<u>Tours Along Trajectories</u> Another efficient and intuitively attractive simplification of (26) becomes possible when the necessary and sufficient condition (10) of D–optimality may be replaced by the necessary and sufficient condition (24), i.e. when observations along every trajectory are independent, Fedorov and Uspensky, 1975.

Let again set T be discrete and contain r elements. Then, accordingly to definition (6) for situations described by (17) optimization problem (28) can be split in r separate simple optimization problems:

$$\max_x \ tr \ m(x)M^{-1}(\xi_s) = \max_{x(t_1),...,x(t_r)} \ \sum_{j=1}^{r} f^T(x(t_j),t_j)M^{-1}(\xi_s)f(x(t_j),t_j) \tag{29}$$

$$= \sum_{j=1}^{r} \max_x f^T(x,t_j)M^{-1}(\xi_s)f(x,t_j) = \sum_{j=1}^{r} \max_x d(x,t_j,\xi_s).$$

Strictly speaking (29) may have a non unique solution (if one would forget about the real computing peculiarities). It occurs usually in problems with the basis functions $f(x,t)$ with some symmetricity properties in $X(t)$.

This non uniqueness allows to take into account various needs which a practitioner has in mind. For instance, when one is looking for the solution with smooth and continuous trajectories, then one has merely to crawl along the ridge $x_s^*(t)$ of function $d(x,t,\xi_s)$:

$$x_s^*(t_j) = \arg \max_{x \epsilon X(t_j)} \ d(x,t_j,\xi_s) \tag{30}$$

and $x_s^*(t_2)$ has to be the nearest to $x_s^*(t_1)$ local maximum of $d(x,t_2,\xi_s)$, then $x_s^*(t_s)$ has to be the nearest to $x_s^*(t_2)$ local maximum of $d(x,t_3,\xi_s)$ and so on.

Unlikely to Cook and Thibodeau ,1980, and Huang and Hsu, 1993, we do not recommend to recalculate ξ_s and subsequently $M(\xi_s), d(x,t,\xi_s)$ after finding a solution of (30) for each t_j. First, there are no theoretical results in favor of that, and second, it makes calculations much more complicated comparatively to the proposed procedure where ξ_s, $M(\xi_s)$ and $d(x,t,\xi_s)$ must be calculated only at the end of the tour along ridge $x_s^*(t_1),...,x_s^*(t_r)$

7 Extension: Linear Functions of the Response

Assume that model (1), (2) properly describes the observed system, but instead of the detailed information (3) an observer has less, namely:

$$\mathbf{z}_i = L\mathbf{y}_i, \tag{31}$$

where L is matrix $q \times r$, $q \leq r$ in the discrete case, and a linear operator mapping $y_i(t)$ to q–dimension space in the continuous case. To avoid any gridlock in inventing further notations, only the discrete case is considered in this section.

The expediency of (31) becomes evident when one can observe only some aggregated characteristics such as an average or total outcome of process $y(t)$ ($L_{1i} = r^{-1}$ or 1), increments ($L_{i,i-1} = -1$, $L_{i,i} = 1$, all other elements equal 0), and so on. Some particular version of (31) were intensively studied (see for instance, Ermakov, 1983 Ch. 7) and we have no intention to discuss the problem in detail trying only to emphasize its connection with the problem of our prime interest.

Combining (1) and (31) one can get similarly to (6) that

$$m(\mathbf{x}_i) = \phi(\mathbf{x}_i)\,\phi^T(\mathbf{x}_i) \tag{32}$$

where

$$\phi(\mathbf{x}) = (LC(\mathbf{x})L^T)^{-1/2}LF(\mathbf{x}).$$

It is obvious that all the results of sections 2 and 3 can be applied in the considered case after replacement of (6) by (32). Unfortunately, there is no similar easy transition for the results considered in Section 5. The reason is that the information matrix (32) for one series of observations (or in this case for one realization of the process) is not generally any more additive with respect to time.

References

Balakrishnan, A. (1971). "Introduction to Optimization Theory in Hilbert Space," *Springer–Verlag*, New York, pp. 152.

Cook, R. D. and Thibodeau, L. A. (1980). "Marginally Restricted D–Optimal Designs," *JASA*, 75, pp. 366–371.

Cook, R. D. and Fedorov, V. V. (1994). "Constrained Optimization of Experimental Design," *Statistics*, (in press).

Ermakov, S. M. (ed) (1983). "Mathematical Theory of the Design of Experiments (in Russian)," Nauka, 386.

Fedorov, V. V. and Atkinson, A. C. (1988). "The Optimum Design of Experiments in the Presence of Uncontrolled Variability and Prior Information," *Optimal Design and Analysis of Experiments,* Y. Dodge, V.V. Fedorov and H.P. Wynn (eds), North–Holland, Amsterdam.

Fedorov, V. V. and Gaivoronsky, A. (1984). "Design of Experiments under Constraints," International Institute of Applied System Analysis, Vienna, WP–84–8, pp. 1–16.

Fedorov, V. V. and Uspensky, A. B. (1975). "Numerical Aspects of Design and Analysis of Experiments (in Russian)," Moscow State University, Moscow, 1968.

Huang, M. L. and Hsu, M. C. (1993). "Marginally Restricted Linear–Optimal Designs," *JASA*, 35, pp. 251–266.

Nachtsheim, C. J. (1989). "On the Design of Experiments in the Presence of Fixed Co-variates," *JSPI*, 4, pp. 339– 364.

Spruil, M. C. and Studden, W. J. (1979). "A Kiefer-Wolfovitz Theorem in a Stochastic Process Setting, " *AMS*, 7, pp. 1329–1332.

Zarrop, M. B. (1979). "Optimal Experimental Design for Dynamic System Identification," Springer-Verlag, New York.

Robust Optimal Designs with Constraints

GRACE MONTEPIEDRA

Department of Applied Statistics and Operations Research,
Bowling Green State University, OH 43403, USA.

1 Introduction

A researcher wants to design an experiment from which he can investigate the relationship between a specified response variable y and several controllable independent variables. The model, notations, and design problem are the same as in Fedorov et. al. (1994), and therefore our introductory part will be very sketchy. Suppose that the true response surface model can be represented as a linear model:

$$y_i = \theta^T f_1(\mathbf{x}_i) + \delta^T f_2(\mathbf{x}_i) + \epsilon_i \qquad (1)$$

where $\mathbf{x}_i = (x_{i1}, \ldots, x_{ik})$ is the vector of values of the k controllable variables for the ith observation, $\theta \in \mathbf{R}^m, \delta \in \mathbf{R}^p$, and the $\{\epsilon_i\}$ $(i = 1, \ldots, N)$ are uncorrelated random errors with zero means and variances σ^2. Here we let N be the number of observations, some of which can be obtained from the same support points. Note that this representation is reasonable in most cases since we can usually approximate even models which are nonlinear in the parameters by some linear function of the form decribed in (1).

In practice, since the true model is typically unknown, the investigator decides to fit a simpler (and, hence, more parsimonious) model, say $\hat{y}_i(\mathbf{x}_i) = \hat{\theta}^T f_1(\mathbf{x}_i)$, to describe the relationship. For instance, this simplified version might be more feasible if the number of support points is not sufficient to handle a more complicated, however more exact, model. The "ignored" part of the true model, $\delta^T f_2(\mathbf{x}_1)$, can be called the contamination function. Inasmuch as using the simpler model might result in a more inferior analysis, it is the investigator's hope that an appropriate optimal design can be used to make the discrepancy as minimal as possible.

An approach which tackles the problem in a very simple and intuitively appealing manner is that which was motivated by the pioneering paper of Box and Draper (1959).

*Kitsos, C.P., and Müller, W.G., Eds., *Proceedings of MODA4*, Physica Verlag, Heidelberg, 1995

16

They introduced a method of obtaining designs by using the mean squared error of the estimates, instead of the standard variance-based criteria which the other methods still retain. Since then, numerous studies on optimal design for this problem have been studied and an extensive reference of the literature is provided in Montepiedra (1994). However, the criteria used in most of these works which are based on the mean squared error are too stringent, making the search for globally optimal designs not feasible in general. Moreover, there has been a general lack of integration of these methods with the already developed body of equivalence theorems. These theorems can be potentially useful mathematical tools for understanding the effect of ignoring the contamination function in the analysis (but not in the design) as well as practical tools for actually finding the optimal designs themselves.

In this paper, we propose a new class of mean squared error-based optimality criteria which is a variation of the class of designs discussed in Fedorov, et. al. (1994). The new class of criteria, which adopts the use of constrained optimization, is introduced in section 2. Its application to the **D**- and **A**- optimal criteria is discussed in section 3, where the specific criteria are defined and equivalent criteria are obtained. Some examples are investigated in section 4. In section 5 we summarize our findings and explore other possible directions for research.

The designs presented here are of the form

$$\xi = \{\mathbf{x}_i, p_i\}_{i=1}^n, p_i = \frac{r_i}{N}, \sum_{i=1}^n r_i = N \tag{2}$$

but they are actually approximate designs in the sense that we are searching over the whole set of possible probability measures in obtaining the optimal design.

Under the assumed model, a reasonable choice of estimator for θ is the least squares estimator $\hat{\theta} = \text{Arg min}_\theta \sum_{i=1}^n (y_i - \theta^T f_1(\mathbf{x}_i))^2$. Standard experimental designs simply utilize the variance-covariance matrix of $\hat{\theta}$, given by $V = E\{[\hat{\theta} - E(\hat{\theta})][\hat{\theta} - E(\hat{\theta})]^T\}$, as the basis for defining optimality criteria. However, since we are using a model which only approximates the true relationship between y and $\mathbf{x}$, then a more appropriate "measure" of "goodness" of the estimate $\hat{\theta}$ would be the matrix of mean squared errors of $\hat{\theta}$: $R = E[(\hat{\theta} - \theta_t)(\hat{\theta} - \theta_t)]$ (where the subscript "t" stands for true value of the indicated parameter). This "measure" takes into account the possible presence of the contamination function which we are ignoring in the analysis. Extending the set of discrete designs denoted by (2) to the set Ξ of all probability measures ξ with support in design region χ, it is straightforward to show that

$$\sigma^{-2}R(\xi) = \mathbf{M}_{11}^{-1}(\xi) + \mathbf{M}_{11}^{-1}(\xi)\mathbf{M}_{12}(\xi)B_t\mathbf{M}_{21}(\xi)\mathbf{M}_{11}^{-1}(\xi)$$

where $B_t = \sigma^{-2}N\delta_t\delta_t^T$ and $\mathbf{M}_{\alpha\beta} = \int_\chi f_\alpha(\mathbf{x})f_\beta^T(\mathbf{x})\xi(d\mathbf{x})$, such that $\alpha, \beta = 1, 2$. Note that $\mathbf{M}_{\alpha\beta}$ can can be recognized as the information submatrices under the true model (1).

Note that the expression above can be expressed as $R(\xi) = V(\xi) + W(\xi)$, where

$$V(\xi) \;=\; E[(\hat{\theta} - E(\hat{\theta}))(\hat{\theta} - E(\hat{\theta}))^T]$$

is the variance term of the matrix, and:

$$W(\xi) = [E(\hat{\theta}) - \theta_t][E(\hat{\theta}) - \theta_t]^T$$

is the corresponding bias term.

2 Convex Criteria with Convex Constraints

Several design criteria have been developed in the literature which deal with $R(\xi)$, or some variation of it. In Fedorov et. al. (1994), a straightforward extension of the standard optimal design framework is done by defining a class of design criteria as any reasonable *convex* function of $R(\xi)$. The $\mathbf{D}_R$-optimal design is introduced in the said paper and defined as $\xi^* = \text{Arg}\min_{\xi \in \Xi} \ln|R(\xi)|$, which is simply the **D**-optimal counterpart of this problem. This is, however, not a convex function in general so that globally optimal designs can only be obtained in a limited number of cases.

The major setback in using the above class of design criteria is the difficulty in finding such meaningful *convex* objective functions, for example, those which are direct counterparts of the standard optimality criteria. The approach in this paper, hence, allows for the feasibility of finding globally optimal designs based on the mean squared error since the simultaneous minimization of the variance term and the bias term of $R(\xi)$ in the strictest sense is not required.

In most cases, whenever $V(\xi)$ is minimized, it is usually at the expense of $W(\xi)$. This is also true vice versa. Suppose there is a greater need to minimize bias in estimation than there is to maximize precision in estimation. Then a logical solution is to find designs which minimize some function of $W(\xi)$, or an expression which reflects the bias, but at the same time ensure that $M_{11}^{-1}(\xi)$ will not be too large. An alternative class of design criteria which makes use of the idea of constrained optimization will thus be proposed.

The inclusion of constraints in other experimental design problems is not a new one and was used, for instance, by Stigler (1971), Studden (1982) and Lee (1987, 1988). Fedorov and Gaivoronski (1984) provided equivalence theorems which deal with the linear constraints. Cook and Fedorov (1994) utilized these ideas and results and on the possibility of linearization of a specified functional $\Psi(\xi)$ near an optimal design (refer to Gaivoronski (1986) and Lee (1988)) to obtain a parallel result for nonlinear convex constraints. This is reproduced below:

Consider the general constrained design problem of finding

$$\xi^* = \text{Arg}\min_{\xi \in \Xi} \Psi(\xi) \tag{3}$$

such that

$$\Phi(\xi) \leq 0, \Phi \in \mathbf{R}^1. \tag{4}$$

Suppose one can make the following assumptions:

(a) χ is compact and $f(\mathbf{x})$ is continuous on χ;

(b) $\Psi(\xi)$ is a convex function;

(c) there exists q such that

$$\{\xi : \Psi(\xi) \leq q \leq \inf, \xi \in \Xi, \Phi(\xi) \leq 0\} = \Xi_q \neq \varnothing;$$

(d) for any $\xi_1 \in \Xi_q$ and $\xi_2 \in \Xi$:

$$\Psi[(1-\alpha)\xi_1 + \alpha\xi_2] = \Psi(\xi_1) + \alpha \int_\chi \psi(\mathbf{x}, \xi_1)\xi_2(d\mathbf{x}) + o(\alpha);$$

(e) $\Phi(\xi)$ is a convex function;

(f) for the same ξ_1 and ξ_2 defined in (d):

$$\Phi[(1-\alpha)\xi_1 + \alpha\xi_2] = \Phi(\xi_1) + \alpha \int_X \phi(\mathbf{x}, \xi_1)\xi_2(d\mathbf{x}) + o(\alpha)$$

where $o(\alpha)/\alpha \to 0$ when $\alpha \to 0$.

It is known that when (a)-(f) hold and constraints are active, then a necessary and sufficient condition for a design ξ^* to be optimal is the existence of a nonnegative real number u^* such that

$$\min_{\mathbf{x} \in X} q(\mathbf{x}, u^*, \xi^*) \geq 0, \tag{5}$$

where $q(\mathbf{x}, u, \xi) = \psi(\mathbf{x}, \xi) + u^T \phi(\mathbf{x}, \xi)$ and $\Phi(\xi^*) = 0$.

This result is the basis of the equivalence theorems given for the newly defined criteria discussed in the next section. In the meantime, we are now ready to introduce a new class of optimality criteria.

Consider the set $\mathbf{R}$ of all $m \times m$ matrices $R(\xi)$. The mean squared error-based optimal (*continuous*) design with variance constraints is defined as

$$\xi^*_{WV} = \text{Arg} \min_{\xi \in \Xi} \Psi[W(\xi)] \text{ such that } \Phi[V(\xi)] \leq C, \tag{6}$$

where C is a user-defined constant.

Ideally, $\Psi[B(\xi)]$ and $\Phi[V(\xi)]$ are chosen to be convex functions of the design measure. This ensures the existence of globally optimal designs and allows the formulation of equivalence theorems by using (5).

The "inverse" formulation of the problem can also be of interest. The mean squared error-based optimal (*continuous*) designs with bias constraints is defined as

$$\xi^*_{VW} = \text{Arg} \min_{\xi \in \Xi} \Phi[V(\xi)] \text{ such that } \Psi[W(\xi)] \leq D, \tag{7}$$

where D is a user-defined constant.

The practitioner can choose to use (6) or (7) as the basis for obtaining a design, depending on whether variance or bias is more important.

3 The D- and A-restricted Minimum Bias Designs

A modification of the $\mathbf{D}_R$-optimality criterion leads to another criterion which allows us to imbed the problem under consideration in the framework of convex design theory.

It can be shown (see Fedorov et. al. (1994)) that the function $\Psi[R(\xi)] = \ln|R(\xi)|$ can be written as $\ln|R(\xi)| = S_1(\xi) + S_2(\xi)$, where $S_1(\xi) = \ln|M_{11}^{-1}(\xi)|$ and $S_2(\xi) = \ln\{1 + \gamma_t^T M_{21}(\xi)M_{11}^{-1}(\xi)M_{12}(\xi)\gamma_t\}$. We have, hence, just been able to represent $\ln|R(\xi)|$ as a sum of two terms: a function of the variance component, $S_1(\xi)$, and a representation of the bias component, $S_2(\xi)$. Note that $S_1(\xi)$ is simply the D-optimal criterion under the assumed model (2).

*In compliance with (6) and (7), the **D**-restricted minimum bias design ξ^* is defined as*

$$\xi^* = \text{Arg}\min_{\xi\in\Xi} \gamma_t^T M_{21}(\xi)M_{11}^{-1}(\xi)M_{12}(\xi)\gamma_t \tag{8}$$

such that

$$\ln|M_{11}^{-1}(\xi)| \le C. \tag{9}$$

We now consider the optimization problem of finding a **D**-restricted minimum bias design. When $\Psi[\xi] = \gamma_t^T M_{21}(\xi)M_{11}^{-1}(\xi)M_{12}(\xi)\gamma_t$, then (a)-(c) are satisfied when χ is compact and $f(\mathbf{x})$ is continuous and linearly independent on this compact set. It is also a well-known fact that $\ln|M_{11}^{-1}(\xi)|$ (the standard **D**-optimal criterion) satisfies (e) and (f). We use these facts to obtain the following result:

Theorem 1 *A design ξ^* is **D**-restricted minimum bias if and only if there exists $u^* \ge 0$ such that*

$$u^* d_1(\mathbf{x},\xi^*) + d_2(\mathbf{x},\xi^*) \le u^* m - \Psi[\xi^*] \tag{10}$$

for all $\mathbf{x} \in \chi$, where $\ln|M_{11}^{-1}(\xi^*)| = C, d_1(\mathbf{x},\xi) = f_1^T(\mathbf{x})M_{11}^{-1}(\xi)f_1(\mathbf{x})$, and $d_2(\mathbf{x},\xi) = f_1^T(\mathbf{x})M_{11}^{-1}(\xi)M_{12}(\xi)\gamma_t\gamma_t^T[M_{21}(\xi)M_{11}^{-1}(\xi)f_1(\mathbf{x}) - 2f_2(\mathbf{x})]$.

The proof is in the Appendix.

Note 1. In applications when nonnormalized variance-covariance matrices are of importance, one should take into account that in (3.2), the constraint becomes: $\ln|M_{11}^{-1}(\xi)| \le C'$ or $\ln|M_{11}^{-1}(\xi)| \le C' + m\ln[N/\sigma^2] = C$, where N is the number of available observations. When the ratio N/σ^2 decreases then, at some N, only a **D**-optimal design must be considered as a solution.

Note 2. It is obvious that the solution to (8) and (9) does not depend on the length of the vector γ but only on its direction, but this solution does depend upon C.

Note 3. Probably for practitioners who have used **D**-optimal designs, the "inverse" version of (8) and (9) will look more familiar:

$$\xi^* = \text{Arg}\min_{\xi\in\Xi} \ln|M_{11}^{-1}(\xi)|$$

such that $\gamma_t^T M_{21}(\xi)M_{11}^{-1}(\xi)M_{12}(\xi)\gamma_t \le B$. When $\gamma_t^T\gamma_t$ is small then the constraint is not active and a standard **D**-optimal design ξ_D^* is a solution. Only when the equality $\gamma_t^T M_{21}(\xi_D^*)M_{11}^{-1}(\xi_D^*)M_{12}(\xi_D^*)\gamma_t = B$ is satisfied does one face a more difficult optimization problem.

Evidently, instead of (10), the fulfillment of the inequality and existence of $u^* \ge 0$ where $d_1(\mathbf{x},\xi^*) + u^* d_2(\mathbf{x},\xi^*) \le m - u^* B$ provides a necessary and sufficient condition for optimality of ξ^*.

In a similar fashion, an **A**-optimal counterpart can also be defined. Under the class of designs considered by Fedorov et. al. (1994) the relevant criterion would be to minimize $\text{tr}[AR(\xi)]$, which is just $\text{tr}[AV(\xi)] + \text{tr}[AW(\xi)]$. Unfortunately, we cannot use this partitioning in the same way that we did for the **D**-criterion. But we can start by strictly replacing A in the second term by $M_{11}(\xi)$. Under the constrained optimization framework of (6), we can let $\Psi[W(\xi)] = \text{tr}[M_{11}(\xi)W(\xi)]$ and $\Phi[V(\xi)] = \text{tr}[AM_{11}^{-1}(\xi)]$. Note that

$\Psi[W(\xi)]$ can be interpreted as a measure (which is the trace in this case) of the bias vector in M_{11} matrix norm, which is a standardized version of the bias factor. We are now ready to define a new criterion:

The A-restricted minimum bias design ξ^* is defined as

$$\xi^* = \text{Arg}\min_{\xi \in \Xi} \gamma_t^T M_{21}(\xi) M_{11}^{-1}(\xi) M_{12}(\xi)\gamma_t \tag{11}$$

such that

$$\text{tr}[AM_{11}^{-1}(\xi)] \leq D, \tag{12}$$

where A is an appropriately chosen matrix.

While this definition is not that natural and closely connected with our old results, it is still appealing for practical purposes.

The corresponding equivalence theorem closely resembles Theorem 3.1:

Theorem 2 *A design ξ^* is A-restricted minimum bias if and only if there exists $u^* \geq 0$ satisfying*

$$u^*d_1(\mathbf{x}, \xi^*) + d_2(\mathbf{x}, \xi^*) \leq u^*\text{tr}[AM_{11}^{-1}(\xi^*)] - \Psi[\xi^*]$$

for all $\mathbf{x} \in \chi$, where $\text{tr}[AM_{11}^{-1}(\xi^*)] = D, d_1(\mathbf{x},\xi) = f_1^T(\mathbf{x})M_{11}^{-1}(\xi^*)AM_{11}^{-1}(\xi^*)f_1(\mathbf{x})$ and $d_2\mathbf{x},\xi)$ is the same as in Theorem 3.1.

The proof is similar to that for Theorem 3.1.

4 Examples

In many relatively simple regression problems, the equivalence theorem allows the construction of optimal designs using semi-intuitive considerations, for example, such as using symmetry or examining the form of the response surface. In the succeeding examples, we guess the number of observations and weights according to intuition. The exact coordinates of the design points which will satisfy the corresponsing necessary and sufficient conditions are consequently determined, thus ensuring an optimal design under the defined criterion.

4.1 THE LINEAR MODEL WITH QUADRATIC (AND OTHER TYPES OF) CONTAMINATION. Suppose we have only one predictor variable x. Let the assumed model be $\theta^T f_1(\mathbf{x}) = \theta_0 + \theta_1 x$ with the contamination function being $\delta^T f_2(x) = \delta x^2$. Design points are chosen in the design region χ which coincides with the interval [-1,1].

First of all, consider all designs which satisfy $|M_{11}^{-1}(\xi)| \leq \kappa^2$ where $\kappa^2 = e^C$. It is evident that instead of the above inequality we may consider the equality $|M_{11}^{-1}(\xi)| = \kappa^2$. ¿From symmetry considerations it is natural to assume that an optimal design has the form:

$$\xi = \left\{ \begin{array}{cc} \bar{x} & -\bar{x} \\ \frac{1}{2} & \frac{1}{2} \end{array} \right\}. \tag{13}$$

When $\kappa^2 \geq 1$, then it is straightforward to show that

$$\bar{x} = \frac{1}{\kappa} \tag{14}$$

satisfies $|M_{11}^{-1}(\xi)| = \kappa^2$ as well as condition (10) if we set $u^* = 2\gamma^2\kappa^{-4}$. Hence, the design described by (13) and (14) is a **D**-restricted minimum bias design.

Note that for $\kappa^2 > 1$, the optimal design is not unique. It can be verified, using a similar argument, that the symmetric three-point design:

$$\xi = \left\{ \begin{array}{ccc} -1 & 0 & 1 \\ \frac{1}{2\kappa^2} & 1 - \frac{1}{\kappa^2} & \frac{1}{2\kappa^2} \end{array} \right\}$$

is also a **D**-restricted minimum bias design.

We find that in this example the optimal design does not depend on γ, but on the constraint κ^2 put on the value of $|M_{11}^{-1}(\xi)|$. This observation differs from Fedorov's et. al (1994) **D**$_R$-optimal design, which depends considerably on γ. Note that the design points for the **D**-restricted design approaches the design region's center as the square root of $|M_{11}^{-1}(\xi^*)|$ increases. The same is observed for the **D**$_R$-optimal design, but this time with increase in the square root of γ.

Now, the design ξ^* given by (13) and (14) also satisfies condition (10) for several other types of contamination functions other than the quadratic form. The following result therefore extends the scope of importance of this same design:

Lemma 1 Let $C = \ln\kappa^2$ and suppose $\chi = [-1, 1]$. For the linear model with contamination function $\gamma^T f_2(x)$ which satisfies:

(i) $\gamma^T f_2(x) = \gamma^T f_2(-x)$ and $\gamma^T f_2(x) \geq z_0 x^2$ for every $x \in \chi$, where $z_0 = \gamma^T f_2(1/\kappa)\kappa^2$;

or

(ii) $\gamma^T f_2(x) = -\gamma^T f_2(-x)$ for every $x \in \chi$, such that $\gamma^T f_2(x)/x > w_0$ whenever $\gamma^T f_2(1/\kappa) > 0$ and $\gamma^T f_2(x)/x < w_0$ whenever $\gamma^T f_2(x) < 0$, where $w_0 = \gamma^T f_2(1/\kappa)\kappa$,

then the design described by (13) and (14) is a **D**-restricted minimum bias design.

A proof is given in the Appendix. As an interesting corollary, this result implies that if a contamination function, say $g(x)$, is an even function such that $g(x) \geq z_0 x^2$ or if $g(x)$ is an odd function such that $g(x)/x > w_0$ whenever $g(1/\kappa) > 0$, then (13) and (14) is an optimal design for the linear model.

For the linear criterion, suppose we choose

$$A = \left[\begin{array}{cc} 1 & 0 \\ 0 & 1 \end{array} \right] = \text{Var}(\theta_0) + \text{Var}(\theta_1)$$

and we want to find an **A**-restricted minimum bias design satisfying $\text{tr}[AM_{11}^{-1}(\xi)] \leq \kappa^2$ where $\kappa^2 = D - 1$. Whenever $D \geq 2$, the design described by (13) and (14) will be optimal.

4.2 THE QUADRATIC MODEL WITH CUBIC CONTAMINATION. Consider the case when we assume a single predictor quadratic model: $\theta^T f_1(\mathbf{x}) = \theta_0 + \theta_1 x + \theta_2 x^2$ with the contamination function $\delta^T f_2(\mathbf{x}) = \delta x^3$. As in the previous example, the design region χ can be chosen to be, without loss of generality, the interval [-1,1].

For a specified value of C, it is clear that we simply need to find a **D**-restricted minimum bias design among designs ξ which satisfy

$$|M_{11}^{-1}(\xi)| = \tau^2$$

22

where, again, $\tau^2 = e^C$.

It can be shown that the three-point symmetric design of the form:

$$\xi = \left\{ \begin{array}{ccc} -1 & 0 & 1 \\ \frac{\bar{p}}{2} & 1-\bar{p} & \frac{\bar{p}}{2} \end{array} \right\} \tag{15}$$

where $\bar{p}$ satisfies

$$p^2(1-p) = \frac{1}{\tau^2} \tag{16}$$

is **D**-restricted minimum bias whenever $\tau^2 \geq 27/4$.

Note that $27/4$ is the smallest possible value of $|M_{11}^{-1}(\xi)|$ among all designs $\xi \in \Xi$ and, of course, the standard **D**-optimal design for a quadratic model (which is design (15) with $p = 2/3$) achieves this minimum value. This design is also **D**-restricted minimum bias when $\tau^2 = 27/4$, since $\bar{p} = 2/3$ satisfies (16).

As in the previous example, note that the optimal design obtained here does not depend on the magnitude of the cubic contamination (signified by $\gamma = \delta N/\sigma^2$), but it does depend on τ^2, the upper limit constraint on the value of $|M_{11}^{-1}(\xi)|$. Upon comparison of this design with the $\mathbf{D}_R$-optimal design for this example (see Fedorov et. al. (1994)), we can see the same relationship that was observed in the earlier example.

5 Appendix: Proofs

PROOF OF THEOREM 3.1: The convexity of $\Psi(\xi)$ follows from the following basic result in matrix theory (see, for example, Fedorov (1972)):

$$[(1-\alpha)A_1 + \alpha A_2][(1-\alpha)B_1 + \alpha B_2]^{-1}[(1-\alpha)A_1^T + \alpha A_2^T] \leq$$
$$(1-\alpha)A_1 B_1^{-1} A_1^T + \alpha A_2 B_2^{-1} A_2^T,$$

if we set $A_1 = \gamma_t^T M_{21}(\xi^*)$, $A_2 = \gamma_t^T M_{21}(\bar{\xi})$, $B_1 = M_{11}(\xi^*)$ and $B_2 = M_{11}(\bar{\xi})$. This result, hence, makes condition (3.3) necessary and sufficient. All we need is to find $\psi(\mathbf{x}, \xi)$ and $\phi(\mathbf{x}, \xi)$ here to obtain the appropriate $q(\mathbf{x}, u, \xi)$. Straightforward differentiation of $\Psi[(1-\alpha)\xi^* + \alpha\bar{\xi}]$ and $\Phi[(1-\alpha)\xi^* + \alpha\bar{\xi}]$ with respect to α, with its subsequent annealing gives $\psi(\mathbf{x}, \xi^*) = -d_2(\mathbf{x}, \xi^*) - \Psi(\xi^*)$ and $\phi(\mathbf{x}, \xi^*) = m - d_1(\mathbf{x}, \xi^*)$, respectively. This completes the proof of the theorem.

PROOF OF LEMMA 4.1: Consider the contamination function which satisfies (i). For the design ξ^* given by (13) and (14) which satisfies $\ln|M_{11}^{-1}(\xi^*)| = C$, it is easy to verify that condition (10) of Theorem 3.1 for this case can be written as follows:

$$2[\gamma^T f_2(\bar{x})]^2 + x^2[u^*\bar{x}^{-2} - 2[\gamma^T f_2(\bar{x})]\frac{\gamma^T f_2(x)}{x^2}] \leq u^* \tag{17}$$

for some nonnegative u^* and for every $x \in [-1,1]$. Now let $u^* = 2[\gamma^T f_2(\bar{x})]^2$. Replacing u^* in (17) and noting that $\gamma^T f_2(x)$ satisfies (i), it is clear that the inequality holds for all $x \in [-1,1]$. A similar argument can be used for the case when $\gamma^T f_2(x)$ satisfies (ii).

References

Box, G.E.P. and Draper, N.R. (1959). A Basis for the Selection of a Response Surface Design, J. Amer. Statist. Assoc. 54, 622-654.

Cook, D.R. and Fedorov, V.V. (1994). Optimal Design Construction with Constraints. *Statistics*. (in press).

Fedorov, V.V. (1972). *Theory of Optimal Experiments*. Academic Press

Fedorov, V.V. and Gaivoronski, A. (1984). Design of Experiments under Constraints, WP-84-8, International Institute of Applied Systems Analysis, Laxenburg, Austria.

Fedorov, V.V., Montepiedra, G. and Nachtsheim, C. (1994). Optimal Design with Finite Model Validity Range. *to be submitted to JASA*.

Gaivoronski, A. (1986). Linearization Methods for Optimization of Functionals which Depend on Probability Measures. *Mathematical Programming Study 28*, 157-181.

Lee, C.S. (1987). Constrained Optimal Designs for Regression Models. *Comm. Statist. Ser. A 16*, 765-783.

Lee, C.S. (1988). Constrained Optimal Designs. *JSPI 3*, 339-364.

Montepiedra, G. (1994). *Optimal Design with Finite Model Validity Range.* Unpublished doctoral thesis at the University of Minnesota, School of Statistics.

Stigler, S. (1971). Optimal Experimental Designs for Polynomial Regression. *J. Amer. Statist. Assoc. 66*, 311-318.

Studden, W.J. (1982). Some Robust-type D-optimal Designs in Polynomial Regression. *J. Amer. Statist. Assoc. 77*, 916-921.

Bayesian Designs for Approximate Normality

MERLISE A. CLYDE

Institute of Statistics and Decision Sciences, Duke University, Durham, NC 27708–0251, USA.

Abstract:

In many experimental design problems, the primary interest is in estimating functions of the parameters and a design is selected according to some optimality criterion. The assumption that parameter estimates are approximately normally distributed is often used to find optimal designs, as well as simplify data analysis. How well this approximation holds for small to moderate sample sizes depends on the intrinsic and parameter-effects curvatures. These measures depend on both the parameterization used as well as the experimental design. For a particular parameterization of interest, these curvatures can be reduced by the choice of the experimental design. A Bayesian approach is taken to find designs that optimize the primary design criterion subject to satisfying constraints based on these curvature measures, with the goal of improving normal approximations. The constrained designs depend on the sample size, but as the sample size increases the constraints are satisfied. A nonlinear regression example is used to illustrate the approach.

1 Introduction

The experimental design problems considered here involve a nonlinear regression model,

$$Y_i = \mu(\theta, x_i) + e_i, \quad i = 1, \ldots, n$$

where $\theta \in \mathbb{R}^p$ are unknown parameters and e_i are assumed to be independent normal random variables with mean 0 and known variance σ^2. The design points $x_1, \ldots, x_n$ which

---

*Kitsos, C.P., and Müller, W.G., Eds., *Proceedings of MODA4*, Physica Verlag, Heidelberg, 1995

make up the design ξ are to be chosen from a specified design region χ to maximize a design criterion ϕ. In nonlinear regression problems, the assumption that parameter estimates are approximately normally distributed is used to find optimal designs (Chaloner 1987, 1992; Chaloner and Larntz 1989) as well as to simplify the data analysis.

Normal approximations for the least squares estimators of the nonlinear regression model are based on the tangent plane approximation to the expectation surface. How well this approximation holds for small to moderate samples depends on the second derivatives of the model function. This array can be partitioned into the intrinsic curvature array, which is independent of the model parameterization, and the parameter effects array which does depend on the model parameterization. Bates and Watts (1980) derived several summaries of the second derivative arrays to indicate when the linear approximation would be suspect. Once the data are collected for a particular experimental design, the parameter effects curvature may be reduced by reparameterizing the model (Bates and Watts 1980; Hougaard 1982, 1985; Kass and Slate 1990; Hills and Smith 1992). One particular parameterization, however, may be the most meaningful to the researcher, in which case, reducing the parameter effects curvature in this parameterization as much as possible by choice of design is desirable. If intrinsic curvature is high, then no matter what parameterization is used, inferences based on the normal distribution may be misleading. Intrinsic curvature can only be reduced by choice of the design points for a given model. In this paper, a Bayesian approach is taken to find designs that optimize a primary design criterion, ϕ, subject to satisfying constraints based on these curvature measures, with the goal of improving normal approximations. In section 2, the Bates and Watts curvature measures are reviewed and in section 3, the constrained design problem is formulated. An example from nonlinear regression is used to illustrate the approach in section 4.

2   Curvature Measures

Let V denote the $n \times p$ matrix of partial derivatives of the mean function, $\dot{\mu}(\theta, x_i)$. Using a QR decomposition, the matrix V can be written as $V = QR$ where Q is a $n \times n$ orthonormal matrix that is partitioned as $Q = (U|N)$ where U is $n \times p$ and N is $n \times (n-p)$, and R is the $n \times p$ matrix

$$R = \left[\begin{array}{c} R_1 \\ 0 \end{array} \right]$$

with R_1 a $p \times p$ upper triangular matrix. So that the curvature measures will be invariant under affine transformations of the parameters, the parameter vector is transformed so that $\varphi = R_1 \theta$. The coordinates of the sample space are rotated so that the first p coordinate vectors are parallel to the tangent space and the last $n - p$ vectors are orthogonal to the tangent space.

Let $D_\theta^2 \mu$ denote the $n \times p \times p$ array of partial derivatives of the mean function with elements,

$$\frac{\partial \mu(\theta, x_i)}{\partial \theta_j\, \partial \theta_k}.$$

After rotating the sample space and reparameterizing the model so that the arrays are invariant to scale and affine transformations of Y and θ, the acceleration array is $A(\varphi, \xi) = \sigma(Q^T \otimes R_1^{-T} \otimes R_1^{-T}) D_\theta^2 \mu$, which corresponds to the second derivatives of the mean vector in the new coordinate system. This array is partitioned so that $A^P(\varphi, \xi) = \sigma(U^T \otimes R_1^{-T} \otimes$

$R_1^{-T})D_\theta^2\mu$ represents the parameter effects array and $A^N(\varphi,\xi) = \sigma(N^T \otimes R_1^{-T} \otimes R_1^{-T})D_\theta^2\mu$ represents the intrinsic curvature array.

The maximum intrinsic curvature and maximum parameter effects curvature (Bates and Watts 1986) are one dimensional summaries of these two arrays and can be expressed as functions of φ (or θ) and the design ξ by,

$$\gamma^{max}(\varphi,\xi) = \sup_{\{d:\ ||d||=1, d\in \mathbb{R}^p\}} ||(I_{(n-p)} \otimes d^T \otimes d^T)A^N(\varphi,\xi)|| \tag{1}$$

$$\omega^{max}(\varphi,\xi) = \sup_{\{d:\ ||d||=1, d\in \mathbb{R}^p\}} ||(I_p \otimes d^T \otimes d^T)A^P(\varphi,\xi)||. \tag{2}$$

Since both tangential and normal components must be small for the normal approximation to be valid, a summary analogous to the above based on the entire second derivative array is

$$\tau^{max}(\varphi,\xi) = \sup_{\{d:\ ||d||=1, d\in \mathbb{R}^p\}} ||(I_n \otimes d^T \otimes d^T)A(\varphi,\xi)||. \tag{3}$$

Clyde (1993) discusses other one-dimensional summaries of the two arrays in the context of design. Hougaard (1982, 1985), Kass (1989) and Smyth (1987) discusses the relationship of the curvature arrays and various summaries to normal approximations for nonlinear regression models.

3   Designs for Approximate Normality

We formulate a constrained design problem to find designs that have small curvature in order to improve normal approximations while still being efficient for parameter estimation. The approach used here is to consider the constrained problem of maximizing a design criterion ϕ, reflecting information gain from the experiment, subject to constraints on posterior normality (or likelihood normality) as measured by the curvature summaries. For these constrained problems, we need some upper bound on the curvature summaries which indicate when the normal approximation is expected to be satisfied.

Using the guidelines of Bates and Watts (1980), we would like to have a design such that $\omega^{max}(\theta,\xi)\sqrt{\chi_{\alpha,p}^2} \le 1$ and/or $\gamma^{max}(\theta,\xi)\sqrt{\chi_{\alpha,p}^2} \le 1$. We use the chi–square rather than the F distribution since we are taking σ known. Since the value of 1 is just a suggested guideline, we can examine the sensitivity of the designs to the parameter effects curvature by instead using some value k for the constraint. For $k = 1$, this corresponds to Bates and Watts' guideline, but we could use lower or higher values of k to evaluate tradeoffs of efficiency versus valid normal approximations.

After the data are collected, the curvature measures are typically evaluated at the maximum likelihood estimates. At the design stage, we may use the prior distribution as the predictive distribution for the estimator and thus are concerned with the behaviour at all potential estimators. We take the approach of finding a design that minimizes the maximum value of the curvature summary over the support of the prior distribution, so that the normal approximation is expected to be valid for any potential estimator. An alternative is to use the expected value of the curvature measure (Clyde, 1993).

The general formulation of constrained design problem is

$$\max_{\xi,\ x_i \in \mathcal{X}} \phi(\xi) \tag{4}$$

28

subject to

$$\max_{\theta \in \Theta} \tau^{\max}(\theta, \xi)\sqrt{\chi^2_{\alpha,p}} \leq k_\tau \qquad (5)$$

$$\max_{\theta \in \Theta} \omega^{\max}(\theta, \xi)\sqrt{\chi^2_{\alpha,p}} \leq k_\omega \qquad (6)$$

$$\max_{\theta \in \Theta} \gamma^{\max}(\theta, \xi)\sqrt{\chi^2_{\alpha,p}} \leq k_\gamma \qquad (7)$$

which will be denoted as the $[\psi, k_\tau, k_\omega, k_\gamma]$ constrained design problem. We can consider a range of possible values of the k's to examine the sensitivity of the designs to the constraints and the change in efficiency. Varying the values of the k's can also be viewed as changing the variance σ^2. Note that the curvature measures depend on the sample size so that the constrained optimal design also will depend on the sample size. This is discussed further in Section 4.

4   Rumford Example

In this section we will consider a simple example to illustrate the constrained design problem. Consider the one parameter nonlinear regression model from Bates and Watts (1986, page 33) for the Rumford data,

$$Y_i = 60 + 70e^{-\theta x_i} + \epsilon_i$$

where ϵ_i are independent normally distributed random variables with mean zero and constant variance. We can instead use the model $\mu(x_i; \theta) = e^{-\theta x_i}$ with error standard deviation σ (after rescaling) for finding the curvature, since the curvature measures are location and scale invariant with respect to the model.

As the primary design criterion, we will use the Bayesian D–optimality criterion, ϕ_D. The expected Fisher information for this problem is

$$I(\theta, \xi) = \sum_{i=1}^n x_i^2 \exp^{-2\theta x_i}/\sigma^2.$$

and the Bayesian D–optimal designs maximize

$$\phi_D(\xi) = E_\theta(\log I(\theta, \xi))$$

with the expectation taken over a prior distribution on θ. Other Bayesian design criteria are discussed in Verdinelli (1992) and Chaloner and Larntz (1989) and could also be used in place of the Bayes D-optimality criterion. Locally optimal designs (Chernoff 1953, Box and Lucas 1959) can be obtained as a special case of Bayesian optimal designs by using degenerate prior distributions. In this problem, the locally optimal designs put all observations at a single point.

4.1   Minimum Curvature Designs

We can examine how the design changes under the curvature criteria. In this problem, the maximum intrinsic curvature given in equation (1) is zero for one point designs. The maximum parameter effects curvature in equation (2) for the class of one point designs ξ_x for $x \in \chi$ simplifies to

$$\omega^{\max}(\theta, \xi_x) = \frac{\sigma}{\sqrt{n}} e^{x\theta}.$$

The design that minimizes this for all values of θ is to take $x = 0$, regardless of the prior distribution, but this design provides no information about θ, as $I(\theta, 0) = 0$.

4.2 Constrained Designs

We will find optimal constrained designs under two different prior distributions.

Prior A

Chaloner (1993) gives necessary conditions on the prior distribution such that the Bayesian D–optimal design is equivalent to a locally optimal design. Let θ^+ denote the $\sup_\Theta \theta$, and θ^- denote the $\inf_\Theta \theta$ where Θ is the support of the prior distribution on θ. For this model, the conditions require that

$$\frac{\theta^+}{\theta^-} < \frac{2 + \sqrt{2}}{2}$$

and that the interval $[\inf_\Theta \theta^{-1}, \sup_\Theta \theta^{-1}]$ is a subset of the design region χ. Under these conditions there is a unique optimal Bayesian design with one support point, and the Bayes D–optimal design takes all observations at $x = [E(\theta)]^{-1}$. The prior distribution A is a two point prior distribution with equal probability at $\theta = 1$ and $\theta = (2 + \sqrt{2})/2$, which satisfies Chaloner's conditions so that Bayes D–optimal design is a one point design at 0.74. For $n = 2, \sigma = .3$ and the Bayes D-optimal design, the maximum parameter effects summary in equation (6) is 1.47 which is greater than the suggested cutoff of $k = 1$. As the constraint (6) is not satisfied, the ϕ_D–optimal design is not optimal for the Bayesian $[\phi_D, k_\tau = \infty, k_\omega = 1, k_\gamma = \infty]$ constrained problem, although it does have zero intrinsic curvature as it has only one support point, so it also satisfies the problem $[\phi_D, k_\tau = \infty, k_\omega = 1, k_\gamma = 0]$.

The additional constraint of zero maximum intrinsic curvature, $(k_\gamma = 0)$, restricts the set of design measures to one–point designs ξ_x for $x \in \chi$. We have that $\omega^{\max}(\theta, \xi_x)$ is a monotone nondecreasing function of x for one point designs ξ_x, and additionally that it is a monotone nondecreasing function in θ. This last part implies that $\max_\Theta \omega^{\max}(\xi, \theta) = \omega^{\max}(\xi, \theta^+)$. The nondecreasing aspect of this function in x then implies if

$$x(k) = \frac{1}{\theta^+} \log \left(\frac{k\sqrt{n}}{1.96\,\sigma} \right)$$

then the design $\xi_{x(k)}$ satisfies the constraint (6) with equality, and for any values of x less than or equal to $x(k)$ the constraint (6) is satisfied. If $x(k)$ is greater than $[E(\theta)]^{-1}$, then the Bayes D–optimal satisfies all constraints and is optimal for the constrained problem (the constraints are nonbinding). Otherwise, if $x(k)$ is contained in χ, then $\xi_{x(k)}$ is the constrained optimal design for $[\phi_D, k_\tau = \infty, k_\omega = k, k_\gamma = 0]$. This follows from the fact that the designs that satisfy both the parameter effects constraint and the intrinsic curvature constraint are the one point designs ξ_x where x is less than or equal to $x(k)$. Since $I(\theta, \xi_x)$ is strictly increasing in x, for x less than $[E(\theta)]^{-1}$, we have that $I(\theta, \xi_x)$ is less than $I(\theta, \xi_{x(k)})$ for all x less than $x(k)$, which implies $\phi_D(\xi_x)$ is less than $\phi_D(\xi_{x(k)})$. Therefore, $\xi_{x(k)}$ is the Bayesian optimal constrained design. For any values of k less than 0.415779, however, no constrained optimal designs exist as $x(0.415779) = 0$. In this problem, the constraints that $k_\gamma = 0$ is nonbinding, since all optimal designs with other values of k_γ have only one support point.

One approach to viewing the tradeoffs involved in the constraints versus the primary optimality criterion is to examine the efficiency of the constrained designs relative to the unconstrained design for different values of the constraint. The ϕ_D efficiency of design ξ relative to ξ^*, the Bayes D–optimal design, is calculated as

$$\exp(\phi_D(I(\theta, \xi)) - \phi_D(I(\theta, \xi^*)))$$

which can be viewed as the fractional number of times the Bayes D–optimal design would have to be replicated in order to have the same value of the ϕ_D criterion as the design ξ. Figure 1 shows the efficiency of the constrained designs relative to the Bayes D–optimal design as the value of the constraint increases. If we require that k_ω be less than or equal to 1, the suggested cutoff, then there is a corresponding cost in terms of efficiency relative to the Bayes D–optimal design of .89. If the constraint cutoff of 1.25 is acceptable then the constrained design has an efficiency of .98. However, reducing the cutoff to values less than one has a much greater impact on the efficiency of the designs. For a constraint cutoff of .75 the efficiency is .63, whereas decreasing the constraint cutoff to .5 drastically reduces the efficiency to .12 for the design with support on the point 0.108. The design with a constraint cutoff of 0.415779, the smallest value possible, corresponds to the design with all observations at 0: the design with zero parameter effects curvature, zero intrinsic curvature, and zero ϕ_D efficiency!

Prior B

If the prior distribution is changed by putting equal probability at $\theta = 1$ and $\theta = 10$, then one-point designs are not optimal with respect to the Bayes D–optimality criterion. We will consider the total curvature $\tau^{\max}$ since the intrinsic curvature will not necessarily be zero. Using $n = 2$ again, the Bayes D–optimal design corresponds to putting one observation at each of $(0.106, 0.999)$, however this design has a maximum total curvature of 1.69 and is therefore not optimal for the $[\phi_D, k_\tau = 1, k_\omega = \infty, k_\gamma = \infty]$ design problem. For $k_\tau = 1$, the constrained optimal design has only one support point at $x = 0.088$. Figure 2 shows the contours of $\sup_\Theta \tau 1.96 = k_\tau$ over the design region (x_1, x_2) for possible two point designs. For values of the constraint greater than 1.08, there are actually two different regions of the design space that satisfy the constraint with one region having optimal designs with only one unique support point and the second region corresponding to optimal designs with two distinct supporting points not far from the Bayes D–optimal design.

As the constraint cutoff is decreased below 1.08, the optimal constrained designs "jump" from having two distinct support points to one support point, corresponding to the two different feasibility regions. At the point $k_\tau = 1.08$, there is a large jump discontinuity in the ϕ_D efficiency curve, with a sharp drop in efficiency for the constrained designs relative to the Bayes D–optimal design for values of k_τ less than 1.08 (efficiencies less than .45) (Figure 2). For constraint cutoffs greater than 1.1, the constrained designs have efficiencies greater than .80.

31

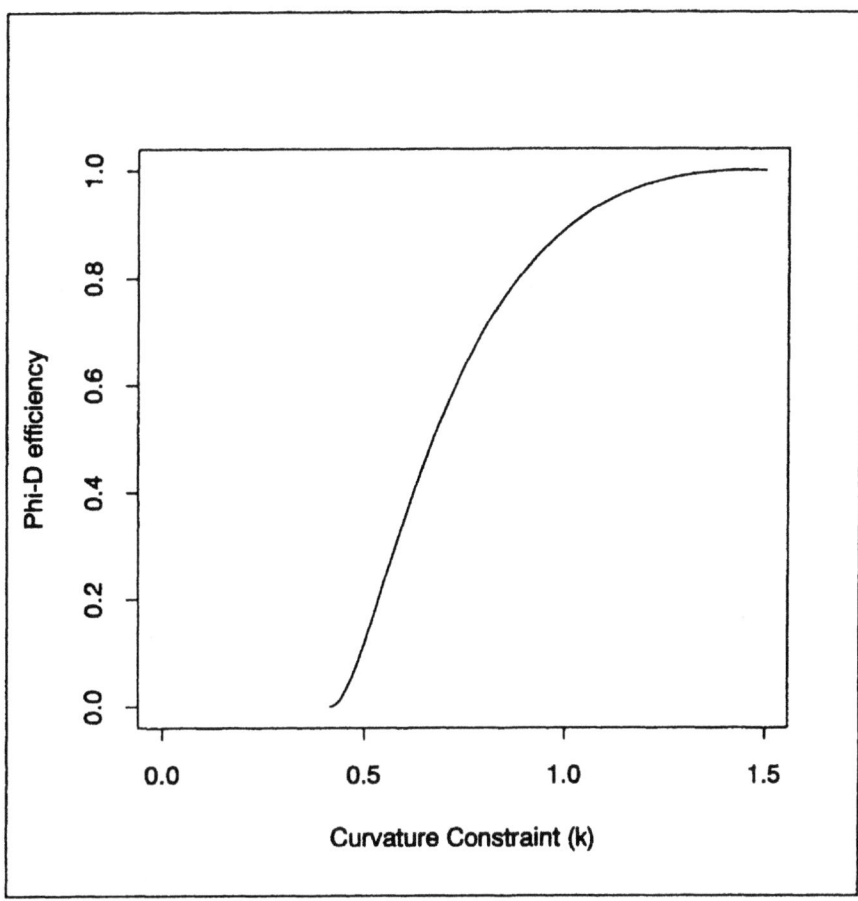

Figure 1: Efficiencies of the maximum total curvature constrained designs relative to the Bayes D–optimal design under prior distribution A. Sample size of 2.

For n greater than or equal to 2, the designs no longer necessarily have zero intrinsic curvature unless they have only one support point or have two support points, but one of the support points is at $x = 0$. Thus there is a corresponding decrease in parameter effect curvature at the expense of a slightly higher intrinsic curvature (for the same value of k_τ).

For this prior distribution, exact optimal designs were found for $n = 2, 3, \ldots, 10$ for the $[\phi_D, k_\tau = 1, k_\omega = \infty, k_\gamma = \infty]$ constrained problem (Table 1). For n greater than or equal to 3, the constrained optimal designs all have two distinct support points, with the same number of points at the low and high point as in the Bayes D–optimal designs (Table 1). For n odd, these designs do not have equal weights on the two support points, but rather have more weight on the lower support point. The distance between the two support points is greater for the constrained designs. Table 1 also gives ϕ_D efficiencies and values of the active constraint function, $1.96 \sup_\Theta \tau(\theta, \xi)$. Designs were also constructed with equal spacing of the x values over the design region, $[.00001, 1]$, as well as taking design points equally spaced on the log scale for this region. Comparing the simple design based on equally spaced observation to the constrained design and the Bayes D–optimal design for each sample-size n, we find that these designs have much higher maximum total curvature and are relatively inefficient compared to the Bayes D–optimal designs. The designs based on equal spacing on the log scale have much lower total curvature than the designs of equal spacing of the x values. For n as large as 25, the total curvature of the

32

(a)

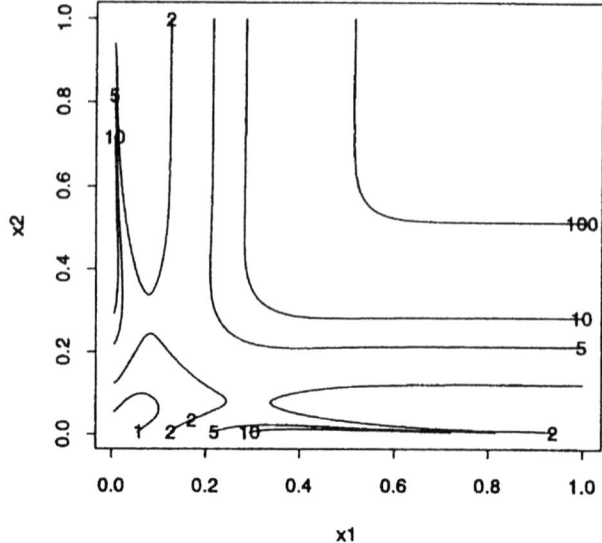

(b)

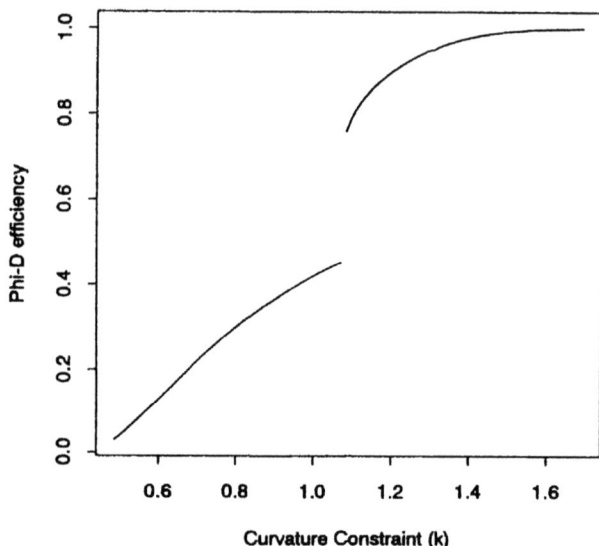

Figure 2: (a) contours of $\sup_\theta \tau^{max}(\theta, \xi)1.96 = k_\tau$ in the design region for two point designs, and (b) efficiencies of the optimal constrained designs relative to the Bayes D-optimal design under prior distribution B for $n = 2$ for the constrained problem $[\phi_D, k_\tau = k, k_\omega = \infty, k_\gamma = \infty]$.

equally spaced design is still above the guideline of $k_\tau = 1$, while it is satisfied for the equally spaced design on the log scale.

The number of times that one would have to replicate a design to reduce k_τ to 1, is simply k_τ^2. For the design with $n = 2$, $k_\tau^2 = (1.69^2) = 2.86$, so that replicating the design 3 times should reduce the curvature to below 1. For n greater than 5 with this prior distribution and standard deviation σ, the maximum total curvature constraint no longer is active so the Bayes D–optimal design is optimal for the constrained criterion as well. Simulation studies indicate that the posterior distributions are more normal, but with more variability, under the constrained designs than with the Bayes D–optimal designs.

5 Discussion

This paper has explored relationships between curvature measures of non–normality and design in nonlinear regression. The curvature summaries are used to find designs that are constrained so that the resulting posterior or likelihood is expected to be well approximated by the normal approximation to the posterior or likelihood. Completely replicating a design, ξ, r times can be viewed as decreasing σ so that the acceleration array for the replicated design is $r^{-1/2}A$. Thus as the sample size increases in this manner, the curvature decreases. These designs, unlike the approximate Bayes D–optimal designs, will depend on the sample size, but as the sample size is increased, the constraints will be satisfied. One use of the curvature summaries in design is to determine an appropriate sample size for valid normal approximations.

The inverse of the observed information matrix often provides a better approximation to the posterior covariance matrix than the inverse of the expected information matrix (Berger 1985, p. 224), but in most design problems expected information is used in the design criterion rather than observed information which depends on Y. In the case of $n = p$ point designs, the intrinsic curvature array A^N is zero, so the observed Fisher information information equals the expected Fisher information information. Let $\tilde{G}$ denote the negative of the observed second derivative matrix of the log posterior. In the case that a uniform prior distribution on θ is used, for a p–point design, $\tilde{G}(\theta) = I_{obs}(\theta) = I(\theta)$. Thus the normal approximations to the posterior distribution based on using the observed second derivative of the log posterior, the observed information matrix, or the expected information matrix to approximate the posterior covariance matrix are equivalent in this situation. It is clear that designs with the same number of support points as model parameters have zero intrinsic curvature. But for a model that is intrinsically nonlinear, Clyde (1993) showed that it is possible to have more than p support points and still have zero intrinsic curvature. However, in those cases the extra points provided no additional information about θ.

While this paper focused on the Bates and Watts maximum curvature summaries, in multiparameter problems there are several alternative summaries that are relevant and which could be used as additional constraints. Clyde (1993) has also developed criteria for normality based on the third derivative of the log of the posterior distribution, which is more meaningful if a Bayesian analysis of the experiment is planned. While in this paper the primary design criterion ϕ_D is maximized subject to constraints on the curvature, one can take the same approach using the design criteria of Hamilton and Watts (1985) or Pazman and Pronzato (1992a, b) as the primary design criterion.

34

Table 1: Comparison of designs for the Rumford example.

n	Design	Design Points (n_i)				ϕ_D Efficiency	$\sup_\Theta \tau(\theta,\xi)1.96$
2	Bayes D–optimal	0.10559	(1)	0.99999	(1)	1.00	1.69
	Constrained	0.08776	(1)	0.08776	(1)	.42	1.00
	Equal Spacing					.0012	12352
	Equal Spacing (Log)					.0012	12352
3	Bayes D–optimal	0.111368	(2)	0.999992	(1)	1.00	1.39
	Constrained	0.086586	(2)	0.635537	(1)	.90	1.00
	Equal Spacing					.08	87.28
	Equal Spacing (Log)					.06	2.91
4	Bayes D–optimal	0.105593	(2)	0.99999	(2)	1.00	1.19
	Constrained	0.087729	(2)	0.92429	(2)	.98	1.00
	Equal Spacing					.24	16.57
	Equal Spacing (Log)					.23	1.59
5	Bayes D–optimal	0.10847	(3)	0.99999	(2)	1.00	1.02
	Constrained	0.10803	(3)	0.97617	(2)	.99	1.00
	Equally Spaced					.37	7.35
	Equal Spacing (Log)					.34	1.57
6	Bayes D–optimal	0.105593	(3)	0.99999	(3)	1.00	0.98
	Constrained	0.105593	(3)	0.99999	(3)	1.00	0.98
	Equal Spacing					.46	4.64
	Equal Spacing (Log)					.34	1.51
7	Bayes D–optimal	0.10751	(4)	.99999	(3)	1.00	.86
	Constrained	0.10751	(4)	.99999	(3)	1.00	.86
	Equal Spacing					.51	3.50
	Equal Spacing (Log)					.30	2.01
8	Bayes D–optimal	0.105593	(4)	0.99999	(4)	1.00	.85
	Constrained	0.105593	(4)	0.99999	(4)	1.00	.85
	Equal Spacing					.55	2.91
	Equal Spacing (Log)					.28	2.17
9	Bayes D–optimal	0.10703	(5)	.99999	(4)	1.00	.77
	Constrained	0.10703	(5)	.99999	(4)	1.00	.77
	Equal Spacing					.57	2.55
	Equal Spacing (Log)					.27	1.96
10	Bayes D–optimal	0.105593	(5)	0.99999	(5)	1.00	.76
	Constrained	0.105593	(5)	0.99999	(5)	1.00	.76
	Equal Spacing					.58	2.32
	Equal Spacing (Log)					.27	1.71

References

- Bates, D.M. and Watts, D.G. 1980. Relative curvature measures of nonlinearity. *J. Royal Statist. Soc. B*, **42**:1-25.

- Bates, D.M. and Watts, D.G. 1986. *Nonlinear regression analysis and its Application*. John Wiley & Sons. New York.

- Berger, J. 1985. *Statistical Decision Theory and Bayesian Analysis.* 2nd edition. Springer–Verlag. New York.

- Box, G.E.P. and H. L. Lucas. 1959. Design of experiments in nonlinear situations. *Biometrika* **46**:77-90.

- Chaloner, K. 1987. An approach to design for generalized linear models. In *Proceedings of the workshop on Model-oriented Data Analysis, Wartburg* – Lecture Notes in Economics and Mathematical Systems, #297. Springer–Verlag, Berlin. 3–12.

- Chaloner, K. and Larntz, K. 1989. Optimal Bayesian design applied to logistic regression experiments. *J. Statistical Planning and Inference,* **21**:191-208.

- Chaloner, K. 1993. A note on optimal Bayesian design for nonlinear problems. *J. Statistical Planning and Inference,* **37**:229-235.

- Clyde, M.A. 1993. *Bayesian Optimal Designs for Approximate Normality.* Ph.D. thesis. University of Minnesota.

- Hills, S.E. and Smith, A.F.M. 1992. Parameterization issues in Bayesian inference. In *Bayesian Statistics 4*, edited by J.M. Bernardo, J.O. Berger, A.P. Dawid, and A.F.M. Smith. Oxford University Press, New York.

- Hougaard, P. 1982. Parameterizations of non–linear models. *J. R. Statist. Soc. B.* **44**:244-252.

- Hougaard, P. 1985. The appropriateness of the asymptotic distribution in a nonlinear regression model in relation to curvature. *J. R. Statist. Soc. B.* **47**:103-114.

- Kass, R.E. 1989. The geometry of asymptotic inference.(with discussion) *Statistical Science* 4:188-234.

- Kass, R.E. and Slate, E.H. 1990. Some diagnostics of maximum likelihood and posterior non-Normality. Technical Report No. 490, Dept. of Statist., Carnegie Mellon University.

- Pazman, A. and Pronzato, L. 1992a. Nonlinear experimental design based on the distribution of estimators. *J. Statist. Planning and Inference* **33**:385–402.

- Pazman, A. and Pronzato, L. 1992b. Nonlinear experimental design for constrained LS estimation. In *Model Oriented Data-Analysis*, V. Fedorov, W. G. Müller, and I. N. Vuchkov editors. Springer–Verlag, New York.

- Smyth, G.K. 1987. Curvature and convergence. *ASA Proc. of Stat'l. Computing Sect.* pages 278-283.

- Verdinelli, I. (1992) Advances in Bayesian experimental design. In: *Bayesian Statistics 4*, editors J.O. Berger, J.M. Bernardo, A.P. Dawid, and A.F.M. Smith. Oxford University Press.

Simulation Approach to One-Stage and Sequential Optimal Design Problems

GIOVANNI PARMIGIANI and PETER MÜLLER

Institute of Statistics and Decision Sciences, Duke University, Durham, NC 27708-0251, USA.

1 Introduction

In this presentation we introduce an algorithm for Bayesian optimal design based on smoothing a scatterplot of observed losses (or utilities) for a Monte Carlo sample of simulated experiments. Denote with d, θ and y the design parameters, the parameter vector and the data, respectively. The Bayesian optimal design problem is to find the design d^* which maximizes the pre-posterior expected utility $\mathcal{U}(d) = \int U(d, y, \theta) dp_d(\theta, y)$, where $p_d(\theta, y)$ is the joint distribution under design d on parameter and sample space, and $U(d, y, \theta)$ is the relevant payoff when the data y is observed under design d and the parameter θ. For example, we might want to estimate the parameters θ in a regression model under squared error loss. The relevant utility would be $U(d, y, \theta) = -(\theta - \bar{\theta})^2$, where $\bar{\theta} = E(\theta|y, d)$ is the posterior mean.

Among the many fine reviews of Bayesian optimal design are Verdinelli (1992) and Chaloner and Verdinelli (1993). Recent results concerning structural properties and numerical strategies for specific classes of models include Chaloner and Larntz (1989), Mukhopadhyay and Haines (1994), Chaloner (1992), Flournoy (1992), Clyde (1993), Clyde, Müller and Parmigiani (1994). Verdinelli and Kadane (1992) and Parmigiani and Polson (1992) suggest alternate utility functions as design criterions.

Sequential design problems add an additional difficulty to the stochastic optimization problem by allowing later stage decisions to depend on earlier observations. Finding the optimal decision amounts to the maximization of a functional whose evaluation requires integration with an integrand which itself requires the optimization of another integral, etc.

---

*Kitsos, C.P., and Müller, W.G., Eds., *Proceedings of MODA4*, Physica Verlag, Heidelberg, 1995

Conventional solution strategies are based on dynamic programming. While dynamic programming offers a very general principle for solving sequential statistical decision problems, the computational burden of it becomes prohibitive even for a moderate number of stages in the decision process, especially when integration is difficult. The literature on circumventing some of the computational difficulties is extensive. See Berger (1985) for a review of the techniques more common in statistical applications. For the fundamentals of Bayesian sequential decision making see DeGroot (1970). Sequential design problems are discussed in Chernoff (1979) and Berry and Fristedt (1985).

This talk summarizes and puts in perspective earlier work in Müller and Parmigiani (1994a), Müller and Parmigiani (1994b) and Clyde, Müller and Parmigiani (1994).

2 One-Stage Design Problems

In many important applications the design criterion $\mathcal{U}(d)$ takes the form of an analytically intractable integral, and also numerical evaluation is typically very computation extensive. However, it is typically easy to evaluate $\mathcal{U}(d)$ by Monte Carlo integration. The joint distribution $p_d(\theta, y)$ is simply the product of prior $p(\theta)$ and likelihood $p_d(y|\theta)$. Both are commonly chosen in a way which allows efficient random variate generation, and hence Monte Carlo integral evaluation $\mathcal{U}(d) \approx \mathcal{U}_m(d) = 1/M \sum U(d, y_i, \theta_i)$, where $\{(y_1, \theta_1), \ldots, (y_M, \theta_M)\}$ is a Monte Carlo sample generated from $p_d(\theta, y) = p(\theta)p_d(y|\theta)$. We are using the subindex d for the likelihood to indicate that the likelihood will typically depend on the design choice d. While conceptually no problem, the prior commonly does not change with d. Assuming that the observed loss $U(d, y, \theta)$ can be evaluated easily this would allow solution of the optimal design problem by application of any global optimization method, replacing the design criterion $\mathcal{U}(d)$ by the approximation $\mathcal{U}_m(d)$. However, this might require enormous computational effort. For each point d considered over the course of the optimization we would need to run a Monte Carlo integration to evaluate $\mathcal{U}_m(d)$.

The outlined large scale Monte Carlo approach includes a certain amount of redundant duplication of computations by simulating similar experiments over and over again, under often only little different joint distributions $p_d(\theta, y)$. It is this observation which motivates us to propose the following numerical optimal design scheme:

1. Select designs d_i, $i = 1, \ldots, N$ from the k-dimensional design space $\mathcal{D}$.
2. Simulate pairs $(\theta_i, y_i) \sim p_{d_i}$. For each simulated experiment (θ_i, y_i) evaluate $u_i = U(d_i, y_i, \theta_i)$.
3. Fit a smooth k-dimensional curve $\tilde{\mathcal{U}}(d)$ to the points (d_i, u_i).
4. Find the optimal design d^* which maximizes the fitted curve $\tilde{\mathcal{U}}(d)$.

The rationale of the proposed algorithm is to evaluate $\tilde{\mathcal{U}}$ at any particular design d by "borrowing strength" from simulated experiments at neighboring designs d'. "Borrowing strength" is formalized by smoothing. Instead of requiring large Monte Carlo sample sizes for each evaluation of $\mathcal{U}(d)$ the scheme exploits continuity of $\mathcal{U}(d)$ to reconstruct the expected utility surface with only a few hundred simulated points.

The choice of the design points d_i in Step 1 could be a meta design problem in itself. However, this would clearly defy the purpose. We suggest instead an evenly spaced grid over a region of reasonable design choices. To find the surface $\tilde{\mathcal{U}}(d)$ any commonly available curve fitting method can be used. Whenever possible we recommend the use of parameterized curves because of (i) the ease in incorporating theoretical constraints, such as monotonicity, asymptotics, or known intercepts; (ii) straightforward implementation; and (iii) simple or possibly even analytic optimization. A parametric curve is used in Example 1. If parameterization is not practicable, standard scatterplot smoothers can be used (Example 3). Another alternative are smoothing models based on Bayesian nonparametric regression models. In Examples 2 and 4 we used a method based on mixture of Dirichlet process models. A full description of this approach would be beyond the scope of this paper. We refer to West, Müller and Escobar (1994) and Müller, Erkanli and West (1994).

3 Examples for One-Stage Optimal Designs

3.1 Example 1: An Information Theoretic Stopping Rule

Byar et al. (1977) report a study comparing placebo, pyridoxine and topical thiotepa in preventing recurrence of stage I bladder cancer. Stage I tumors tend to recur in around 50% of the patients within two years after removal. Instillations of thiotepa and regular ingestion of pyridoxine are hoped to retard the growth of tumors, or cure them altogether.

In this example we find the optimal number of patients for a hypothetical second stage. We assume patients arrive at random time, are randomized to one of the three groups, and followed until the end of the study. For each patient we record the time they enter the study, an indicator of the assigned treatment ($X_1 = 1$ for pyridoxine, $X_2 = 1$ for thiotepa), size (X_3) and number (X_4) of initial bladder cancer occurences, and up to nine recurrence times t_j per patient. Using a Weibull proportional hazard rates model for the recurrence times, the hazard function $h(t_{ij})$ is given by: $\log(h(t_{ij})) = \log\alpha + (\alpha - 1)\log t_{ij} + \beta_0 + \beta_1 X_{1i} + \beta_2 X_{2i} + \beta_3 X_{3i} + \beta_4 X_{4i}$, $i = 1, \ldots, n$, $j = 1, \ldots, n_i$. We used the data from Byar et al. (1977) to construct a multivariate normal prior distribution $(\alpha, \beta_0, \ldots, \beta_e) \sim N(\mu, 10\Sigma)$, where μ and Σ are the posterior moments conditional on the data reported there. The arrival times when patients enter the study are simulated from the empirical distribution of the arrival times reported in Byar et al. (1977).

Purpose of the study is to learn about the relative effects of the two treatments. We therefore consider as design criterion the Kullback-Liebler divergence between the marginal prior and marginal posterior on the treatment effects (β_1, β_2):

$$
\begin{aligned}
\mathcal{U}(n) &= \int_y \mathrm{KL}\{p(\beta_1, \beta_2), p_n(\beta_1, \beta_2|y)\} m_n(y) dy \\
&= \int_{y, \Omega} \log(p_n(\beta_1, \beta_2|y)/p(\beta_1, \beta_2)) dp_n(\beta_1, \beta_2, y),
\end{aligned}
$$

where $m_n(y)$ denotes the marginal distribution of the data. See DeGroot (1984) or Carlin and Polson (1992) for a discussion of Kullback-Liebler divergence as design criterion.

To find the optimal design n^* we selected a grid of $N = 120$ design points $\{d_1 = d_2 = \ldots d_5 = 5, d_6 = \ldots = d_{10} = 10, \ldots, d_{116} = \ldots = d_{120} = 100\}$. For each d_i we simulated an experiment (θ_i, y_i) and evaluated $u_i = \log\{p(\beta_{i1}, \beta_{i2}|y_i)/p(\beta_{i1}, \beta_{i2})$. We compute u_i by a Monte Carlo integral approximation (Carlin and Polson 1992):

40

$p(y|\beta_1, \beta_2) \approx 1/M \sum_j p(y|\eta_j, \beta_1, \beta_2)$ and $p(y) \approx 1/M \sum_j p(y|\theta_j)$, where $\{\eta_1, \dots, \eta_M\}$ is a Monte Carlo sample from the conditional prior $p(\eta|\beta_1, \beta_2)$ and $\{\theta_1, \dots, \theta_M\}$ is a simulated prior Monte Carlo sample. We chose Monte Carlo sample size $M = 100$. Figure 1 shows the simulated points and a fitted second degree polynomial.

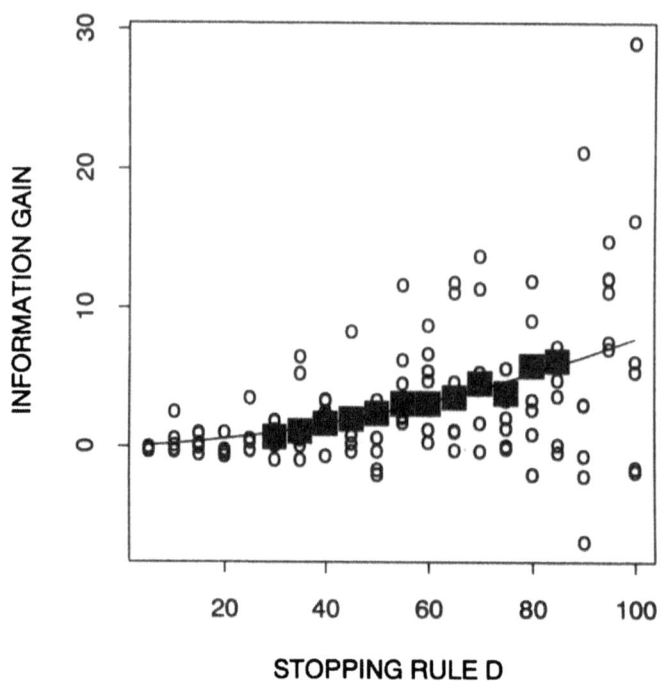

Figure 1: Bladder cancer study: Estimated expected KL divergence as a function of n. Black squares indicate the result of pointwise Monte Carlo integration. The design criterion does not include a sampling cost leading to a unique optimal design. Instead we can consider goal design – finding the minimum n to achieve a targeted expected information gain $\mathcal{U}(n)$.

3.2   Example 2: Timing of Medical Exams

We consider the problem of optimal screening schedules for breast cancer. Design parameters are the age α at which to begin regular screening and the frequency δ of screenings. The problem is the subject of an ongoing debate in the literature, see for example American Medical Association (1988). Zelen (1993), Parmigiani (1993) and Parmigiani and Kamlet (1993) define a four-state semi Markov process to describe the history of a chronic disease. The four states are "disease is absent or present but not detectable", "detectable pre-clinical", "clinical" and "death". We use the transition densities given in Parmigiani and Kamlet (1993), except for the transition from "absent" to "pre-clinical" which we simulate by direct generation from an empirical distribution.

To find the optimal policy $d^* = (\alpha^*, \delta^*)$ we chose $N = 240$ designs d_i on a grid over $(30, 65) \times (0.8, 3.5)$ and simulated patients (i.e. transition times θ and outcomes y) for each design. For each simulated patient we evaluated an observed utility $U(d, y, \theta)$

determined by a payoff of 1000 units for early breast cancer detection, and a cost of 1 unit per screening.

Figure 2 shows the smooth surface $\tilde{\mathcal{U}}(D)$ fitted to the simulated experiments. To reduce the enormous range of $U(d, y, \theta)$ we actually simulated $M = 1000$ patients for each design choice d_i and replaced u_i by $\bar{u}_i = 1/M \sum_j U(d_i, y_{ij}, \theta_{ij})$. The surface $\tilde{\mathcal{U}}(d)$ was fitted by a mixture of Dirichlet process based smoothing model. See Müller, Erkanli and West (1994) or West, Müller and Escobar (1994) for a description. Alternatively, local regression models could be used.

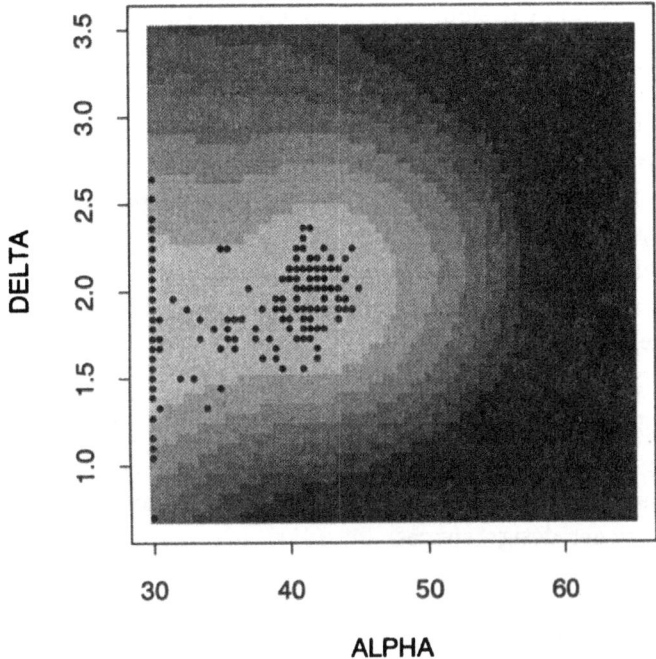

Figure 2: Breast cancer screening schedules: The grey shades show the estimated expected utilities $\tilde{\mathcal{U}}(D)$. The maximum occurs around $\alpha^* = 42$, $\delta^* = 1.96$. The surface falls off steeply for large δ, large α and small δ, corresponding to missed detections and unnecessarily frequent screenings. However the very soft decline towards smaller values of α indicates that the optimal choice of α could be very sensitive to the specific choice of utility function. The plotted points represent draws from the posterior distribution on the point of maximum $(\alpha^*, \delta^*) = d^*(\theta)$. Each point corresponds to an imputed parameter vector θ for the MDP model. Conditional on θ, the optimal design parameters are computed by a simple grid search. This requires no additional computation effort since the grid is anyway required to compute $\hat{\mathcal{U}} = E_\theta \mathcal{U}$.

3.3 Example 3: Optimal Design for Heart Defibrillators

Clyde, Müller and Parmigiani (1994) consider an application of Bayesian optimal design to the problem of designing testing strengths when setting the energy level for implantable heart defibrillators (IHD). A recent reference for more background discussion about this important application is Malkin et al. (1993).

From a decision theoretic viewpoint, the problem amounts to choosing test strengths $x_i, i = 1, \ldots, I$ for which we observe a binary response y_i with

$$Pr(y_i = 1|\theta, x_i) = (1 + exp\{-\beta(x_i - \lambda) - log(.95/.05))^{-1},$$

where x_i are the test strenghts at which the physician tries to stop artificially triggered heart arrhythmia during implantation, y_i are indicators for successful defibrillation, and $\theta = (\beta, \lambda)$ are unknown parameters, slope and LD95, describing the patient. The parameters vary from patient to patient. We are concerned with finding λ, the LD95, i.e. the level at which defibrillation will be successfull with probability 95%.

We consider eight point designs $x_1, \ldots, x_8$ on a fixed grid characterized by the center of the grid d_1 and the spacing d_2. As loss function we choose $U(d, y, \theta) = (L - \bar{L})^2$, where $L = \log(\lambda)$ is the log LD95, and $\bar{L} = E(L|y)$ is the posterior mean conditional on data y. We chose this quadratic estimation loss in log LD95 to reflect the asymmetry of the decision problem. Underestimation of λ is almost certainly fatal because of failed defibrillation. Overestimation is to be avoided because of possible collateral damage. The optimal policy d^* is found by maximizing $\mathcal{U}(d) = \int \int [-\{\log(\lambda) - E[\log(\lambda)|y, d]\}^2 - C(d, y)] \, dp_d(\theta|y) \, dp_d(y) = \int(-Var(\log \lambda|y, d) - c\, n(d, y))dp_d(y)$, where $p_d(y)$ is the marginal distribution on the sample space under design d, $n(d, y)$ is the realized number of defibrillations, and $c = 0.02$ is a trade-off parameter.

Figure 3 shows the estimated expected utility surface for different choices of testing strengths. The optimum is found at $d_1^* = 6.6$ and $d_2^* = 0.8$, i.e. for a grid centered at 6.6 with a spacing of 0.8. Müller and Parmigiani (1994b) discuss alternative schemes to evaluate the optimal design in this setup. Clyde, Müller and Parmigiani (1994) describe optimal design on richer design spaces.

4 Sequential Problems

4.1 Sequential Optimal Design by Smoothing of Monte Carlo Experiments

We propose an approach to sequential design based on an extension of the methodology we suggest for non-sequential stochastic optimization. The sequence of integrations and optimizations required for the solution of the sequential decision problem is replaced by a sequence of response surface estimations and optimizations of profiles of the estimated surfaces.

The method discussed in Section 2 lends itself nicely to application in the sequential context. For illustration, we consider a two-stage design problem. The methods apply to an arbitrary number of stages, with the obvious modifications in the notation. The sequence of observations/decisions is (1) Choose design d_1; (2) observe the experiment y_1 with probability model $p_{d_1}(y_1|\theta)$; (3) Choose design d_2; (4) observe the experiment y_2 with probability model $f_{d_2}(y_2|\theta)$; and (5) Choose terminal decision conditional on y_1, y_2, based on the payoff function $U(d_1, d_2, y_1, y_2, \theta)$. The optimal first stage design d_1^* is determined by maximizing over d_1 the expected utility

$$\mathcal{U}(d_1) = \int \max_{d_2} \left\{ \int U(d_1, d_2, y_1, y_2, \theta)dp_d(y_2|\theta, y_1) \right\} dp_d(y_1, \theta).$$

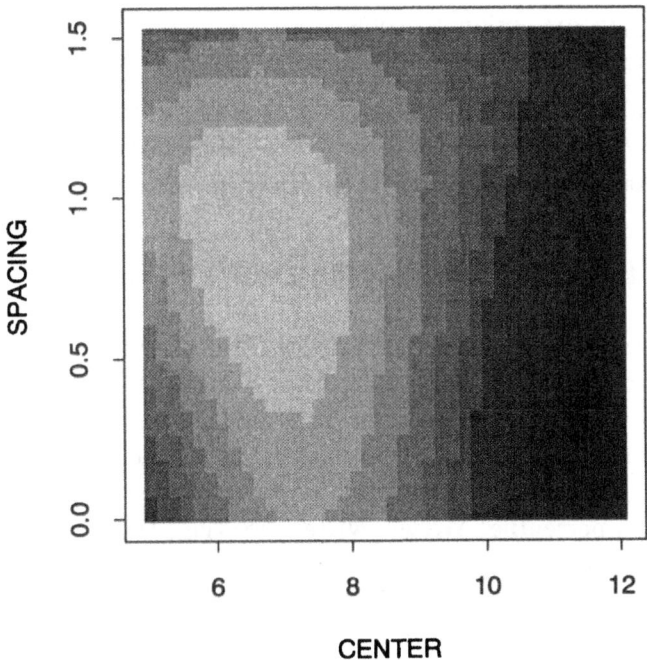

Figure 3: Implantable heart defibrillator design: Estimated expected utilities (i.e. negative pre-posterior expected variance on the log LD95). The relatively flat expected utility surface is typical for many design problems.

where $d = (d_1, d_2)$. Without loss of generality, y_1, y_2 are the sufficient statistics. The following is a curve fitting based implementation of the backward induction algorithm.

1. Select $d_i = (d_{i1}, d_{i2}) \in \mathcal{D}_1 \times \mathcal{D}_2$, $i = 1, \ldots, N$;
2. Draw N points θ_i, y_{i1}, y_{i2} from $p_{d_i}(\theta, y_1, y_2)$. Compute $u_{i2} = U(d_i, y_i, \theta_i)$ and record the Monte Carlo sample points $(d_{i1}, y_{i1}, d_{i2}, u_{i2})$.
3. Fit a curve $\tilde{\mathcal{U}}_2(d_1, d_2, y_1)$ to the points $(d_{i1}, y_{i1}, d_{i2}, u_{i2})$.
4. Deterministically find the maximum over d_2 of $\tilde{\mathcal{U}}_2(d_1, d_2, y_1)$. Replace the Monte Carlo points by the pairs (d_{i1}, u_{i1}), where $u_{i1} = max_{d_2} \tilde{\mathcal{U}}_2(d_{i1}, y_{i1}, d_2)$; The value of d_2 at which the maximum is reached is the solution to the second stage design problems and it depends on d_1 and y_1;
5. Fit a curve $\tilde{\mathcal{U}}(d_1)$ to the points (d_{i1}, u_{i1}).
6. Deterministically find the maximum over d_1 of $\tilde{\mathcal{U}}(d_1)$, yielding the optimal first stage design.

For sequential problems with $q > 2$ stages, Steps 3/4 would need to be repeated $q - 1$ times, each time replacing the subvector $(y_{i,s-1}, d_{i,s}, u_{i,s})$ in the Monte Carlo sample points by $u_{i,s-1}$. Incidentally, Steps 5/6 are tantamount to repeating Steps 3/4 for $s = 1$.

The curve fitting in Steps 3 and 5 corresponds to taking expectations with respect to y_2 and y_1 respectively. Depending on the dimensionality of the problem, different implementations of the curve fitting will be appropriate. The smoothing problem in Step 5 can be approached by nonparametric regression or any common smoothing algorithm.

For the possibly high dimensional problem in Step 3 we used the method proposed in Müller, Erkanli and West (1994), based on a mixture of Dirichlet process model.

This solution strategy can be combined successfully with any of the available methodologies for circumventing the so-called curse of dimensionality, such as rolling horizon techniques, or n-step look-ahead procedures (Berger, 1985).

4.2   Example 4: Optimal Sequential Design for a Logistic Growth Model

In this example we apply the ideas outlined in the previous section, to find optimal sequential designs for a nonlinear regression model, using squared error loss in a specific inference problem. In particular, we consider a logistic growth curve $f(X, \beta, \gamma) = 1/\{1 + exp(\beta - \gamma X)\}$. The model is completed by assuming a binomial likelihood $y_i \sim Bin(p_i, n)$ with $p_i = f(X_i, \beta, \gamma)$ and fixed $n = 20$. As design criterion we will use squared error loss in estimating the 80-th percentile. We will write L for the LD80. We optimize over the class of all two point sequential designs, i.e. all experiments of the following form: (1) Observe data y_1 for $X = x_1$, (2) select a second design point x_2 to observe $y_2 \sim Bin(f(x_2, \beta, \gamma), N)$. The choice of the second stage design point x_2 can depend on the outcome y_1 of the first stage.

The prior distribution on the parameters β and γ was chosen as $\alpha \sim Gamma(a, b)$ and $\beta \sim Gamma(c, d)$, with a, b, c and d fixed such that prior expectation and S.D. for β are 3.0 and 1.0, and prior expectation and S.D. for γ are 1.0 and 0.2 respectively.

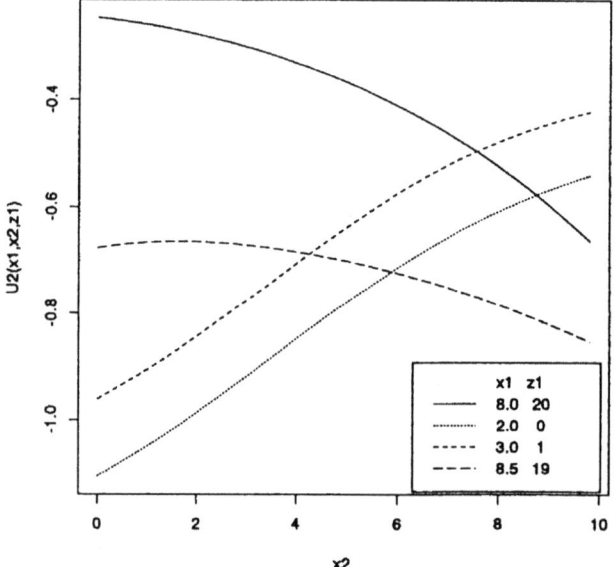

Figure 4: Logistic growth model: $\bar{\mathcal{U}}_2(x_1, y_1, x_2)$ for fixed values of (x_1, y_1). The specific values are given in the legend (y_1 is denoted as z_1). For each pair (x_1, y_1) the point of maximum on the shown curve gives the optimal second stage design point $x_2 = r_2(x_1, y_1)$.

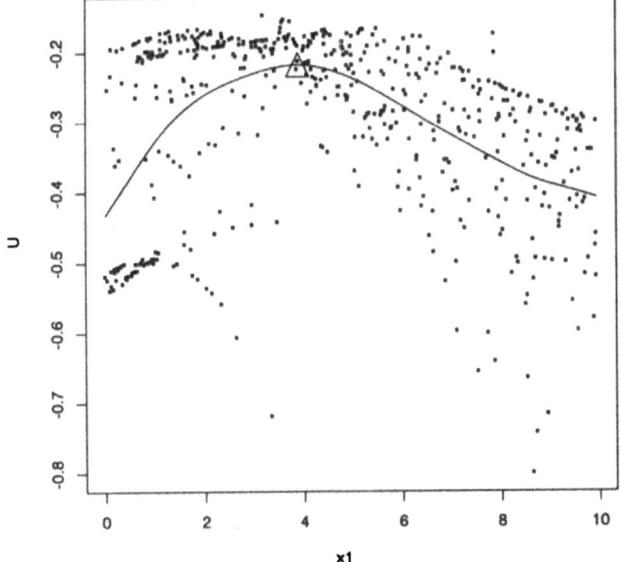

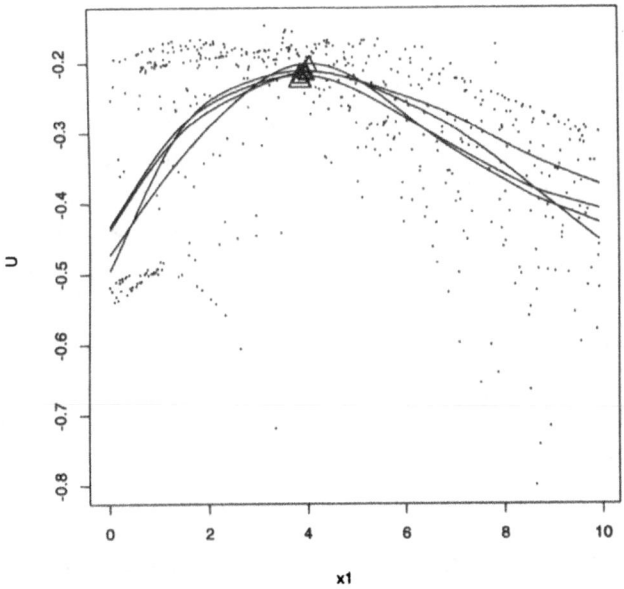

Figure 5: Logistic growth model: Panel (a) shows maximum expected utilities $\tilde{\mathcal{U}}_2(x_1, y_1, r_2(x_1, y_1))$ plotted against x_1. The line approximates the expectation $\mathcal{U}(x_1) = \int \tilde{\mathcal{U}}_2(x_1, y_1, r_2(x_1, y_1)) \, dP(y_1)$. The triangle marks the optimal first stage design at $x_1 = 3.88$. Panel (b) plots estimated curves $\tilde{\mathcal{U}}(x_1)$ in four runs of the simulation experiment. The points show the simulated points from the last run of the simulation experiment. The triangles indicate the maxima for each run. While there are slight differences in the fitted curves, the final point of maximum, and the corresponding expected utilities are very close.

Following the suggested procedure, for 500 randomly chosen pairs x_{i1}, x_{i2} we simulated quadruples $(\beta_i, \gamma_i, y_{i1}, y_{i2})$ from the joint distribution of parameter vector and data. Then we recorded the observed payoffs $u_{i2} = -(\bar{L} - L)^2$. Here $\bar{L}$ is the posterior mean of the 80-th percentile, conditional on the data (x_1, y_1, x_2, y_2), and L is the true 80-th quantile, computed from the simulated values for β and γ. The posterior means $\bar{L}$ were computed by Monte Carlo integration with importance sampling, employing the prior distribution as importance sampling density. This importance sampling step could be replaced by appropriate quadrature methods in more involved examples.

Based on the simulated Monte Carlo sample we fit a response surface $\tilde{\mathcal{U}}_2(x_1, y_1, x_2)$ to the quadruples $(x_{i1}, y_{i1}, x_{i2}, u_{i2})$. This is a smoothing problem in a four dimensional data set, making application of commonly used smoothing algorithms difficult. We used an implementation of the multivariate smoothing procedure based on mixture of Dirichlet process models (Müller, Erkanli and West 1994).

Fixing (x_1, y_1) in the fitted model shows expected utility as a function of x_2. Figure 4 shows such curves for different pairs of x_1 and y_1. One of the lines, for example, shows expected utility as a function of x_2 if in the first stage of the experiment a response of $y_1 = 19$ was observed at $x_1 = 8.5$. The curve indicates that the optimal second stage design in this experiment is around $x_2 = 3.5$. The set of all such maxima defines $r_2(x_1, y_1)$, the optimal second stage rule as a function of the first stage outcome.

Having estimated the surface $\tilde{\mathcal{U}}_2$, the original sequential problem can be reduced to a format like the one stage design problems. Figure 5 plots maximum expected utilities $u_{1i} = \tilde{\mathcal{U}}_2(x_{1i}, y_{1i}, r_2(x_{1i}, y_{1i}))$ against x_{1i}. The still visible randomness is due to y_1. The optimal design is now found by integrating over y_1 and locating the maximum of the remaining function $\mathcal{U}(x_1)$. Using the strategy outlined in Section 2. we replaced the integration by another curve fitting problem. The curve $\tilde{\mathcal{U}}(x_1)$ shown in the plot was obtained by fitting a cubic smoothing spline to the bivariate data set (x_{1i}, u_{1i}). From the curve $\tilde{\mathcal{U}}$, the estimated optimal first stage design can be read off as $x_1^* = 3.88$.

Acknowledgement

The authors were partially supported by NSF under grants DMS-9305699, DMS-9403818 and DMS-9404151.

References

American Medical Association; Council on Scientific Affairs, (1989). Mammographic Screening in asymptomatic women aged 40 years and older. *Journal of the American Medical Association* 261, pp. 2535–2542.

Berger, J.O. (1985). *Statistical Decision Theory and Bayesian Analysis.* Springer Verlag, New York.

Berry, D.A. and Fristedt, B. (1985). *Bandit Problems: Sequential Allocation of Experiments.* Chapman and Hall, London.

Byar, D.P., Blackard, C., and VACURG (1977). Comparison of Placebo, Pyridoxine and Topical Thiotepa in Preventing Recurrence of Stage I Bladder Cancer. *Urology* 10, pp. 556–561.

Carlin, B.P and Polson, N.G. (1992). An expected utility approach to influence diagnostics. *Journal of the American Statistical Association* 87, pp. 1013–1021.

Chaloner, K. and Larntz, K. (1989). Optimal Bayesian design applied to logistic regression experiments. *Journal of Statitical Planing and Inference* 21, pp. 191-208.

Chaloner, K. (1992). A note on optimal Bayesian design for nonlinear problems. Technical Report 572, School of Statistics, University of Minnesota.

Chaloner, K. and Verdinelli, I. (1993). Bayesian experimental design: a review. Technical Report, Department of Statistics, Carnegie Mellon University.

Chernoff, H. (1979). *Sequential analysis and optimal design.* SIAM, Philadelphia .

Clyde, M. (1993). An object-oriented system for Bayesian nonlinear design using XLISP-STAT. Technical Report 587, School of Statistics, University of Minnesota.

Clyde, M., Müller, P. and Parmigiani, G.: (1994) "Optimal Design for Heart Defibrillators" in *Case Studies in Bayesian Statistics, II* (C. Gatsonis, J. Hodges, R. E. Kass, N. Singpurwalla eds.), Springer, NY. forthcoming.

DeGroot, M.H. (1970). *Optimal Statistical Decisions.* McGraw Hill, New York.

DeGroot, M.H. (1984). Changes in utility as information. *Theory and Decision* 17, pp. 287-303.

Flournoy, N. (1992). A clinical experiment in bone marrow transplantation: Estimating a percentage point of a quantal response function. In *Bayesian Statistics in Science and Technology: Case Studies,* ed. C. Gatsonis, J. Hodges, R. E. Kass, N. Singpurwalla eds., pp. 324-337.

Malkin, R.A., Pilkington, T. C., Burdick, D. S., Swanson, D. K., Johnson, E. E., and Ideker, R. E. (1993). Estimating the 95 effective defibrillation dose. *IEEE Trans on EMBS* 40(3), pp. 256-265.

Mueller, P., Erkanli, A., and West, M. (1994). Bayesian curve fitting using multivariate normal mixtures. ISDS Discussion Paper, Duke University.

Mueller, P. and Parmigiani, G. (1994a). Numerical evaluation of information theoretic measures. In *Bayesian Statistics and Econometrics: Essays in Honor of A. Zellner,* ed. Berry D.A., Chaloner K.M., Geweke J.F.

Mueller, P. and Parmigiani, G. (1994b). Optimal design via curve fitting of Monte Carlo experiments. Discussion paper, ISDS, Duke University.

Mukhopadhyay, S. and Haines, L. (1994). Bayesian D-optimal design for exponential growth model. *Journal of Statistical Planing and Inference,* to appear.

Parmigiani, G. and Polson N. G. (1992). Bayesian design for random walk barriers. In *Bayesian Statistics IV,* ed. J. M. Bernardo, J. O. Berger, A. P. Dawid and A. F. M. Smith eds., pp. 715-72.

Parmigiani, G. and Kamlet, M. S.: (1993) "Cost-Utility Analysis of Alternative Strategies in Screening for Breast Cancer" in *Case Studies in Bayesian Statistics* (C. Gatsonis, J. Hodges, R. E. Kass, N. Singpurwalla eds.), Springer, NY, 390-402.

Parmigiani, G.: (1993) "On Optimal Screening Ages", *Journal of the American Statistical Association,* **88** 622-628.

Tierney, L. (1990). *LISP-STAT: An object-oriented environment for statistical computing and dynamic graphics.* J. Wiley, New York.

Tsutakawa, R. (1980). Selection of dose levels for estimating a percentage point on a logistic quantal response curve. *Applied Statistics* 29, pp. 25-33.

Verdinelli, I. and Kadane, J. (1992). Bayesian designs for maximizing information and outcome. *Journal of the American Statistical Association* 86, pp. 510-515.

Verdinelli, I. (1992). Advances in Bayesina experimental design. In *Bayesian Statistics 4,* ed. J. M. Bernardo, J. O. Berger, A. P. Dawid and A. F. M. Smith eds., pp. 467-48.

West, M., Mueller, P., and Escobar, M.D. (1994). Hierarchical priors and mixture models, with application in regression and density estimation. In *Aspects of Uncertainty: A tribute to D. V. Lindley,* ed. A.F.M. Smith and P. Freeman.

One Bound for the Mean Duration of Sequential Testing Homogeneity

LEONID I. GALTCHOUK and MICHAIL B. MALJUTOV

UFR de Mathematique et d'Informatique, Louis Pasteur Universite, 7, Rue Rene Descartes, 67084 Strasbourg Cedex, France
and
Lomonosov Moscow State University, Faculty of Mechanics and Mathematics, 119899, Moscow, Russia.

Abstract:

A lower bound is proved for the mean duration of any sequential strategy for testing homogeneity under the alternative formulated in terms of maximal distance in variation between $m \geq 2$ populations.

1 Introduction and Main Results

Let independent observations may be taken from $m \geq 2$ populations with distributions $P_{\theta_1}, \ldots, P_{\theta_m}, \theta_i \in \Theta, i = 1, \ldots, m$ defined on a measure space $(Y, \mathcal{Y})$, where Θ is a parametric set which we assume to be a metric space. All the distributions are supposed to be mutually absolutely continuous and possessing densities $p_{\theta_i}(\cdot)$ with respect to a σ-finite measure μ. We consider two natural types of distances between densities:

D 1. $d_v(P_\theta, P_\phi) = \int |p_\theta - p_\phi| d\mu$.

D 2. *Assume that the measures P_θ on $(\mathbb{R}^d, \mathcal{B})$ under consideration differ only by their location parameters $\theta \in \mathbb{E}^d$ (where $\mathbb{E}^d$ is the d-dimensional euclidean space with the arithmetic norm $|\cdot|$) and*

$$d_l(P_\theta, P_\phi) = |\theta - \phi|,$$

---

*Kitsos, C.P., and Müller, W.G., Eds., *Proceedings of MODA4*, Physica Verlag, Heidelberg, 1995

For both distances we consider sequential testing

$$H_0 = \{P_{\theta_1} = \ldots = P_{\theta_m}\}$$

against

$$H_1 = \{\max d(P_{\theta_i}, P_{\theta_j}) \geq \Delta\}$$

in case D 1 and

$$H_1 = \{\sum_{i=1}^{m} d(P_{\theta_i}, \bar{P})^2 \geq \Delta^2\}, m\bar{P} := \sum_{i=1}^{m} P_{\theta_i},$$

in case D 2 for some specified $\Delta > 0$.(This condition is similar to the positiveness of the non-centrality parameter in testing constancy of effects in ANOVA).

The corresponding m-tuples of parameters $t \in \Theta^m := T$ constitute the sets T_0 and T_1. Other m-tuples of parameters are treated as an indifference zone I (i.e. no decision is considered as false under $t \in I$).

Our observations are controlled, i. e. we can stop measurements at any Markov stopping time N and for any $n < N$ choose the observation from $u(n)$-th population, where $u(n) : Y^{n-1} \to [m] := \{1, \ldots, m\}$ are $\mathcal{Y}^{n-1}$-measurable. The stopping time N and a sequence of functions $u(n), n = 1, \ldots, N-1$ determine a *sequential experimental design* (SED) π. A *strategy* s consists of a design and a measurable decision $\delta : Y^N \to \{0, 1\}, \delta = r, r = 0, 1$ means that H_r is accepted. Formalization, generalization and the fundamental properties of sequential strategies are outlined below in Section 2.

We denote $A(t) = T_{1-r}$ if $t \in T_r$ and consider class $G(\alpha)$ of strategies s satisfying

Condition $G(\alpha)$: $\max_{r=0,1} \max_{t \in T_r} P_t^s(\delta = 1 - r) \leq \alpha$.

The expression under the maximum sign in condition $G(\alpha)$ which will be formally introduced in Section 2, is called the *error probability* (EP) of strategy s under parameter t.

In Galtchouk and Maljutov (1995) we described a lower bound for the *mean duration* (MD) $E_t^s N$ of strategies from $G(\alpha)$ in framework D 2 which was nonasymptotic for $t \in T_0 \cup T_1$ and asymptotic (when $\alpha \to 0$) for $t \in I$. We evaluated this bound for gaussian distributions P_θ and found it to be asymptotically equivalent to the respective mean duration of the strategy constructed there for any $t \in T$. Although the generalization of calculations for an arbitrary family of equivalent distributions of populations differing by their location parameters seems complicated, the numerical evaluation of the corresponding bounds is feasible. It seems also possible to generalize the proof of asymptotic optimality of that bound. We mention only one of numerous possible applications of that theory when studying the minimal mean duration of separation of significant variables of nonparametric additive model studied in Maljutov and Wynn (1994), which was the principal part of the search algorithm there.

This paper starts the similar program for the more difficult case of distance D 1. We evaluate the maximum of the general lower bound for MD of strategies from G_α over $T_i, i = 0, 1$, and discuss the problems of construction of asymptotically optimal strategies.

Early papers on various sequential testing problems are reviewed in Bechhofer et al (1968), Chernoff (1972) and Shiryaev (1978). A lower bound for the mean duration of a sequential test for homogeneity in our first setting D 1 , $m = 2$, was evaluated in

Volodin (1979b) as a particular case of his general lower bound (Volodin (1979a)). Note that the optimal control is trivial in case $m = 2$ consisting of the uniform allocation of observations between populations. The uniform improvement of the Volodin's general bound is described in Maljutov (1983a, 1983b). We outline these general bounds in Section 2 and evaluate the minimal value of one of them for distance D 1 in Section 3 .

Formalization of Sequential Testing Homogeneity

In this section we formalize and generalize the model outlined above to include e. g. randomized controls and introduce the notions used further. Our framework follows (Maljutov, 1983a, 1983b).

Let $(\mathbf{Y}, \mathcal{Y})$ be a measurable space, $(\Omega, \mathcal{F})$ be its Descartes product with $([0, 1], \mathcal{B})$, (reserved for randomizations of a control at each step) where $\mathcal{B}$ is its Borel σ-algebra, $(\Omega^\infty, \mathcal{F}^\infty)$ be the infinite product of the identical copies $(\Omega_n = \mathbf{Y_n} \times [0, 1], \mathcal{F_n} = \mathcal{Y_n} \times \mathcal{B_n})$ of the above space, $Y_1(\omega), Y_2(\omega), ...$, where $Y_n(\omega)$ be $\mathcal{Y}_n$-measurable function with values in $\mathbb{E}^d$ for any $n \geq 1$ and $\omega = (\omega_1, \omega_2, ...) \in \Omega^\infty$.

Let $\mathbb{F} = (\mathcal{F}^k = \prod_1^k \mathcal{F}_n, k = 1, ...)$ be the filtration of the σ -algebras and we put $\mathcal{F}_0 = \{\Omega, \emptyset\}$.

A *sequential design* (SED) is a pair $\pi = (\mathcal{U}, N)$ consisting of a predictable process $\mathcal{U} = (u^*(n), n = 1, ...)$ of randomized controls with values in $[m]^*$, where $[m]^*$ is the set of probability distributions over $[m] = \{1, ..., m\}$, such that $u^*(n)$ is F^{n-1}-measurable and a stopping time N with respect to the filtration $\mathcal{F}^n$. We denote by $u(n)$ the random realization of the control in n-th experiment from the distribution $u(n)^*$ by the following rule: if $[0, 1] = \cup_1^m A_i, \lambda(A_i) = u_i^*(n)$, then $u(n) = r$ if the uniformly distributed random point belongs to A_r.

By Ionescu-Tulcea lifting Theorem (Neveu, 1971, Theorem 1.V.1) SED π and values $u(1) \in [m], t \in \Theta^m$, uniquely determine the measure P_t^π on $(\Omega, \mathbb{F})$ such that for any $B \in \mathcal{Y}, n = 1, ...$

$$P_t^\pi(y_n \in B | \mathcal{F}^{n-1} \times \mathcal{B}_n) = P_{\theta_{u(n)}}(B), n \leq N.$$

Therefore, SED π generates uniquely the statistical experiment $\mathcal{E} = (\Omega^\infty, \mathbb{F}^N, (P_t^\pi, t \in T))$. Below we shall consider the testing problem for such experiments. A strategy s consists of a design π and a binary-valued decision δ which is $\mathcal{F}^N$-measurable. The definition of the measure

$$P_t^s(A, \{r\}) = P_t^\pi(A, \delta = r), A \in \mathcal{F}^N, r = 0, 1,$$

corresponding to the strategy s is now straightforward. Since the measure corresponding to a SED is marginal with respect to the distribution of the strategy we may substitute the latter in all expectations without further notice.

Now we introduce the static projection (SP) $\pi_t \in [m]^*$ of a SED π depending on $t \in \Theta^m$:

$$\pi_t(r) = \sum_{i=1}^N P_t^\pi(u(i) = r, i \leq N)/E_t^\pi N, r \in [m].$$

52

SP shows how often the SED uses observations from each population. Its importance for the theory of SED follows from the following direct generalisation of the Wald identity

Proposition 1 (Malyutov, 1983a, 1983b).*For any SED $\pi = (\mathcal{U}, N)$ with $\sup E_t^\pi N < \infty$ and for any integrable function $g^u(y)$ it holds:*

$$E_t^\pi \sum_{i=1}^N g^{u(i)}(y_i) = E_t^\pi N \int \bar{g}_t^u \pi_t(du), \tag{1}$$

where $\bar{g}_t^u = \int g^u(y) P_{\theta_u}(dy)$.

Proposition shows that as far as expectations of additive functions are of concern, SED is equivalent to the non-homogeineous independent statistical experiment described by its SP. This fact enables generalization to SED of many well-known results derived for independent observations (Maljutov, 1983a, 1983b).

Definition. We call a SED π nondegenerate if $P_t^\pi(\sum_{n=1}^N 1\ (u(n) = r, n \le N) > 0) = 1$ for all $r \in [m], t \in \Theta^m$.

The following result describes derivatives of measures P_t^π with respect to each other for a given nondegenerate design π, being a direct consequence of Proposition 1 in (Malyutov, 1983a, 1983b).

Proposition 2.*If measures $P_\theta, \theta \in \Theta$ are mutually equivalent and design π is nondegenerate, $P_t^\pi(N < \infty) = 1$ for all t, then measures P_t^π and $P_{t'}^\pi$ on $\mathcal{F}^N$ are also equivalent and their density is*

$$dP_t^\pi / dP_{t'}^\pi = \prod_{n=1}^N (p_{\theta_{u(n)}} / p_{\theta'_{u(n)}}),$$

where $t' = (\theta'_1, \ldots, \theta'_m)$.

It follows from proposition 2 that the loglikelihood is an additive function.

Its expectation with respect to P_t^π is called the Kullback information $K^\pi(t, t')$ of SED π (known also under other titles such as the Kullback divergence, the Kullback-Leibler information, relative entropy etc.) It follows from (1) that omitting all nonsignificant indices we have:

$$K^\pi = E^\pi N \int k^u(\theta, \theta') \pi_t(du),$$

where

$$k^u(\theta, \theta') \int [\log p_{\theta_u} / p_{\theta'_u}](y) p_{\theta_u}(y) \mu(dy)$$

is the Kullback information related to a particular observation from the u-th population. Its integral

$$k^{\pi_t} = \int k^u \pi_t(du)$$

with respect to SP π_t describing the average Kullback information in experiments of SED π plays the major role further.

As an easy consequence of the above expression for K^π and of Jensen's inequality (guaranteeing that K^π is not less than the Kullback information for the final decision's distributions $K_\delta(\alpha)$, where

$$K_\delta(\alpha) = \alpha \log \alpha / (1 - \alpha) + (1 - \alpha) \log (1 - \alpha) / \alpha))$$

the following lower bound for MD of any SED π from $G(\alpha)$ with a uniformly bounded mean duration was found in (Malyutov, 1983a, 1983b) as a particular case of a much more general inequality:

Proposition 3. *For any $t \in T_0 \cup T_1, \alpha < 1/2$, any SED π from $G(\alpha)$ we have*

$$E_t^\pi N \geq K_\delta(\alpha)/k(t), \tag{2}$$

where

$$k(t) = \max_{\rho \in [m]^\bullet} \min_{t' \in A(t)} k^\rho(t, t').$$

A similar reasoning yields a lower bound for the *maximal over T_i mean duration m_i under the same conditions as in Proposition 3*

$$m_i \geq K_\delta(\alpha)/\tilde{k}_i,$$

where

$$\tilde{k}_i = \max_{\rho \in [m]^\bullet} \min_{t \in T_i} \min_{t' \in A(t)} k^\rho(t, t').$$

2 Maximum of the Lower Bound (2) for the Case D 1.

Let $\mathcal{P}$ be the class of mutually absolutely continuous probability distributions with densities $p(\cdot)$ (which we call admissible below) with respect to a sigma-finite nonatomic measure μ on $(\mathbf{Y}, \mathcal{Y})$. For a m-tuple $t = (P_1, \ldots, P_m)$ of elements of $\mathcal{P}$ let us denote $< t >= \max_{1 \leq i,j \leq m} d_v(P_i, P_j)$, $d_v(\cdot, \cdot)$ is introduced in the definition of D 1. As was indicated above, the sets of m-tuples T_0, T_1, I correspond to the following conditions:

$$< t >= 0, < t >\geq \Delta, 0 << t >< \Delta.$$

Assume that T_1 is not void.

Denote $k_r = \min_{t \in T_r} k_r(t)$,

$$\Delta_1 = -2^{-1} \sum_{k=-1}^{1} \log(1 + k\Delta) > 0, \Delta_2 = 2^{-1} \sum_{k=-1}^{1} (1 + k\Delta)\log(1 + k\Delta) > 0.$$

Theorem. $k_0 = 2\Delta_1/m, k_1 = \Delta_2.$

Remark.1. The natural extension of the lower bound in Volodin (1979b) for $m > 2$ gives the same expression of k_1 and $2/m$ times less lower bound for MD under H_0.

2.In proving this theorem we use unpublished calculations of G. N. Semenov who has been a student of M. Maljutov. The tragic death of G. N. Semenov was a great loss for science.

Proof of the theorem is based on the following lemmas

Lemma 1. $t \in T_1 \Leftrightarrow \min_{P \in \mathcal{P}} \max_{i,j \in [m]} \max(d_v(P_i, P), d_v(P_j, P)) \geq \Delta/2.$

Proof is straightforward.

Choose an arbitrary $P_0 \in \mathcal{P}$ and a set A such that $P_0(A) = 1/2$. Let the extremal populations in Lemma 1 be the first and second ones for definiteness. In this section instead of the previous notation $k(\theta, \theta')$ we use $k(P, P')$ because the measure itself serves as a parameter of the distribution.

Lemma 2. *Let $P_i \in \mathcal{P}, i = 0, 1, 2, d_v(P_1, P_2) \geq \Delta$. Then*

$$a) k(P_0, P_1) + k(P_0, P_2) \geq 2\Delta_1.$$

$$b) \max(k(P_0, P_1), k(P_0, P_2)) \geq \Delta_1.$$

$$c) k(P_1, P_0) + k(P_2, P_0) \geq 2\Delta_2.$$

$$d) \max(k(P_1, P_0), k(P_2, P_0)) \geq \Delta_2.$$

Proof. It is clear that a) implies b), c) implies d). Hence we prove only a) and c). Let p_i be density of P_i with respect to $P_0, i = 1, 2$. We have

$$k(P_0, P_i) = - \int \log p_i dP_0, k(P_i, P_0) = \int p_i \log p_i dP_0, i = 1, 2.$$

Let $p^+ = \max(p_1, p_2), p^- = \min(p_1, p_2)$. Then it is straightforward that

$$\int (p^+ - p^-) dP_0 \geq 2\Delta, \int (p^+ + p^-) dP_0 = 2. \tag{3}$$

Let $p_k^* = 1 + (-1)^{k+1} \Delta 1_A + (-1)^k \Delta 1_{Y \backslash A}, k = 1, 2$, where 1_A is the indicator of the set A. It is clear that

$$p^{*+} \equiv 1 + \Delta, p^{*-} \equiv 1 - \Delta.$$

Hence $\int (p^{*+} - p^{*-}) dP_0 = 2\Delta$. Let $t^* = (P_1^*, P_2^*, P_0, \ldots, P_0), t_0 = (P_0, \ldots, P_0)$. It is easy to verify that

$$k(t_0, t^*) = - \int (\log p_1^* + \log p_2^*) dP_0 = 2\Delta_1,$$

$$k(t^*, t_0) = \int (p_1^* \log p_1^* + p_2^* \log p_2^*) dP_0 = 2\Delta_2.$$

To prove a) and c) it remains to establish that the above m-tuples are the extremal ones. For this aim first notice that for any admissible p_1, p_2 satisfying conditions of Lemma 2 we have from (3):

$$\int (p^+ - (1 + \Delta)) dP_0 = - \int (p^- - (1 - \Delta)) dP_0 \geq 0. \tag{4}$$

Lemma 3. *For any admissible densities $p_1, p_2, d_v(P_1, P_2) \geq \Delta$, it holds*

$$\int [\log p_1 + \log p_2 - (\log p_1^* + \log p_2^*)] dP_0 \leq 0.$$

Proof. We have from (4)

$$(1 + \Delta)^{-1} \int (p^+ - (1 + \Delta)) dP_0 \leq (1 - \Delta)^{-1} \int ((1 - \Delta) - p^-) dP_0,$$

which implies

$$\int (p^+ / p^{*+} - 1 + p^- / p^{*-} - 1) dP_0 \leq 0.$$

Using the inequality $\log x \leq x + 1$ we get from here

$$\int (\log(p^+/p^{*+}) + \log(p^-/p^{*-})) \leq 0$$

which is equivalent to the statement of the Lemma.

Lemma 3.*For any admissible densities $p_1, p_2, d_v(P_1, P_2) \geq \Delta$, it holds*

$$\int [p_1 \log p_1 + p_2 \log p_2 - (p_1^* \log p_1^* + p_2^* \log p_2^*)]dP_0 \geq 0.$$

Proof. We have from (4)

$$\log(1 + \Delta) \int [p^+ - (1 + \Delta)]dP_0 \geq \log(1 - \Delta) \int [(1 - \Delta) - p^-]dP_0,$$

which implies

$$\int [(p^+ - p^{*+}) \log p^{*+} + (p^- - p^{*-}) \log p^{*-}]dP_0 \geq 0. \tag{5}$$

By the same inequality for logarithms as above we have

$$\int [p^+ \log(p^{*+}/p^+) + p^- \log(p^{*-}/p^-)]dP_0 \leq \int [p^+(p*+/p^+ - 1) + p^-(p^{*-}/p^- - 1)]dP_0 = 0.$$

Subtracting the last inequality from (5) we get the statement of the Lemma after a straightforward transformation.

The proof of the Theorem is now completed by the indication of the optimal mixed strategy of the statistician. The argument similar to that of Galtchouk and Maljutov (1995) for the case of univariate gaussian location parameters shows that the optimal ρ in (2) assigns equal weights to the extremal populations when the distributions differ in variation not less than by Δ and it is uniform under any null hypothesis t_0. The saddle strategy of Nature in $A(t_0)$ is the uniform mixture of all permutations of t^*. The appearence of mixed saddle strategy in $A(t_0)$ is explained by its non-convexity: $A(t_0)$ is the complement to the convex set. Bearing in mind these maximin strategies the statement of the theorem follows from the straightforward counting.

The first type of asymptotically optimal strategies for testing two composite hypotheses in case of finite parameter and finite control sets was proposed by H.Chernoff in his pioneer paper of 1959. It is based on studying the vector of likelihoods of the current ML-estimate of parameters $\hat{t}_n$ against all alternatives from $A(t)$. It is reproduced in (Chernoff, 1972) with a discussion of many generalizations and extensions. Probably, the most important of them at that moment were those of (Kiefer and Sacks, 1963). Assuming compactnes of parametric sets T_0, T_1 they proved particularly asymptotic optimality of the following two-step strategy. The first step of their procedure should end with a consistent estimate of all parameters of the distribution . The next stage uses the design which would be optimal if this consistent estimate of parameters were its true value. In our case it is possible to construct consistent estimates of all underlying distributions and of their variation distances and use the design corresponding to this preliminary estimate which was indicated above. It remains to prove its asymptotic optimality which is not implied by the existing theory because even the local compactness of the parametric sets does not hold.

56

Acknowledgement

This paper was written while the second author was a guest of UFR de Mathematique et d'Informatique, Universite Louis Pasteur de Strasbourg. He is grateful to the authorities of UFR and ISF, grant MCF 000, for their support.

References

Bechhofer, R. E., Kiefer, J. and Sobel, M. (1968). Sequential Identification and Ranking Procedures. Univ. Chicago Press, Chicago.

Chernoff, H. (1972). Sequential Analysis and Optimal Design, SIAM, Philadelphia.

Galtchouk, L.I. and Maljutov, M.B. (1995). Asymptotically Optimal Testing Coincidence of Location Parameters,in: Statistics and Control of Stochastic Processes (Festschrift for A. N. Shiryaev), TVP, Moscow.

Kiefer, J. and Sacks, J. (1963). Asymptotically Optimal Sequential Inference and Design,Ann. Math. Statist., 34, 705-750.

Malyutov, M, B. (1983a). Lower Bounds for an Average Number of Sequentially Designed Experiments. Springer Lect. Notes Math. 1021, 419-435

Malyutov, M. B. (1983b). Lower Bounds for Mean Duration of a Sequentially Designed Experiment, Soviet Math. (Iz. VUZ) , 27, No 11, 21-46 of English edition by Allerton Press.

Maljutov, M. B. and Wynn, H. P. (1994). Sequential Screening of Significant Variables of an Additive Model. In: Festschrift for E.B.Dynkin, Birkhäuser, Boston.

Neveu, J.(1971). Bases Mathematiques du Calcul des Probabilites, 2-eme edition, Masson, Paris.

Shiryaev, A. N. (1978). Optimal Stopping Rules. Springer-Verlag, New York.

Volodin, I. N. (1979a). Lower Bounds for Average Sample Size and Efficiency of Statistical Inference Procedures, Theory Probab. Appl. 24, 120-129

Volodin, I. N. (1979b). Lower Bounds for Mean Sample Size in Goodness-of-Fit and Homogeneity Tests, Theory Probab. Appl., 24, 640-648.

MV-optimization in Simple Linear Regression

BEN TORSNEY and JESUS LÓPEZ-FIDALGO

Department of Statistics, University of Glasgow, G12 8QW, U.K.
and
Department of Pure and Applied Mathematics, University of Salamanca, 37008 Spain.

Abstract:

MV-optimality is a potentially difficult criterion because of its nondifferentiability at equal variance designs. However in many cases such designs can be easily determined. In this paper MV-optimum designs for simple linear regression are found. The equivalence theorem of and the directional derivative of the MV-criterion derived by Ford I., (3), have been used for this purpose. It turns out that for simple linear regression there exist an MV-optimal design with a support of at most two points. Such designs could be of a wide ranging practical value.

1 Introduction

MV-optimality has rarely been treated in the literature. There are a few papers which consider it. Elfving G., (2), uses minimax designs, that is, E-optimal designs, aiming to minimize the maximum of the variances of any unitary linear combination of the parameters. Here unitary is understood with respect to the standard norm in euclidean space. Dette H. and Studden W.J., (1), generalize the Minimax criterion to any norm over euclidean space,

$$\Phi_{|\cdot|}[M(\xi)] = \max\{c^t M^{-1}(\xi)c : c \in R^m, |c| = 1\}$$

In particular, for the ℓ_1-norm,

$$|c|_1 = \sum_{i=1}^{m} |c_i|$$

*Kitsos, C.P., and Müller, W.G., Eds., *Proceedings of MODA4*, Physica Verlag, Heidelberg, 1995

58

we have the MV-optimality criterion,

$$\Phi_{|\cdot|_1}[M(\xi)] = \max_i e_i^t M^{-1}(\xi) e_i = \max_i Var_\xi(\alpha_i)$$

where ξ is a design over a compact set X and $\alpha_i, i = 1, 2, \ldots, m$, are the parameters of the model. We will denote it by Φ_{MV}.

Ford I., (3), proves some general results for this criterion. In fact, he establishes the following properties of the criterion function,

1. Φ_{MV} is a decreasing function over the positive definite symmetric matrices in the sense that,

$$\Phi_{MV}(M_1 + M_2) \leq \Phi_{MV}(M_1)$$

being M_1 a positive definite symetric matrix and M_2 a positive semi-definite symetric matrix.

2. Φ_{MV} is a convex function over the positive definite symetric matrices.

3. Φ_{MV} can be not differentiable at points in $\mathcal{M}$ at which the function may be minimized.

4. The directional derivative at M in the direction of N is given by,

$$\partial \Phi_{MV}(M, N) = \{M^{-1} - M^{-1} N M^{-1}\}_{ss}$$

where $\{M^{-1}\}_{ss}$ is the biggest diagonal element of M^{-1}. If there are r coincident biggest diagonal elements $ss_1, \ldots, ss_r$, say, then we choose s such that,

$$\{M^{-1} N M^{-1}\}_{ss} = \min_{ss_i}\{M^{-1} N M^{-1}\}_{ss_i}$$

We will say that a matrix N has the ss_j property if,

$$\partial \Phi_{MV}(M, N) = \{M^{-1} - M^{-1} N M^{-1}\}_{ss_j}$$

Ford I., (3), denotes by $\mathcal{M}'$ the convex sub-set of $\frac{1}{2}m(m + 1)$ dimensional space whose points have as components the elements of the upper triangular parts of the matrices of which $\mathcal{M}$ is composed.

5. Let N_1, $N_2 \in \mathcal{M}$ have the ss_j property, then $\lambda N_1 + (1 - \lambda) N_2$ has the ss_j property for $\lambda \in [0, 1]$. This means that the members of $\mathcal{M}$ which have the ss_j property, for a given M, form a convex sub-set of $\mathcal{M}$. We shall denote this sub-set by $\mathcal{M}_{ss_j}$, or similarly $\mathcal{M}'_{ss_j}$ as a sub-set of $\mathcal{M}'$.

6. Let $N \in \mathcal{M}'$ have the ss_j property, then σN has the ss_j property for $\sigma > 0$. This means that the sets $\mathcal{M}'_{ss_j}$ are formed by intersections of convex cones with $\mathcal{M}'$. And because M has all the ss_j properties, $j = 1, \ldots, m$, then M lies in the intersection,

$$\mathcal{M}'_{ss_1} \cap \mathcal{M}'_{ss_2} \cap \ldots \cap \mathcal{M}'_{ss_m}$$

7. $\partial \Phi_{MV}(M, \lambda M_1 + (1-\lambda) M_2) = \lambda \partial \Phi_{MV}(M, M_1) + (1-\lambda) \partial \Phi_{MV}(M, M_2), \lambda \in [0, 1], M_1, M_2 \in \mathcal{M}_{ss_j}, j = 1, \ldots, r$. In particular, if $M_2 = M$ then $\partial \Phi_{MV}(M, \lambda M_1 + (1 - \lambda) M) = \lambda \partial \Phi_{MV}(M, M_1), \lambda \in [0, 1]$, that is, the sign of $\partial \Phi_{MV}(M, \lambda M_1 + (1 - \lambda) M)$ and the sign of $\lambda \partial \Phi_{MV}(M, M_1)$ are the same.

8. In general $\Phi(aM) < \Phi(M), M \in \mathcal{M}, a > 1$.

This tells us that if we think of $\mathcal{M}$ as being surrounded by the minimal convex cone which contains it, then the only feasible points in $\mathcal{M}'$, corresponding to possible optimal M lie on the upper surface of $\mathcal{M}'$ not hidden by the cone.

Moreover, the equivalence theorem says, among other things, that the following statements are equivalent,

(a) $\xi^\star$ is Φ-optimum,

(b) $\partial\Phi[M(\xi^\star), M(\xi)] \geq 0, \xi \in \Xi$.

Taking these properties into account and also the equivalence theorem we see that to check optimality of a matrix M, where Φ_{MV} is not differentiable at M, we do not need to calculate $\partial\Phi_{MV}[M,N]\forall N \in \mathcal{M}$, but only for the N which are the generators of the convex sets $\mathcal{M}'_{ss_j}, j = 1, \ldots, m$, and which are feasible optimal solutions.

Takeuchi K., (11), used this criterion for balanced weighing designs. Later Jacroux M. ((5), (6), (7), (8), e.g.) called it MV-optimality. Hotelling, (4), also proved implicity some results for this criterion.

We have given the gradient of this criterion function on every interior set of a partition of the information matrix space determined by the equality of the variances in every possible combination of the parameters (see (9)). This allows the possibility of using algorithms for differentiable criterion functions on each of these sets. We called the criterion V(β)-optimality, but in deference to Jacroux M. we revert to use of the name of MV-optimality. Ford I., (3), did not invoke a special name for the criterion.

The model we are concerned with is that of simple linear regression,

$$E(y) = \gamma + \delta x, \sigma(x) = 1, x \in [a,b].$$

The design problem for the MV- criterion is to choose a probability measure $p(\cdot)$ on $[a,b]$ in order to find,

$$\min_{p(\cdot)} \max\{Var_{p(\cdot)}(\hat{\gamma}), Var_{p(\cdot)}(\hat{\delta}\}).$$

We aim to solve this problem for all $[a,b]$. Let M be the information matrix associated with the design $p(\cdot)$ and let X denote a random variable with probability measure $p(\cdot)$. Then:

$$M = \begin{pmatrix} 1 & E(X) \\ E(X) & E(X^2) \end{pmatrix}$$

hence,

$$\det M = E(X^2) - [E(X)]^2 = Var(X)$$

$$Var_{p(\cdot)}(\hat{\gamma}\} = \frac{E(X^2)}{Var(X)} = c_1^t M^{-1} c_1, \text{ where } c_1^t = (1,0)$$

$$Var_{p(\cdot)}(\hat{\delta}\} = \frac{1}{Var(X)} = c_2^t M^{-1} c_2, \text{ where } c_2^t = (0,1)$$

thus,

$$\Phi_{MV}(M) = \max\left\{ \frac{E(X^2)}{Var(X)}, \frac{1}{Var(X)} \right\}$$

We note also that the correlation beetwen $\hat{\gamma}$ and $\hat{\delta}$ is ,

$$\rho = Corr(\hat{\gamma}, \hat{\delta}) = \frac{-E(X)}{\sqrt{E(X^2)}}$$

and hence,

$$c_1^t M^{-1} c_1 = (1 - \rho^2)^{-1}$$
$$c_2^t M^{-1} c_2 = [E(X^2)(1 - \rho^2)]^{-1}$$

2 Solution for Different Cases

2.1 Case 1

Now we will study the different cases for the interval extremes a and b. Firstly we assume that $[a, b] \subseteq [-1, 1]$. This means $E(X^2) < 1$ always. Hence the criterion in this case is equivalent to $c_2^t M^{-1} c_2$ which in fact is equivalent to that of D-optimality since,

$$\Phi_{MV}(M) = \max\left\{ \frac{E(X^2)}{Var(X)}, \frac{1}{Var(X)} \right\} = \frac{1}{Var(X)} = \frac{1}{\det M}.$$

It is well-known that the D-optimal design in this case is,

$$\begin{array}{ccc} x & a & b \\ p_\delta^\star(\cdot) & 1/2 & 1/2 \end{array}$$

and the value of the function at this design is $4/(b - a)^2$, whatever the values of a and b. The design also maximises $E(X^2)(1 - \rho^2)$.

2.2 Case 2

A second trivial case is when $[a, b] \subseteq (-\infty, 1)$ or $[a, b] \subseteq (1, \infty)$, that means $E(X^2) > 1$ always. The criterion in this case is equivalent to that of c_1-optimality,

$$\Phi_{MV}(M) = \frac{E(X^2)}{Var(X)} = c_1^t M^{-1} c_1$$

and the c_1-optimal design is the MV-optimal design. It is known (see for instance (10), corollary 1 of section 3) that the c_1-optimal design is,

$$\begin{array}{ccc} x & a & b \\ p_\gamma^\star(\cdot) & \frac{|b|}{|a|+|b|} & \frac{|a|}{|a|+|b|} \end{array}$$

We are interested in computing the value of the function Φ_{MV} at this design for any a and b. First we note that,

$$E(X) = \begin{cases} 0 & \text{if } a < 0 < b \\ \frac{2ab}{a+b} & \text{otherwise} \end{cases}$$

$$E(X^2) = |a||b| = |ab|$$

Hence,

$$\Phi_{MV}(M) = \begin{cases} 1 & \text{if } a < 0 < b \\ \frac{(a+b)^2}{(b-a)^2} & \text{otherwise} \end{cases}$$

Note that this design maximises $1 - \rho^2$ and hence minimises ρ^2.

2.3 Case 3

The remaining case to consider is when $-1 \in [a, b]$ or $1 \in [a, b]$. We have the following result:

Proposition: If there is a design, $p^{\star\star}$, such that $E(X) = 0$ and $E(X^2) = 1$ then this is a MV-optimal design and c_1-optimal design.

Proof: This is true because,

$$\frac{E(X^2)}{E(X^2) - [E(X)]^2} \geq 1$$

for any distribution for X, that is, $\Phi_{MV}(M) \geq 1$ for every information matrix M. Thus, a design such that $\Phi_{MV}(M) = 1$ should be MV-optimal and c_1- optimal.

Now we must prove the existence of such designs. The condition $E(X) = 0$ is satisfied only if $a < 0 < b$. Assuming this, let us consider three point designs,

$$
\begin{array}{cccc}
x & a & z & b \\
p^{\star}(\cdot) & p_1 & p_2 & p_3
\end{array}
$$

with $a < z < b, p_i \geq 0, \sum_i p_i = 1, E(X) = 0, E(X^2) = 1$. These conditions are equivalent to the linear system,

$$
\begin{pmatrix}
1 & 1 & 1 \\
a & z & b \\
a^2 & z^2 & b^2
\end{pmatrix}
\begin{pmatrix}
p_1 \\
p_2 \\
p_3
\end{pmatrix}
=
\begin{pmatrix}
1 \\
0 \\
1
\end{pmatrix}
$$

whose solution is,

$$p_1 = \frac{bz + 1}{(b - a)(z - a)}, p_2 = \frac{ab + 1}{(a - z)(b - z)}, p_3 = \frac{az + 1}{(a - b)(z - b)}$$

Since $a < z < b$ and $p_1, p_2, p_3 \geq 0$ then we must have $ab < -1$ and $bz, az > -1$. The key condition for finding an optimal design $p^{\star\star}$ is $ab < -1$. If this condition is satisfied then there is a point z such that $bz, az > -1$, e.g. $z = 1/2|a|$. We have assumed that $a < 0 < b$, but notice that if $ab < -1$ and $a < b$ then $a < 0 < b$ automatically. Thus, if $ab < -1$ then these are many c_1-optimal designs and there are all MV-optimal designs.

For example, given $ab < -1$ it is possible to find c, d such that $a < c < 0 < d < b$, $cd < -1$ and then to find z such that $dz, cz > -1$. Thus c, z, d satisfy the same conditions as a, z, b above. Effectively a and b have been replaced by c and d. In consequence another optimal design is,

$$
\begin{array}{cccc}
x & c & z & d \\
p^{\star}(\cdot) & p_1 & p_2 & p_3
\end{array}
$$

where,

$$p_1 = \frac{dz + 1}{(d - c)(z - c)}, p_2 = \frac{cd + 1}{(c - z)(d - z)}, p_3 = \frac{cz + 1}{(c - d)(z - d)}$$

If $a < 0 < b$ then one optimal design is,

$$
\begin{array}{ccc}
x & a & b \\
p^{\star}(\cdot) & \frac{|b|}{|a| + |b|} & \frac{|a|}{|a| + |b|}
\end{array}
$$

62

Note that 2 and 3 point designs and designs with bigger supports, even continuous designs can be optimal. An example of an optimal continuous design on $[-\sqrt{2}, \sqrt{2}]$ is,

$$p(x) = \frac{|x|}{2}, -\sqrt{2} \le x \le \sqrt{2}$$

Note also that since $Corr(\hat{\gamma}, \hat{\delta}) = \frac{-E(X)}{\sqrt{E(X^2)}}$, $\hat{\gamma}$ and $\hat{\delta}$ are uncorrelated under the above design.

We summarize our results so far diagramatically in Figure 1, which considers ploting b versus a $(b \ge a)$. We have established the MV-optimal design for subregions of this infinite triangle. These are depicted in the figure.

It remains to consider without loss of generality the case $-1 < a < 1, b > 1, ab > -1$. The case $a < -1, -1 < b < 1, ab > -1$ is similar. In this region we must determine whether the MV-optimal design is an equal variance design or a c_1-optimal or a D-optimal design. We first establish the best equal variance design, i.e., one satisfying $E[X^2] = 1$. We denote such a design by p_E^*. Note that

$$\Phi_{MV}(p_E) = \frac{1}{1 - [E(X)]^2}$$

Thus a best p_E-design should minimise $|E(X)|$ (and hence minimises the numerical correlation between $\hat{\gamma}$ and $\hat{\delta}$ subject to their variances being equal).

Now we shall look for the MV-optimal design in the remaining regions. Ford I., (3), in Appendix 2 proves that Φ_{MV} is decreasing, therefore the minimum is reached on the boundary of $\mathcal{M}$. From Caratheodory theorem 3 points should be enough for an MV-optimal design. Therefore a 2 points MV-optimal design can be found. We consider first a two points the design,

$$\begin{array}{ccc} x & a' & b' \\ p(\cdot) & p & 1-p \end{array}$$

with $0 \le p \le 1$ and $a \le a' < b' \le b$. It must satisfy the condition,

$$E(X^2) = a'^2 p + b'^2 (1-p) = 1$$

that is,

$$0 \le p = \frac{b'^2 - 1}{b'^2 - a'^2} \le 1$$

For this last condition to hold some constraints are necessary on a' and b'. On the one hand if $b' > 1$ then $b'^2 > a'^2$ and $a'^2 \le 1$, that is, $b' > 1$ and $-1 < a' < 1$. Since $ab > -1$ then $a'b' > -1$ in this case. This is clearly true in the case $a' > 0$; otherwise $a \le a' < 0, b' \le b \rightarrow a'b' \ge ab' \ge ab > -1$. On the other hand the case $b' < 1$ is not possible, otherwise it must be $a' \ge 1$, that is in contradiction with $a' < b'$.

Consequently a' and b' must satisfy the same conditions as a and b. Therefore,

$$E(X) = a'p + b'(1-p) = a'\frac{b'^2 - 1}{b'^2 - a'^2} + b'\frac{1 - a'^2}{b'^2 - a'^2} = \frac{a'b' + 1}{b' + a'}$$

$$Var(X) = \frac{(b'^2 - 1)(1 - a'^2)}{(a' + b')^2}$$

63

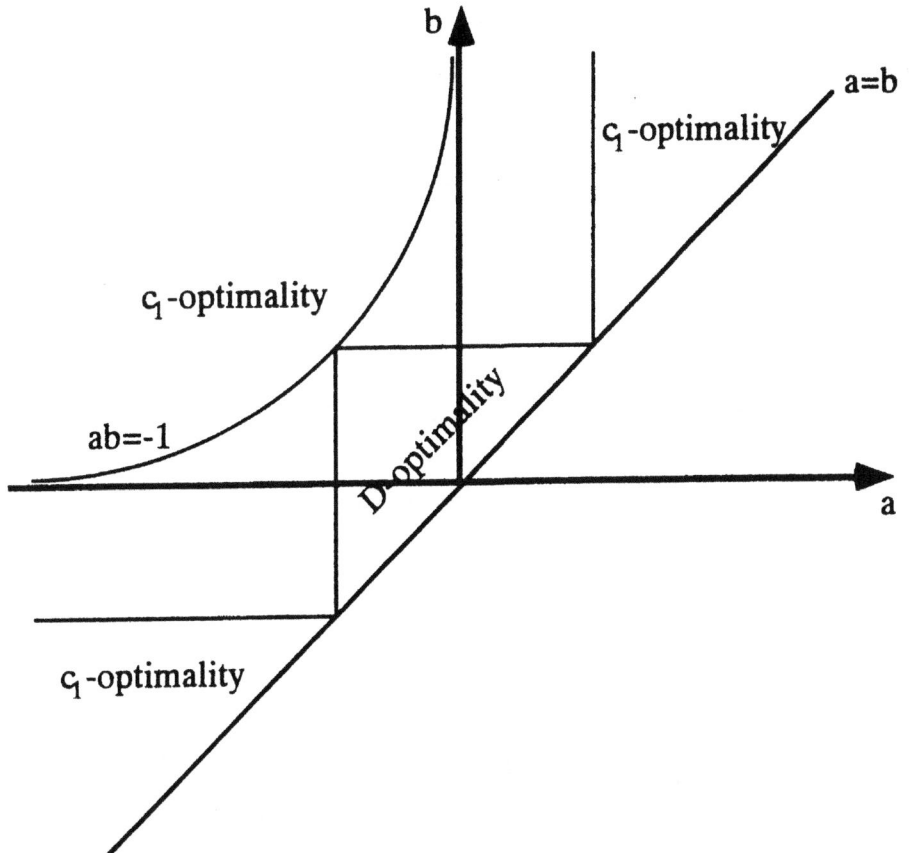

Figure 1: First approach

thus,

$$\Phi_{MV}(M) = \frac{(a'+b')^2}{(b'^2-1)(1-a'^2)}$$

Differenciating with respect to a' and b',

$$\frac{\partial \Phi_{MV}}{\partial a'} = \frac{-2(a'+b')(1+a'b')}{(b'^2-1)(1-a'^2)^2)} < 0$$

$$\frac{\partial \Phi_{MV}}{\partial b'} = \frac{2(a'+b')(1+a'b')}{(b'^2-1)^2(1-a'^2)} > 0$$

we can see that Φ_{MV} is increasing in a' and decreasing in b'. Therefore the best two point equal variance design in this case and with the constraint $E(X^2) = 1$ has to be,

$$
\begin{array}{ccc}
x & a & b \\
p_E^\star(\cdot) & \frac{b^2-1}{b^2-a^2} & \frac{1-a^2}{b^2-a^2}
\end{array}
$$

Consider now a three point design,

$$
\begin{array}{cccc}
x & a & z & b \\
p(x) & p_1 & p_2 & p_3
\end{array}
$$

where $a < z < b$. Then $E(X^2) = 1$ implies,

$$p_1 = \frac{[(b^2 - 1) - (b^2 - z^2)p_2]}{(b^2 - a^2)}$$

$$p_3 = \frac{[(1 - a^2) - (z^2 - a^2)p_2]}{(b^2 - a^2)}$$

and,

$$E(X) = \frac{ab + 1}{a + b} + \frac{(b - z)(z - a)}{a + b}p_2$$

Thus $E(X)$ is an increasing fuction of p_2 and is positive at $p_2 = 0$. Hence $|E(X)|$ and hence Φ_{MV} is minimised at $p_2 = 0$. Thus the best equal varianze design on [a,b] put weight only at a and b as given above.

We can now write down the Φ_{MV}-values of the $p_\delta^\star$, $p_\gamma^\star$ and $p_E^\star$ designs,

$$\Phi_{MV}[M(p_\delta^\star)] = \max\left\{ \frac{4}{(b-a)^2},\ \frac{2(a^2+b-2)}{(b-a)^2} \right\} = \begin{cases} \frac{4}{(b-a)^2} & \text{if } a^2 + b^2 < 2 \\ \frac{2(a^2+b-2)}{(b-a)^2} & \text{if } a^2 + b^2 > 2 \end{cases}$$

If $a < 0 < b$ then,

$$\Phi_{MV}[M(p_\gamma^\star)] = \max\left\{ 1,\ \frac{1}{|a||b|} \right\} = \begin{cases} 1 & \text{if } ab < -1 \\ \frac{1}{|a||b|} & \text{if } ab > -1 \end{cases}$$

If $a > 0$ or $b < 0$ then,

$$\Phi_{MV}[M(p_\gamma^\star)] = \max\left\{ \frac{(a+b)^2}{ab(b-a)^2},\ \frac{(a+b)^2}{(b-a)^2} \right\} = \begin{cases} \frac{(a+b)^2}{(b-a)^2} & \text{if } ab > 1 \\ \frac{(a+b)^2}{ab(b-a)^2} & \text{if } ab < 1 \end{cases}$$

$$\Phi_{MV}[M(p_E^\star)] = \frac{(a + b)^2}{(b^2 - 1)(1 - a^2)}$$

3 Conclusion

Compirasons between these values enable us to draw the conclusion that D-optimal and c_1-optimal designs are MV-optimal in extensions of the regions of Figure 1. These are depicted in Figure 2. The D-optimal design is MV-optimal if $a^2 + b^2 < 2$. The c_1-optimal design is MV-optimal if $|ab| > 1$. The implication is then that the equal variance design is MV-optimal in the two unlabelled regions ($|ab| < 1, a^2 + b^2 > 2$). For completeness we appeal to the general equivalence theorem and Ford's directional derivative formula (see (3)) in the following composite theorem.

In the two unlabelled regions the optimal design would be $p_E^\star$. To prove this we will use the general equivalence theorem and the derivative of Φ_{MV} given by Ford I., (3), p 126 (see section 1).

Theorem: Assume the model,

$$E(y) = \gamma + \delta x, \sigma(x) = 1, x \in [a, b].$$

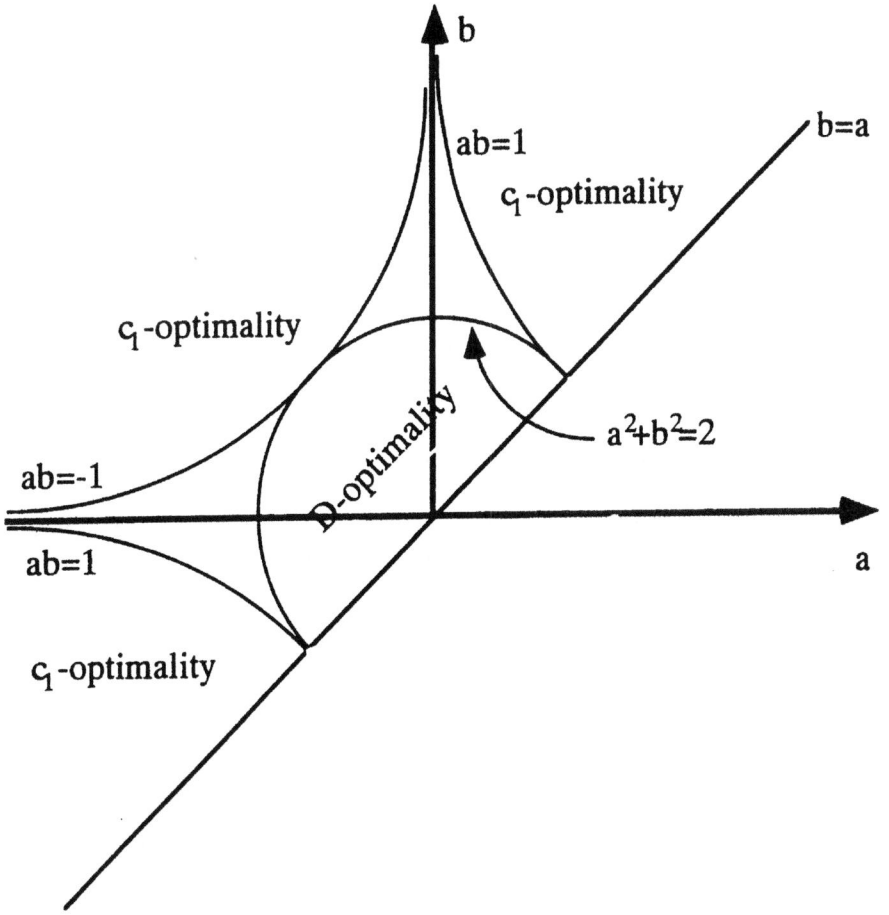

Figure 2: Second approach

1. If $ab < -1$ then the MV-optimal design is any c_1-optimal design, namely any design such that $E(X) = 0$ and $E(X^2) = 1$.

2. If $ab > 1$ then the MV-optimal design is the unique c_1-optimal design, which is the design,

x	a	b
$p^\star(\cdot)$	$\frac{|b|}{|a|+|b|}$	$\frac{|a|}{|a|+|b|}$

and the value of the function at this design is,

$$\Phi_{MV}(M) = \frac{(a+b)^2}{(b-a)^2}.$$

3. If $a^2 + b^2 < 2$ then MV-optimal design is the D-optimal design, namely

x	a	b
$p^\star_\delta(\cdot)$	$1/2$	$1/2$

and the value of the function at this design is

$$\Phi_{MV}(M) = \frac{4}{(b-a)^2}.$$

4. If $a^2 + b^2 > 2$, $ab < 1$ and $ab > -1$ then the MV-optimal design is the following equal variance design,

$$
\begin{array}{ccc}
x & a & b \\
p_E^\star(\cdot) & \frac{b^2-1}{b^2-a^2} & \frac{1-a^2}{b^2-a^2}
\end{array}
$$

and the value of the function at this design is

$$\Phi_{MV}[M(p_E^\star)] = \frac{(a+b)^2}{(b^2-1)(1-a^2)}.$$

Proof: The theorem has been proved for some regions. For the remaining regions the results of section 1 shall be used. Thus, for checking the optimality on M we must compute the sign of the directional derivative in the direction of one point designs if Φ_{MV} is differentiable at M. If Φ_{MV} is not differentiable at M then we must check also in the direction of the design matrices N_A or N_B corresponding to the points A and B in Figure 3. Figure 3 shows the region $\mathcal{M}'$ considered as a sub-set of the two-dimensional space for the last two components. Taking into acount that the first component is always equal to one we can see the two sets $\mathcal{M}'_{ss_1}$ (lower region) and $\mathcal{M}'_{ss_2}$ (upper region) and the points A and B representing matrices N_A and N_B. These matrices may correspond to two point designs on $\{a, b\}$

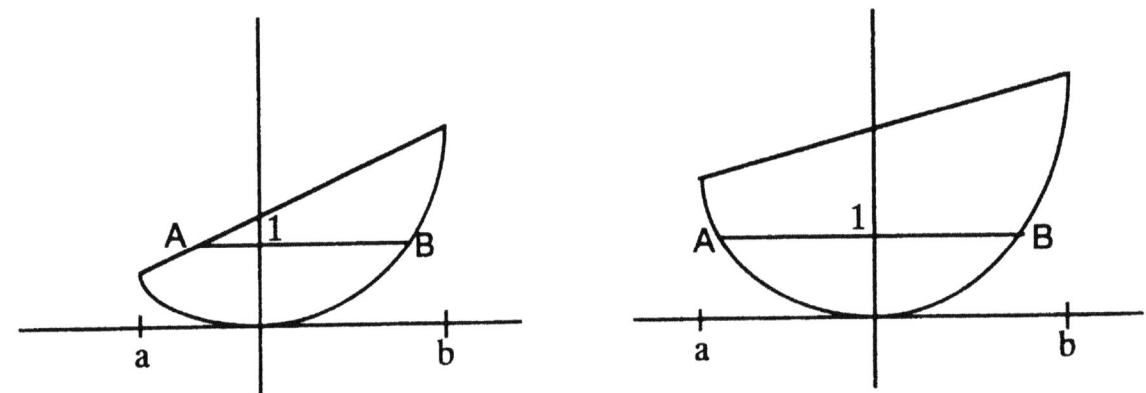

Figure 3: The two cases

Summarizing the last paragraph we can say that,

1. If Φ_{MV} is differentiable at M we need only calculate $\partial\Phi_{MV}(M, N_x)$ for an one point design information matrix N_x.

2. If Φ_{MV} is not differentiable at M, that is, M has equal variances, we need also to calculate $\partial\Phi_{MV}(M, N_A)$ if $|a| < 1$ or $\partial\Phi_{MV}(M, N_B)$ if $|b| < 1$.

The first part of this theorem has been proved before. Now we prove the second part. The case $a > 1$ or $b < -1$ has been proved. It remains to check the case $0 < a < 1$, $ab > 1$ (and similarly $-1 < b < 0$, $ab > 1$). The information matrix of $p_\gamma^\star$ and its inverse are,

$$
M(p_\gamma^\star) = \begin{pmatrix} 1 & \frac{2ab}{a+b} \\ \frac{2ab}{a+b} & ab \end{pmatrix}, \quad M^{-1}(p_\gamma^\star) = \begin{pmatrix} \frac{(a+b)^2}{(b-a)^2} & \frac{-2(a+b)}{(b-a)^2} \\ \frac{-2(a+b)}{(b-a)^2} & \frac{(a+b)^2}{ab(b-a)^2} \end{pmatrix}
$$

Since Φ_{MV} is differentiable at this matrix (different diagonal elements), then we need only check at one point designs,

$$N_x = \begin{pmatrix} 1 & x \\ x & x^2 \end{pmatrix}$$

Moreover $ab > 1$ means that biggest diagonal element in $M = M^{-1}(p_\gamma^\star)$ is $ss = 11$. Therefore,

$$\partial\Phi_{MV}(M, N_x) = (M^{-1} - M^{-1}N_xM^{-1})_{11} = \frac{-(a+b)^2(2b-2x)(2a-2x)}{(b-a)^4} > 0, x \in (a,b)$$

In the third part of the theorem, the case $-1 < a < b < 1$ has been solved. Let us consider the case $a^2 + b^2 < 2$, $b > 1$ (the case $a^2 + b^2 > 2$, $a < -1$ is analogous).The information matrix of design,

$$\begin{array}{ccc} x & a & b \\ p_\delta^\star(\cdot) & 1/2 & 1/2 \end{array}$$

and its inverse are,

$$M(p_\delta^\star) = \begin{pmatrix} 1 & \frac{a+b}{2} \\ \frac{a+b}{2} & \frac{a^2+b^2}{2} \end{pmatrix}, M^{-1}(p_\delta^\star) = \begin{pmatrix} \frac{2(a^2+b^2)}{(b-a)^2} & \frac{-2(a+b)}{(b-a)^2} \\ \frac{-2(a+b)}{(b-a)^2} & \frac{4}{(b-a)^2} \end{pmatrix}$$

Since $a^2 + b^2 < 2$ then the maximum diagonal element of the inverse is for $ss = 22$. We need only check at one point designs because Φ_{MV} is differentiable at $M = M(p_\delta^\star)$ (different diagonal elements). Therefore,

$$\partial\Phi_{MV}(M, N_x) = (M^{-1} - M^{-1}N_xM^{-1})_{22} = \frac{-4(2x-2a)(2x-2b)}{(b-a)^4} > 0, x \in (a,b)$$

It remain to prove the last part of the therorem. The information matrix associated with design,

$$\begin{array}{ccc} x & a & b \\ p_E^\star(\cdot) & \frac{b^2-1}{b^2-a^2} & \frac{1-a^2}{b^2-a^2} \end{array}$$

and its inverse are,

$$M(p_E^\star) = \begin{pmatrix} 1 & \frac{ab+1}{a+b} \\ \frac{ab+1}{a+b} & 1 \end{pmatrix}, M^{-1}(p_E^\star) = \begin{pmatrix} \frac{(a+b)^2}{(b^2-1)(1-a^2)} & \frac{-(a+b)(1+ab)}{(b^2-1)(1-a^2)} \\ \frac{-(a+b)(1+ab)}{(b^2-1)(1-a^2)} & \frac{(a+b)^2}{(b^2-1)(1-a^2)} \end{pmatrix}$$

We have to check at one point designs and at point A (see Figure 3) because Φ_{MV} is not differentiable at $M = M(p_E^\star)$ (equal diagonal elements). But $M = N_A$, thus,

$$\partial\Phi_{MV}(M, N_A) = 0$$

It remains to check at N_x,

$$\partial\Phi_{MV}(M, N_x) = \begin{cases} \frac{-(a+b)2(1+ab)[(1+ab)-2(a+b)x+(1+ab)x2]}{(b2-1)2(1-a2)2} & \text{if } x^2 > 1 \\ \frac{-(a+b)2[(1+ab)2-(b2-1)(1-a2)-2(a+b)(1+ab)x+(a+b)2x2]}{(b2-1)2(1-a2)2} & \text{if } x^2 < 1 \end{cases}$$

There are two cases for $a^2 + b^2 > 2$, $ab < 1$ and $ab > -1$. One is for $-1 < a < 1$, $b > 1$ and the other one is for $-1 < b < 1$, $a < -1$. We will consider just the first one without loss of generality. If $x^2 > 1$ the directional derivative is positive if $(1+ab) - 2(a+b)x + (1+ab)x^2 < 0$ for $1 < x < b$. Since it is a parabola with the coefficient of x^2 positive it is enough to prove the last inequality at points 1 and b. Let us see,

$$(1 + ab) - 2(a + b) + (1 + ab) = 2[1 + ab - (a + b)] < 0$$

$$(1 + ab) - 2(a + b)b + (1 + ab)b^2 = 2[(1 - b^2)(1 - ab)] < 0$$

If $x^2 < 1$ the directional derivative is positive if $(1 + ab)^2 - (b^2 - 1)(1 - a^2) - 2(a + b)(1 + ab)x + (a + b)^2 x^2 < 0$ for $a < x < 1$. With the same reasoning as before we must check the last inequation at points $x = a$ and $x = 1$,

$$(1 + ab)^2 - (b^2 - 1)(1 - a^2) - 2(a + b)(1 + ab)a + (a + b)^2 a^2 = (a^2 + b^2 - 2)(a^2 - 1) < 0$$

$$(1 + ab)^2 - (b^2 - 1)(1 - a^2) - 2(a + b)(1 + ab) + (a + b)^2 = (ab + 1)[ab + 1 - (a + b)] < 0$$

4 A Note on Correlation

We have made various references in the text to the implications of MV-optimal designs for the correlation ρ between $\hat{\gamma}$ and $\hat{\delta}$. In some cases the designs set this to zero; in others they minimise it while the remainder minimises it in some constraint manner. These references were in response to a referee's suggestion that MV-optimal designs can lead to highly correlated estimates. Certainly the MV-criterion would seem to ignore correlation structure. However the above observation suggest that ρ is being held down. Moreover we wonder if this is a problem peculiar to MV-optimality.

Note that the D-optimal design above is somewhat restrained in minimising ρ^2. More formal judgements could be made given that if $a < 0 < b$ then any design such $E(X) = 0$ minimises ρ^2, while otherwise ρ^2 is minimised by the two-point design,

x	a	b
$p(x)$	$\frac{\lvert b \rvert}{\lvert a \rvert + \lvert b \rvert}$	$\frac{\lvert a \rvert}{\lvert a \rvert + \lvert b \rvert}$

See Torsney B., (12), for motivation of this result and Torsney, B. and Alahmadi, A.M., (13), for more general results on minimising squared covariances and correlations.

References

DETTE H. and STUDDEN W.J. (1993). Minimax Designs in Linear Regression Models. Preprint.

ELFVING G. (1959). Design of Linear Experiments. In Probability and Statistics. The Harald Cramér Volume (U. Grenander, ed.). Stockholm: Almagrist & Wiksell, pp 58-74.

FORD I. (1976). Ph.D. Thesis. University of Glasgow.

HOTELLING H. (1944). Some Improvements in Weighing and other experimental Techniques. Ann. Math. Statist., Vol. 15, pp 297–306.

JACROUX M. (1983a). Some Minimum Variance Block Designs for Estimating Treatment Differences. J. R. Statist. Soc. B, Vol. 45, pp 70–76.

JACROUX M. (1983b). On the MV-optimality of Chemical Balance Weighing Designs. Calcutta Statist. Assoc. Bull., Vol. 32, pp 143–151.

JACROUX M. (1987). On the Determination and Construction of MV- optimal Block Designs for Comparing Test Treatments with a Standard Treatment. J. Statist. Plann. Inference, Vol. 15, pp 205–225.

JACROUX M. (1988). Some further Results on the MV-optimality of Block Designs for Comparing Test Treatments to a Standard Treatment. J. Statist. Plann. Inference, Vol. 20, pp 201–214.

LOPEZ-FIDALGO J. (1992). Minimizing the Largest of the Parameters Variances. $V(\beta)$-optimality. Proceedings of the 3rd International Workshop MODA 3 in Petrodvorets, Russia. Physica-Verlag. New York.

PUKELSHEIM F. and TORSNEY B. (1991). Optimal Weights for Experimental Designs on Linearly Independent Support Points. Ann. statics. Vol. 18, No. 3, pp 1614–1625.

TAKEUCHI K. (1961). On the Optimality of Cetain Type of PBIB designs. Rep. Stat. Appl. Res. Un. Japan Sci. Engrs., Vol. 8(3), pp 140–145.

TORSNEY B. (1986). Moment Inequalities via Optimal Design Theory. Lin. Alg. and Its Appl. Vol. 82, pp 237–253

TORSNEY B. and ALAHMADI, A.M. (1995). Designing for Minimal Dependent Observations. Statistica Sinica (to appear).

On the Support Points of D-Optimal Nonlinear Experimental Designs for Chemical Kinetics

CHRISTOS P. KITSOS

Department of Statistics, University of Business and Economics, Athens 10434, Greece.

Abstract:

In principle the nonlinear experiment design problem is distinguished from the linear one on the fact that the design points depend on the unknown parameter under investigation. Moreovere if the sequential principle is adopted there are not general convergence theorems neither for the sequence of estimators nor for the sequence of design measures. Fortunately, in practice, the nonlinear models involved are not too complicated and the support points can be evaluated, for the particular class of nonlinear models from chemical kinetics, so that the experimenter can built the local optimum experiment design.

1 Introduction

The main target of this paper is to provide the support points for experiments where the underlying model describing the physical phenomenon is assumed to be a nonlinear one. A table of the most commonly used nonlinear response functions from chemical kinetics appeared at the review paper of Ford, Kitsos and Titterington (1989), where for each model references are cited. We shall refer to these models provided their support points. For the binary response models, Torsney and Musrati (1992) provide the support points considering different weighting functions as well.

*Kitsos, C.P., and Müller, W.G., Eds., *Proceedings of MODA4*, Physica Verlag, Heidelberg, 1995

2 Background

When an experiment is performed it is usually assumed that the response vector y going with the covariate u, form the experimental region $U \subseteq R^k$ is linked with the parameter of interest θ, from the parameter space $\Theta \subseteq R^p$ through a deterministic function $f(u, \theta)$ and a stochastic part e, known as error through the relation

$$y = f(u, \theta) + e \tag{1}$$

with

$$\eta = E(y \mid u) = f(u, \theta)$$

The function $f : U \times \Theta \to R$ known as "regression" function is assumed to be twice continuously differentiable, while for the i.i.d errors it is assumed $E(e) = 0$ and $V(e) = \sigma^2$. In general it is $\sigma^2 = \sigma^2(\theta, u)$. Any design is representable by the probability measure ξ, say, defined on U. Moreover exact designs can be embedded in a set of continuous designs Ξ, say, the set of all probability measures on U. The set of u values providing positive weight by ξ is known as the support of the design. The average per observation information matrix $M = M(\theta, \xi)$ and a convex decreasing function ϕ on the set of nonnegative definite matrices play an important role on defining optimal experiment design, see Pukelsheim (1993, pg 135), Kitsos (1992) among others. There are cases where interest is focused on a subset s of the p parameters. Without loss of generality assume that these are the first $s < p$ and then the information matrix is reduced to

$$M_s(\theta, \xi) = M_{11} - M_{12} \, M_{22}^- \, M_{12}^T \tag{2}$$

where

$$M(\theta, \xi) = \begin{pmatrix} M_{11} & M_{12} \\ M_{12}^T & M_{22} \end{pmatrix}$$

see Ford et al(1989), for details.

In practice the most widely used ctiterion is D and D_s-optimality, Fedorov (1972), and the most common procedure to solve the problem is linearization of the model $f(u, \theta)$ about θ_0, a known value of θ. That is if we let $X = X(\theta) = \nabla f(u, \theta)$ and $\Delta(\xi, \theta) = \det(X'X)$ then the D-optimal design ξ_D, say, is

$$\xi_D = \text{Arg}[\ max\ [\Delta(\xi, \theta_0),\ \xi \in \Xi]\ ] \tag{3}$$

Because D-optimality minimizes the anticipated volume of any ellipsoidal confidence region of the parameter vector θ for the linearized deferministic portion $f(u, \theta)$ about θ_0, appears an aesthetic appeal in practical problems, see Logothetis and Wynn (1989) as well as for the different weighing experiments, this paper restricts attention on D-optimality. For nonlinear applications on c-optimality see Kitsos, Titterington and Torsney (1988), Kitsos (1992b) while for robust estimation of non-linear aspects see Kitsos and Müller (1994 a and b).

3 Support Points for D-optimal Designs

It was discussed by Ford, Kitsos and Titterington (1989) the different lines of inference (Bayesian, Pure Likelihood, Repeated Sampling) for the two main procedures (Static or Sequential designs) as well as the main frame for the locally optimal designs. The target of this paper is to discuss the support points for the local D-optimal design. On these points either a static or a sequential design, batch or fully, can be built up. It was also presented by Ford et al (1989) all the nonlinear models, response functions from chemical kinetics and related fields. It is easy to see (their Table 1) that in practice when Θ is a subset of R^p, usually assumed closed with regard of the topology of R^p, p runs from 1 to 4. In some cases when we fit the model a nine-term model might appear, Bates and Watts (1988 pg 88).

Now, let the nonlinear model be of the form

$$\eta = f(u, \theta) = f_1(u)\phi_1^T + f_2(u, \phi_2) \qquad (4)$$

i.e the sum of a linear and nonlinear model with $\phi_1 = (\theta_1, \ldots, \theta_s)$, $\phi_2 = (\theta_{s+1}, \ldots, \theta_p)$ and f_1, f_2 twice continuously differentiable functions. Then the D and D_s- optimal design depends only on ϕ_2.

Indeed: $M(\theta, \xi) = M(\phi_2, \xi)$ as trivially $\triangledown$ does not depend on ϕ_1. Therefore the D-optimal design depends only on ϕ_2. Moreover $M_{22} = M_{22}(\phi_2, \xi)$, recall (2.2), and thus the ratio $det(M)/det(M_{22})$ is a function of ϕ_2 only. Therefore the D_s- optimal design depends on ϕ_2 as well.

Nonlinear models are called partially nonlinear if

$$\triangledown = A(\phi_1)\, h\, (\phi_2, u) \qquad (5)$$

with $A(\phi_1)$ a nonsigular matrix and ϕ_1 the subset of s-terms appear in linear mode in the model and $h(\phi_2, u)$ a vector, with ϕ_2 defined as above. Then the D-optimal design depends on ϕ_2 and conditions for D_s-optimal designs are discussed by Khuri (1984).

As an example consider the Mitscherlish equation $\eta = \theta_1 + \theta_2 exp(\theta_3 u)$. Where η is the expected amount of growth, θ_1 the hypothetical growth from an infinite amount of fertilizer and $\theta_1 + \theta_2$ measures the rate at which additional increment of fertilizer decreases. Due to (3.1) the D-optimal design depends on $\phi_2 = (\theta_2, \theta_3)$. Because the $f_2(u, \phi_2) = \theta_2 exp(\theta_3 u)$ is partially nonlinear with $A(\theta_2) = diag(1, \theta_2)$ and $h(\theta_3, u) = (exp(\theta_3 u), u\, exp(\theta_3 u))^T$ the three term design depends only on θ_3. This is an approach to justify the results derived by Box and Lucas (1959).

A second example might be Michaelis - Menden model with $\eta = \frac{\theta_1 u}{\theta_2 + u}$. Then $A(\theta_1) = diag(1, \theta_1)$ and $h(u, \theta_2) = (u(\theta_2 + u)^{-1}, u(\theta_2 + u)^{-2})^T$. Therefore the design depends only on θ_2. As far as D_1-optimality concerned if can be easily varified that Khuri's (1984) condition does not hold.

Therefore the gain on D-optimal designs is substantial and it seems that only for the intrinsic nonlinear terms prior knowledge is needed. Recall that only when one parameter is involved different criteria as A-, c-optimality coincide with D-optimality. We shall refer, see Table 1 below, to the initial vector as $\theta_0 = (\theta_1, \theta_{20})$, say, if prior knowledge only for the θ_2 term is needed. When $U \subseteq R$ we let $U = [\kappa, \lambda]$. If the experimenters have other initial values than the values of Table 1 they can perform the experiment in the first stage

Table 1

Support points for D-Optimal Designs in nonlinear models

No.	Model, η (1)	Initial θ_0-Restrictions (2)	Support points (3)		
			u_1	u_2	u_3
1	$\theta_1 + \theta_2 exp(\theta_3 u)$	$\theta_0 = (\theta_1, \theta_2, \theta_{03})$	κ		
			$-\dfrac{1}{\theta_{03}} + \dfrac{\kappa exp(\theta_{03}\kappa) - \lambda exp(\theta_{03}\lambda)}{exp(\theta_{03}\kappa) - exp(\theta_{03}\lambda)}$		
2	$\theta_1 \{1 - exp(\theta_2 u)\}$	$\theta_0 = (\theta_1, \theta_{02})$	λ		
			$-\dfrac{1}{\theta_{02}} - \dfrac{\lambda exp(\theta_{02}\lambda)}{1 - exp(\theta_{02}\lambda)}$		
3	$\theta_1 - exp(-\theta_2 u)$	$\theta_0 = (\theta_1, \theta_{02})$ (i) $\frac{1}{\theta_{02}} \notin [\kappa, \lambda]$	(i) κ λ		
		(ii) $\frac{1}{\theta_{02}} \in (\kappa, \lambda)$	(ii) $\frac{1}{\theta_{02}}$ $min \left(\frac{\partial \eta}{\partial \theta_2} \big|_{\theta_{02}}^{u=\kappa} , \frac{\partial \eta}{\partial \theta_2} \big|_{\theta_{02}}^{u=\lambda} \right)$		
4	$\theta_1 exp(\theta_2 u)$	$\theta_0 = (\theta_1, \theta_{02})$ (i) $\theta_{02} < 0$ (ii) $\theta_{02} > 0$	(i) κ $\kappa - \frac{1}{\theta_{02}}$ (ii) $\lambda - \frac{1}{\theta_{02}}$ λ		
5	$\dfrac{\theta_1 u}{\theta_2 + u}$	$\theta_0 = (\theta_1, \theta_{02})$	λ $\dfrac{0.5\theta_{20}\lambda}{\theta_{20} + 0.5\lambda}$		
6	$\dfrac{\theta_1}{\theta_1 - \theta_2} \{exp(-\theta_1 u) - exp(-\theta_2 u)\}$	$\theta_0 = (0.7, 0.2)$ (i) estimate both (ii) estimate θ_1 (iii) estimate θ_2	(i) 1.229 6.858 (ii) 1.17 7.74 (iii) 3.52 5.64		
7	$\dfrac{\theta_1}{\theta_1 + \theta_2 - \theta_3} \{exp(-\theta_3 u) - exp(-(\theta_1 + \theta_2)u)\}$	$\theta_0 = (0.25, 0.5, 0.5)$ (i) estimate θ (iiα) estimate (θ_1, θ_2) (iiβ) estimate (θ_2, θ_3) (iiγ) estimate (θ_1, θ_3) (iiiα) estimate θ_1 (iiiβ) estimate θ_2 (iiiγ) estimate θ_3	(i) 0.4,1.3,2.0 (iiα) 0.4,1.2,2.0 (iiβ) 0.4,1.2,2.0 (iiγ) 0.4,1.25,2.0 (iiiα) 0.4,1.4,2.0 (iiiβ) 0.3,1.3,2.0 (iiiγ) 0.3,1.3,2.0		
8	$\dfrac{\theta_2\theta_3 u}{(1-u)\{1 + (\theta_2 - 1)u\}}$	(i) $\theta_2 = 2$ (ii) $\theta_2 = 10, 40, 80, 100, 1000$	(i) 0.13 0.30 (ii) 0.05 0.30		
9	$exp\left\{ -\theta_1 u_1 exp\left(-\theta_2 \left(-\frac{1}{u_2} - \frac{1}{T_0} \right)\right)\right\}$	$\theta_0 = (1.0, 16000)$ $T_0 = 400$	0.419 8.209	420 380	
10	$exp(-\theta_1 u_1 exp(-\theta_2 u_2))$	$\theta_0 = (\theta_{01}, \theta_{02})$	$\frac{1}{\theta_{01}} exp(\theta_{02}\kappa)$	$\frac{1}{\theta_{01}} exp(\theta_{02}\lambda)$	
11	As 6 with $\theta_i = \delta_i exp\left(-E_i \left(\frac{1}{u_2} - \frac{1}{T_0} \right)\right)$	(i) $\theta_0 = (0.7, 0.2)$ $\delta = (0.015, 0.003)$ $E = (5000, 10000)$ (ii) $\theta_i \sim N(\theta_{i0}, w_i^2)$	(i) 48.864 21.988 8304.5 319.42 (ii) 1.15 1.15	600 600 450 450	
12	$\dfrac{\theta_1 \theta_3 u_1}{1 + \theta_1 u_1 + \theta_2 u_2}$	$\theta_0 = (2.9, 12.2, 0.69)$ (i) estimate θ (ii) estimate (θ_1, θ_2) when $\theta_3 = 0.69$	(i) 0.2 3.0 3.0 (ii) 0.345 3.0 3.0	0.0 0.0 1.0 0 0.795	
13	$\dfrac{\theta_1 \theta_3(u_2 - u_1/1.632)}{1 + \theta_2 u_1 + \theta_3 u_2 + \theta_4 u_3}$	$\theta_0 = (38.6, 0.044, .022, .107)$ $\beta = .72$ (exponent of power transformation)	471 107 107 107	294 294 69 294	11 (*) 121 (*) 11 (**) 11 (·)
14	$\dfrac{\theta_1^2 exp(-\theta_3^2 u_3) + \theta_2^2 exp(-\theta_4^2 u_3)u_1 u_2}{\theta_1^2 exp(-\theta_3^2 u_3)u_1 + u_4 \theta_2^2 exp(-\theta_4^2 u_3)u_2}$	$\theta_0 = (0.001123, 0.00527, 110, 107)$ $\beta = -0.218$	1. 4. 16. 1. 16. 7. 1. 4.	673 673 623 623	5.75 (*) 5.75 (*) 5.75 (**) 5.75 (·)

(*): no. of runs : 6 , (**): no. of runs : 7, (·): no. of runs : 8.

at the suggested points, devoting a percentage of the data. Then at the second stage they can adjust the estimators and the design points, i.e. a two stage design, Kitsos (1992b).

Suppose, now, that the sequential principle is adopted and the initial design built up on the support points of a D-optimal one. Then, provided that the stochastic approximation scheme is applied, with batch size $b = dim\ \theta$, then the limiting design will be a D-optimal one, Kitsos (1989). In Table 1 below the main nonlinear models were considered and the support points for D-optimal designs where presented. For references see the review paper of Ford, Kitsos and Titterington (1988) for each model in particular.

We comment that variable u is usually the time of reaction, i.e $u = t$. In models No 9 and No 11 it is $u_2 = T$, the absolute temperature, while in models No 8 and No 12, the explanatory variable u presents pressure, $u = u_1 = P_1$ and $u_2 = P_2$, say. For a discussion of model No 5 see Kitsos (1992a) while for model No 4 see Kitsos (1989) for a discussion on the sequential nature of the design.

From Table 1 it is rather easy to see that at least for D-optimality and for reaction kinetic data the support points can be easily evaluated. Now if the design is static the D-optimal experiment can be produced by replication at these support points. If the sequential principle is adopted the situation is rather more complicated and there is no global theoretical background i.e converging limiting theorems for the sequence of design measures. But if the stochastic approximation is applied, under certain conditions, Kitsos (1989), the limiting design is a D-optimal one.

References

[1] Atkinson, A.C., Donev, A.N. (1992). Optimum Experimental Design. Clarendon Press.

[2] Bates, D.M., Watts, G.D. (1988). Nonlinear Regression Analysis and its Applications. Wiley.

[3] Box, G.E.P., Lucas, H.L. (1959). Design of experiments in Nonlinear Situation. Biometrica, 49, 77-90.

[4] Fedorov, V.V. (1972). Theory of Optimal Experiments. Academic Press.

[5] Ford, I. Kitsos, C.P. , Titterington, D.M. (1989). Recent Advances in Nonlinear Experimental Design. Technometrics, 31, 49-60.

[6] Khuri, A.I. (1984). A Note on D-optimal Designs for partially Nonliear Regression Models. Technometrics, 26, 59-61.

[7] Kitsos, C.P., Titterington, D.M. and Torsney, B. (1988). An Optimal design problem in rhythmonetry. Biometrics, 44, 657-671.

[8] Kitsos, C.P. (1989). Fully sequential procedures in nonlinear design problems. Comp. Stat. and Data Analysis 8, 13-19.

[9] Kitsos, C.P. (1992a). Adopting sequential procedures for Biological Experiments. In Model-Oriented Data Analysis, editors: Müller, Wynn, Zhigljavsky, 9-13. Physica-Verlag.

76

[10] Kitsos, C.P. (1992b). Quasi-sequential procedures for the Calibration problem. In COMPSTAT 1992, Vol. 2, Y. Dodge and J. Whittaker (eds), 227-231. Physica - Verlag.

[11] Kitsos, C.P., Müller, Ch.H. (1994a). Robust estimation of non- linear aspects. Submitted.

[12] Kitsos, C.P., Müller, Ch.H. (1994b). Robust linear Calibration. Submitted.

[13] Logothetis, N., Wynn, H.R. (1989). Quality Through Design. Oxford Science Publications.

[14] Pazman, A. (1993). Nonliear Statistical Models. Kluwer Academic Publications.

[15] Pukelsheim, F. (1993). Optimal Design of Experiments. Wiley.

[16] Torsney, B., Musrati, A.K. (1992). On the Construction of Optimal Designs with Applications to Binary Response and to Weighted Regression Models. In Model - Oriented Data Analysis, editors: Müller, Wynn, Zhigljavsky, 37-52. Physica-Verlag.

Designing Experiments for Additive Nonlinear Models

RAINER SCHWABE

Freie Universität, 1. Mathematisches Institut, Arnimallee 2–6, 14 195 Berlin, Germany.

Abstract:

According to a general result by Schwabe and Wierich (1993) product designs are D-optimal in additive linear models. In the present note this result is extended to a nonlinear setting. Local optimality is considered as well as minimax approaches and weight functions on the parameters. In particular, optimal designs can be constructed as a product of those designs which are optimal in the corresponding single-factor models. The results are obtained for the whole parameter vector and for the parameters associated with the single factors.

1   Introduction

We consider situations in which the outcome of an experiment may depend on two or more factors of influence. This can be described by a functional relationship $X(t) = \eta(t;\beta) + Z$ for the observation $X(t)$ at the setting $t = (t_1, \ldots, t_K)$ of the K factors, where Z denotes random noise. The mean response η is assumed to be a known function of t up to a p-dimensional vector β of unknown parameters. However, η may be nonlinear in β. Under suitable assumptions, e.g. independent observations with equal variances the underlying model is completely described by the response function $E(X(t)) = \eta(t;\beta)$, where t may be chosen from a design region T.

As a measure of performance of an experiment we investigate the asymptotic information matrix $\mathbf{I}(\delta;\beta) := (\int \frac{\partial \eta(t;\beta)}{\partial \beta_i} \frac{\partial \eta(t;\beta)}{\partial \beta_j} \delta(dt))_{i,j=1,\ldots,p}$ for generalized designs δ. The latter are defined as probability measures on the design region T. The use of such an information matrix is justified by the fact that (in the regular case) its inverse $\mathbf{I}(\delta;\beta)^{-1}$ is

---

*Kitsos, C.P., and Müller, W.G., Eds., *Proceedings of MODA4*, Physica Verlag, Heidelberg, 1995

proportional to the asymptotic covariance of the estimators for β if an increasing number of observations is made according to a design scheme converging to δ.

In contrast to the linear theory the asymptotic information matrix $\mathbf{I}(\delta;\beta)$ is a function of β and hence the performance of the design δ heavily depends on the unknown value of the vector of parameters β. A method for determining designs which are locally optimal at a particular value β has been developed already by Chernoff (1953). Later on minimax considerations and weighting functions have been considered to overcome the problem of the parameter dependence. For general remarks on the design of experiments in nonlinear settings we refer to the monographs by Fedorov (1972), Bandemer et al. (1977), Silvey (1980), and Atkinson and Donev (1992) and to the survey article by Ford, Titterington and Kitsos (1989).

Because in most cases it is impossible to find a generalized design that results in a uniformly maximal information matrix we consider the criterion of minimizing the generalized variance or - equivalently - of maximizing the determinant of the information matrix: δ^* is D-optimal at β if δ^* maximizes $\det(\mathbf{I}(\delta;\beta))$. Similarly δ^* is D-optimal at β for a subsystem β' of the parameters if δ^* maximizes $\det(\mathbf{I}_{\beta'}(\delta;\beta))$ where $\mathbf{I}_{\beta'}(\delta;\beta)$ is the information matrix for the subsystem β'. In particular, for a one-dimensional subsystem β_0 the information matrix $\mathbf{I}_{\beta_0}(\delta;\beta)$ reduces to a real number and a design δ^* will be called optimal for β_0 at β if $\mathbf{I}_{\beta_0}(\delta;\beta)$ becomes maximal.

If only one factor of influence is involved the optimization problems of finding D-optimal designs have been solved for a variety of models and powerful algorithms can be employed. In the setup of different factors of influence which is considered here the optimization becomes substantially more difficult. Some progress may be expected by the method of using a canonical form which was proposed by Ford, Torsney and Wu (1992) for one dimension. However, in the present paper we consider the situation of additive nonlinear effects $E(X(t_1,\ldots,t_K)) = \beta_0 + \sum_{k=1}^{K} \eta_k(t_k;\beta_k)$, $t_k \in T_k$, $k = 1,\ldots,K$. This means that no interactions do occur between the factors.

The main tool of our investigations is a complete class result presented in section 2 which allows a reduction of the problems to the much more tractable marginal models $E(X^{(k)}(t_k)) = \beta_0 + \eta_k(t_k;\beta_k)$, $t_k \in T_k$, in which only one factor is involved. In particular, D-optimal designs for the whole parameter vector β as well as for the parameter vectors β_k and $\beta^{(k)} = (\beta_0,\beta_k)$ corresponding to the single factors can be constructed as products of designs which are optimal in the marginal models and hence known or easier to determine. In the subsequent sections results for local optimality, minimax optimality and weighted optimality are obtained. In a final section further applications of the concept are exhibited.

The present paper extends ideas of Kurotschka (1984), Wierich (1986), Schwabe and Wierich (1993) and related results by Rafajłowicz and Myszka (1992) on linear additive models to a nonlinear setting. Generalizations to other optimality criteria can be obtained following the lines indicated in Schwabe (1994).

2 Additive Nonlinear Models

We consider the rather general case of an additive nonlinear model including explicitly a constant term. Under this assumption a two-factor additive model can be written as

$$E(X(t_1, t_2)) = \eta(t_1, t_2; \beta) = \beta_0 + \eta_1(t_1; \beta_1) + \eta_2(t_2; \beta_2), \qquad (1)$$

$t_1 \in T_1$, $t_2 \in T_2$, such that $\eta_k(t_k; \beta_k)$ is the effect of the k-th factor of influence and $\beta = (\beta_0, \beta_1, \beta_2)$ is a p-dimensional vector of unknown parameters with $(p_k - 1)$-dimensional components β_k related to the k-th factor, $k = 1, 2$, $p = p_1 + p_2 - 1$. The corresponding marginal models also include a constant term each:

$$E(X^{(k)}(t_k)) = \eta^{(k)}(t_k; \beta^{(k)}) = \beta_0 + \eta_k(t_k; \beta_k), \qquad (2)$$

$t_k \in T_k$, such that $\beta^{(k)} = (\beta_0, \beta_k)$ is p_k-dimensional, $k = 1, 2$. Note that it is natural to consider a product-type design region $T = T_1 \times T_2$ since there are no interactions assumed between the factors. By sub- und superscripts k and (k) we denote designs, parameters, etc. belonging to the k-th marginal model. For example $\mathbf{I}_k(\delta_k; \beta^{(k)}) = (\int \frac{\partial \eta^{(k)}(t_k; \beta^{(k)})}{\partial \beta_i^{(k)}} \frac{\partial \eta^{(k)}(t_k; \beta^{(k)})}{\partial \beta_j^{(k)}} \delta_k(dt_k))_{i,j=1,\ldots,p_k}$ is the information matrix at $\beta^{(k)}$, $\mathbf{I}_{k,\beta_k}(\delta_k; \beta^{(k)})$ the information matrix for β_k at $\beta^{(k)}$, and $\mathbf{I}_{k,\beta_0}(\delta_k; \beta^{(k)})$ the information for β_0 at $\beta^{(k)}$ in the k-th marginal model.

Some particular one-dimensional models have been considered in the literature which include a constant term β_0. The most popular is the exponential growth curve

$$E(X(t)) = \eta(t; \beta) = \beta_0 + \beta_1 e^{\beta_2 t}$$

associated with the well-known Mitscherlisch law of diminishing returns (see Box and Lucas (1959)). In this case $\frac{\partial \eta(t; \beta)}{\partial \beta} = (1, e^{\beta_2 t}, t e^{\beta_2 t})$. A two-dimensional version of the Mitscherlisch law is consequently given by

$$E(X(t_1, t_2)) = \eta((t_1, t_2); \beta) = \beta_0 + \beta_{11} e^{\beta_{12} t_1} + \beta_{21} e^{\beta_{22} t_2}.$$

Further nonlinear growth curves including a constant term have been used e.g. by Rasch (1990) and Rasch et. al. (1992):

(i)   the tanh-curve $\eta(t; \beta) = \beta_0 + \beta_1 \tanh(\beta_3(t - \beta_2))$;

(ii)   the arctan-curve $\eta(t; \beta) = \beta_0 + \frac{2}{\pi} \beta_1 \arctan(\beta_3(t - \beta_2))$;

(iii) the Janoschek curve $\eta(t; \beta) = \beta_0 + \beta_1 \exp(\beta_2 t^{\beta_3})$.

3 A Complete Class Result

The main tool in the present investigation is the following result which establishes the fact that for every design δ its performance is dominated by that of the corresponding product design $\delta_1 \otimes \delta_2$, where δ_1 and δ_2 are the marginals of δ.

Lemma 1

For every design δ with marginals δ_1 and δ_2 and every $\beta = (\beta_0, \beta_1, \beta_2)$

(i) $\det(\mathbf{I}(\delta; \beta)) \le \det(\mathbf{I}_1(\delta_1; \beta^{(1)})) \det(\mathbf{I}_2(\delta_2; \beta^{(2)})) = \det(\mathbf{I}(\delta_1 \otimes \delta_2; \beta))$;

(ii) $\det(\mathbf{I}_{\beta^{(1)}}(\delta; \beta)) \le \det(\mathbf{I}_1(\delta_1; \beta^{(1)})) \mathbf{I}_{2,\beta_0}(\delta_2; \beta^{(2)}) = \det(\mathbf{I}_{\beta^{(1)}}(\delta_1 \otimes \delta_2; \beta))$;

(iii) $\det(\mathbf{I}_{\beta_1}(\delta; \beta)) \le \det(\mathbf{I}_{1,\beta_1}(\delta_1; \beta^{(1)})) = \det(\mathbf{I}_{\beta_1}(\delta_1 \otimes \delta_2; \beta))$.

80

Proof

For β fixed the linearizations are given by $\eta(t_1, t_2; \beta) \approx \beta_0 + f_1(t_1)'\beta_1 + f_2(t_2)'\beta_2$ in case of the additive model and $\eta^{(k)}(t_k; \beta^{(k)}) \approx \beta_0 + f_k(t_k)'\beta_k$, $k = 1, 2$, for the marginal models respectively where $f_k(t_k) = (\frac{\partial \eta_k(t_k;\beta_k)}{\partial \beta_{ki}})_{i=1,\dots,p_k-1}$. Hence $\mathbf{I}(\delta; \beta)$ and $\mathbf{I}_k(\delta_k; \beta^{(k)})$ etc. can be regarded as the information matrices in the corresponding linearized models and the results follow directly from the Lemma in Schwabe and Wierich (1993).

For the readers' convenience we sketch a slightly different proof here which is based on the following representation given by Schwabe (1994):

For every product design $\delta_1 \otimes \delta_2$ the matrices

$$\mathbf{I}(\delta_1 \otimes \delta_2; \beta)^- = \begin{pmatrix} 1 + \gamma_1 + \gamma_2 & -\int f_1' d\delta_1 J_1^- & -\int f_2' d\delta_2 J_2^- \\ -J_1^- \int f_1 d\delta_1 & J_1^- & 0 \\ -J_2^- \int f_2 d\delta_2 & 0 & J_2^- \end{pmatrix} \quad (3)$$

and

$$\mathbf{I}_k(\delta_k; \beta^{(k)})^- = \begin{pmatrix} 1 + \gamma_k & -\int f_k' d\delta_k J_k^- \\ -J_k^- \int f_k d\delta_k & J_k^- \end{pmatrix} \quad (4)$$

are generalized inverses of $\mathbf{I}(\delta_1 \otimes \delta_2; \beta)$ in the additive model resp. of $\mathbf{I}_k(\delta_k; \beta^{(k)})$ in the marginal models, where $J_k = \int f_k f_k' d\delta_k - \int f_k d\delta_k \int f_k' d\delta_k$ and $\gamma_k = \int f_k' d\delta_k J_k^- \int f_k d\delta_k$, which can be proved by direct verification of $\mathbf{I}(\delta_1 \otimes \delta_2; \beta)\mathbf{I}(\delta_1 \otimes \delta_2; \beta)^- \mathbf{I}(\delta_1 \otimes \delta_2; \beta) = \mathbf{I}(\delta_1 \otimes \delta_2; \beta)$ and $\mathbf{I}_k(\delta_k; \beta^{(k)})\mathbf{I}_k(\delta_k; \beta^{(k)})^- \mathbf{I}_k(\delta_k; \beta^{(k)}) = \mathbf{I}_k(\delta_k; \beta^{(k)})$ respectively.

(i) From the formula for the determinant of a partitioned matrix it is known that

$$\det(\mathbf{I}(\delta; \beta)) = \det(\mathbf{I}_{(\beta_1,\beta_2)}(\delta; \beta)) \leq \det(\mathbf{I}_{\beta_1}(\delta; \beta)) \det(\mathbf{I}_{\beta_2}(\delta; \beta)).$$

Since the marginal models (2) are submodels of the full model (1) we obtain $\det(\mathbf{I}_{\beta_k}(\delta; \beta)) \leq \det(\mathbf{I}_{k,\beta_k}(\delta_k; \beta^{(k)}))$ by the usual refinement argument. Now $J_k = \mathbf{I}_{k,\beta_k}(\delta_k; \beta^{(k)})$ by (4) and hence $\det(\mathbf{I}_{1,\beta_1}(\delta_1; \beta^{(1)})) \det(\mathbf{I}_{2,\beta_2}(\delta_2; \beta^{(2)})) = \det(\mathbf{I}(\delta_1 \otimes \delta_2; \beta))$ in view of (3).

(ii) Again by the formula for the determinant of a partitioned matrix and a refinement argument we have

$$\begin{aligned} \det(\mathbf{I}_{\beta^{(1)}}(\delta; \beta)) &\leq \left(1 - \int f_2' d\delta_2 (\int f_2 f_2' d\delta_2)^- \int f_2 d\delta_2\right) \det(\mathbf{I}_1(\delta_1; \beta^{(1)})) \\ &= \mathbf{I}_{2,\beta_0}(\delta_2; \beta^{(2)}) \det(\mathbf{I}_1(\delta_1; \beta^{(1)})) \\ &= \det(\mathbf{I}_{\beta^{(1)}}(\delta_1 \otimes \delta_2; \beta)) \end{aligned}$$

and the last equality follows from the representation (3).

(iii) The result is straightforward by the refinement argument $\det(\mathbf{I}_{\beta_1}(\delta; \beta)) \leq J_1$ and the representation (3). $\qquad\square$

4 Locally Optimal Designs

In Schwabe and Wierich (1993) it has been shown that in additive linear models there are optimal product designs and that these optimal designs can be directly constructed from optimal designs for the corresponding marginal models. As a direct consequence of Lemma 1 this result is also true for local optimality in additive nonlinear models as follows

Theorem 1

Let $\beta = (\beta_0, \beta_1, \beta_2)$ and $\beta^{(k)} = (\beta_0, \beta_k)$, $k = 1, 2$.

(i) If δ_1^* and δ_2^* are D-optimal at $\beta^{(1)}$ and $\beta^{(2)}$ respectively in the marginal models, then $\delta^* := \delta_1^* \otimes \delta_2^*$ is D-optimal at β in the additive model.

(ii) If δ_1^* is D-optimal at $\beta^{(1)}$ in the first marginal model and δ_2^* is optimal for β_0 at $\beta^{(2)}$ in the second marginal model, then $\delta^* := \delta_1^* \otimes \delta_2^*$ is D-optimal for $\beta^{(1)}$ at β in the additive model.

(iii) If δ_1^* is D-optimal at $\beta^{(1)}$ in the first marginal model, then $\delta^* := \delta_1^* \otimes \delta_2$ is D-optimal for β_1 at β in the additive model for every marginal design δ_2 on T_2.

Proof

In view of Lemma 1 the criterion functions factorize into their marginal counterparts and maximization can be done componentwise. For (iii) note that the criterion functions of D-optimality for β_1 and that of D-optimality for the whole parameter vector at the same $\beta^{(1)}$ coincide in the marginal model (cf the Auxiliary Result in Schwabe and Wierich (1993)). □

Of course, the roles of the first and second factor can be interchanged.

5 Minimax Optimal Designs

In the general nonlinear case the above solutions heavily depend on the unknown vector of parameters β. A compromise is to look for a minimax design δ which minimizes the maximal generalized variance or equivalently maximizes the minimal determinant of the information where β might vary over some region of interest. Note that $\mathbf{I}(\delta; \beta)$ and $\mathbf{I}_k(\delta_k; \beta^{(k)})$ do not depend on the value of the additive constant β_0 because the effect of β_0 is linear. δ^* is minimax D-optimal on B if δ^* maximizes $\min_{(\beta_1, \beta_2) \in B} \det(\mathbf{I}(\delta; \beta))$. Analogously δ_k^* is minimax D-optimal on B_k in the k-th marginal model if δ_k^* maximizes $\min_{\beta_k \in B_k} \det(\mathbf{I}_k(\delta_k; \beta^{(k)}))$. Similarly minimax D-optimality for parts of the parameter vector are defined, e.g. δ^* is minimax D-optimal for β_1 on B if δ^* maximizes $\min_{(\beta_1, \beta_2) \in B} \det(\mathbf{I}_{\beta_1}(\delta; \beta))$. In case of a rectangular region $B = B_1 \times B_2$ of interest the product of optimal designs is again optimal

Theorem 2

(i) If δ_1^* and δ_2^* are minimax D-optimal on B_1 and B_2 respectively in the marginal models, then $\delta^* := \delta_1^* \otimes \delta_2^*$ is minimax D-optimal on $B = B_1 \times B_2$ in the additive model.

(ii) If δ_1^* is minimax D-optimal on B_1 in the first marginal model and δ_2^* is minimax optimal for β_0 on B_2 in the second marginal model, then $\delta^* := \delta_1^* \otimes \delta_2^*$ is minimax D-optimal for $\beta^{(1)}$ on $B = B_1 \times B_2$ in the additive model.

(iii) If δ_1^* is minimax D-optimal on B_1 in the first marginal model, then $\delta^* := \delta_1^* \otimes \delta_2$ is minimax D-optimal for β_1 on B in the additive model for every marginal design δ_2 on T_2, where B_1 is the projection of B onto its first component.

Proof

By Lemma 1 we obtain for every design δ

(i) $\min_{(\beta_1,\beta_2)\in B} \det(\mathbf{I}(\delta;\beta)) \;\leq\; \min_{\beta_1\in B_1} \det(\mathbf{I}_1(\delta_1;\beta^{(1)})) \min_{\beta_2\in B_2} \det(\mathbf{I}_2(\delta_2;\beta^{(2)}))$

$\leq\; \min_{\beta_1\in B_1} \det(\mathbf{I}_1(\delta_1^*;\beta^{(1)})) \min_{\beta_2\in B_2} \det(\mathbf{I}_2(\delta_2^*;\beta^{(2)}))$

$=\; \min_{(\beta_1,\beta_2)\in B} \det(\mathbf{I}(\delta_1^* \otimes \delta_2^*;\beta));$

(ii) $\min_{(\beta_1,\beta_2)\in B} \det(\mathbf{I}_{\beta^{(1)}}(\delta;\beta)) \;\leq\; \min_{\beta_1\in B_1} \det(\mathbf{I}_1(\delta_1;\beta^{(1)})) \min_{\beta_2\in B_2} \mathbf{I}_{2,\beta_0}(\delta_2;\beta^{(2)})$

$\leq\; \min_{\beta_1\in B_1} \det(\mathbf{I}_1(\delta_1^*;\beta^{(1)})) \min_{\beta_2\in B_2} \mathbf{I}_{2,\beta_0}(\delta_2^*;\beta^{(2)})$

$=\; \min_{(\beta_1,\beta_2)\in B} \det(\mathbf{I}_{\beta^{(1)}}(\delta_1^* \otimes \delta_2^*;\beta));$

(iii) $\min_{(\beta_1,\beta_2)\in B} \det(\mathbf{I}_{\beta_1}(\delta;\beta)) \;\leq\; \min_{\beta_1\in B_1} \det(\mathbf{I}_{1,\beta_1}(\delta_1;\beta^{(1)}))$

$\leq\; \min_{\beta_1\in B_1} \det(\mathbf{I}_{1,\beta_1}(\delta_1^*;\beta^{(1)}))$

$=\; \min_{(\beta_1,\beta_2)\in B} \det(\mathbf{I}_{\beta_1}(\delta_1^* \otimes \delta_2;\beta)). \qquad\qquad \square$

Alternatively a minimum regret approach tries to maximize the minimal value for the efficiency $\mathrm{eff}(\delta;\beta) = \det(\mathbf{I}(\delta;\beta))/\det(\mathbf{I}(\delta^*(\beta);\beta))$ where $\delta^*(\beta)$ is the locally D-optimal design at β. Hence δ^* is maximal D-efficient on B if δ^* maximizes $\min_{(\beta_1,\beta_2)\in B} \mathrm{eff}(\delta;\beta)$ and, analogously, δ_k^* is maximal D-efficient on B_k if δ_k^* maximizes the minimal efficiency $\min_{\beta_k\in B_k} \det(\mathbf{I}(\delta_k;\beta^{(k)}))/\det(\mathbf{I}(\delta_k^*(\beta^{(k)});\beta^{(k)}))$, where $\delta_k^*(\beta^{(k)})$ is the locally D-optimal design at $\beta^{(k)}$ in the k-th marginal model, etc. The efficiencies for subsystems of the parameters and there marginal counterparts are defined similarly.

Theorem 3

(i) If δ_1^* and δ_2^* are maximal D-efficient on B_1 and B_2 respectively in the marginal models, then $\delta^* := \delta_1^* \otimes \delta_2^*$ is maximal D-efficient on $B = B_1 \times B_2$ in the additive model.

(ii) If δ_1^* is maximal D-efficient on B_1 in the first marginal model and δ_2^* is maximal efficient for β_0 on B_2 in the second marginal model, then $\delta^* := \delta_1^* \otimes \delta_2^*$ is maximal D-efficent for $\beta^{(1)}$ on $B = B_1 \times B_2$ in the additive model.

(iii) If δ_1^* is maximal D-efficient on B_1 in the first marginal model, then $\delta^* := \delta_1^* \otimes \delta_2$ is maximal D-efficient for β_1 on B in the additive model for every marginal design δ_2 on T_2, where B_1 is the projection of B onto its first component.

Proof

Based on the results of Theorem 1 that for every β the products of locally optimal designs are themselves locally optimal the present proof parallels that of Theorem 2. $\quad\square$

6 Weighted Optimality

In contrast to the worst case consideration of the previous section we introduce here a (normalized) weighting measure ν on the parameter region which describes our interest in the different parameter values. There are some alternative possibilities of weighting the information (see e.g. Atkinson an Donev (1992), p. 214). The most reasonable one is apparently the weighted generalized variance, i.e. δ^* is D-optimal with respect to ν if δ^* minimizes $\int \det(\mathbf{I}(\delta;\beta))^{-1}\nu(d(\beta_1,\beta_2))$, δ_k^* is D-optimal with respect to ν_k if δ_k^* minimizes $\int \det(\mathbf{I}_k(\delta_k;\beta^{(k)}))^{-1}\nu_k(d\beta_k)$ etc. In case of a product-type weighting the products of optimal designs are optimal again.

Theorem 4

(i) If δ_1^* and δ_2^* are D-optimal with respect to ν_1 and ν_2 respectively in the marginal models, then $\delta^* := \delta_1^* \otimes \delta_2^*$ is D-optimal with respect to $\nu = \nu_1 \otimes \nu_2$ in the additive model.

(ii) If δ_1^* is D-optimal with respect to ν_1 in the first marginal model and δ_2^* is optimal for β_0 with respect to ν_2 in the second marginal model, then $\delta^* := \delta_1^* \otimes \delta_2^*$ is D-optimal for $\beta^{(1)}$ with respect to $\nu = \nu_1 \otimes \nu_2$ in the additive model.

(iii) If δ_1^* is D-optimal with respect to ν_1 in the first marginal model, then $\delta^* := \delta_1^* \otimes \delta_2$ is D-optimal for β_1 with respect to ν in the additive model for every marginal design δ_2 on T_2, where ν_1 is the first marginal of ν.

Proof

As in the proof of Theorem 2 we obtain by Lemma 1 for every design δ

$$
\begin{aligned}
\text{(i)} \quad \int \det(\mathbf{I}(\delta;\beta))^{-1} d\nu
&\geq \int \det(\mathbf{I}_1(\delta_1;\beta^{(1)}))^{-1} d\nu_1 \int \det(\mathbf{I}_2(\delta_2;\beta^{(2)}))^{-1} d\nu_2 \\
&\geq \int \det(\mathbf{I}_1(\delta_1^*;\beta^{(1)}))^{-1} d\nu_1 \int \det(\mathbf{I}_2(\delta_2^*;\beta^{(2)}))^{-1} d\nu_2 \\
&= \int \det(\mathbf{I}(\delta_1^* \otimes \delta_2^*;\beta))^{-1} d\nu
\end{aligned}
$$

$$
\begin{aligned}
\text{(ii)} \quad \int \det(\mathbf{I}_{\beta^{(1)}}(\delta;\beta))^{-1} d\nu
&\geq \int \det(\mathbf{I}_1(\delta_1;\beta^{(1)}))^{-1} d\nu_1 \int \mathbf{I}_{2,\beta_0}(\delta_2;\beta^{(2)})^{-1} d\nu_2 \\
&\geq \int \det(\mathbf{I}_1(\delta_1^*;\beta^{(1)}))^{-1} d\nu_1 \int \mathbf{I}_{2,\beta_0}(\delta_2^*;\beta^{(2)})^{-1} d\nu_2 \\
&= \int \det(\mathbf{I}_{\beta^{(1)}}(\delta_1^* \otimes \delta_2^*;\beta))^{-1} d\nu
\end{aligned}
$$

$$
\begin{aligned}
\text{(iii)} \quad \int \det(\mathbf{I}_{\beta_1}(\delta;\beta))^{-1} d\nu
&\geq \int \det(\mathbf{I}_{1,\beta_1}(\delta_1;\beta^{(1)}))^{-1} d\nu_1 \\
&\geq \int \det(\mathbf{I}_{1,\beta_1}(\delta_1^*;\beta^{(1)}))^{-1} d\nu_1 \\
&= \int \det(\mathbf{I}_{\beta_1}(\delta_1^* \otimes \delta_2;\beta))^{-1} d\nu \qquad\qquad \square
\end{aligned}
$$

Similar results hold if we consider the problem of maximizing the weighted logarithmic information $\int \ln \det(\mathbf{I}(\delta;\beta))\nu(d(\beta_1,\beta_2))$ as proposed by Läuter (1974) or the directly weighted information $\int \det(\mathbf{I}(\delta;\beta))\nu(d(\beta_1,\beta_2))$.

7 Applications

In a typical situation the influence of the first factor is a direct group effect α_i, $i = 1,\ldots,I$, and the second factor adds a nonliear contribution

$$
\eta(i,t_2;\beta) = \alpha_i + \eta_2(t_2;\beta_2), \tag{5}
$$

$i = 1,\ldots,I$, $t_2 \in T_2$. Hence the marginal model $\eta^{(1)}(i;\alpha) = \alpha_i$ associated with the first factor is linear and describes a one-way layout with I groups. A constant term can explicitly be involved by means of a linear reparametrization. As such a linear reparametrization of the linear part does not affect the D-optimality this shows

Corollary 1

Let δ_1^* be uniform on $\{1,\ldots,I\}$, i.e. $\delta_1^*(\{i\}) = I^{-1}$.

(i) If δ_2^* is D-optimal at $\beta^{(2)}$ (minimax D-optimal on B_2, maximal D-efficient on B_2, D-optimal with respect to ν_2 resp.) in the second marginal model, then $\delta^* := \delta_1^* \otimes \delta_2^*$ is D-optimal at β (minimax D-optimal on $B = B_1 \times B_2$, maximal D-efficient on $B = B_1 \times B_2$, D-optimal with respect to $\nu = \nu_1 \otimes \nu_2$ resp.) in the additive model.

84

(ii) If δ_2^* is optimal at $\beta^{(2)}$ (minimax optimal on B_2, maximal efficient on B_2, optimal with respect to ν_2 resp.) for β_0 in the second marginal model, then $\delta^* := \delta_1^* \otimes \delta_2^*$ is D-optimal at β (minimax D-optimal on $B = B_1 \times B_2$, maximal D-efficient on $B = B_1 \times B_2$, D-optimal with respect to $\nu = \delta_1^* \otimes \nu_2$ resp.) for α in the additive model.

(iii) If δ_2^* is D-optimal at $\beta^{(2)}$ (minimax D-optimal on B_2, maximal D-efficient on B_2, D-optimal with respect to ν_2 resp.) in the second marginal model, then $\delta^* := \delta_1 \otimes \delta_2^*$ is D-optimal at β (minimax D-optimal on B, maximal D-efficient on B, D-optimal with respect to ν resp. where B_2 is the projection of B onto its second component and ν_2 is the second marginal of ν) for β_2 in the additive model for every marginal design δ_1 on T_1.

Note that the second marginal model is again of the general form (2) $\eta^{(2)}(t_2; \beta^{(2)}) = \beta_0 + \eta_2(t_2; \beta_2)$. An example for the above model (5) is the Mitscherlisch law with different asymptotic levels but common diminishing return $\eta((i, t_2); \beta) = \alpha_i + \beta_{21} e^{\beta_{22} t_2}$. Finally we mention that all results can be extended to K factors as indicated in Schwabe and Wierich (1993) for the linear case, e.g.:

Corollary 2

If $\delta_1^*, \ldots, \delta_K^*$ are D-optimal at $\beta^{(k)}$ (minimax D-optimal on B_k, maximal D-efficient on B_k, D-optimal with respect to ν_k resp.) in the marginal models $\eta^{(k)}(t_k; \beta^{(k)}) = \beta_0 + \eta_k(t_k; \beta_k)$, $t_k \in T_k$, $k = 1, \ldots, K$, then $\delta^* := \delta_1^* \otimes \ldots \otimes \delta_K^*$ is D-optimal at β (minimax D-optimal on $B = B_1 \times \ldots \times B_K$, maximal D-efficient on $B = B_1 \times \ldots \times B_K$, D-optimal with respect to $\nu = \nu_1 \otimes \ldots \otimes \nu_K$ resp.) in the K-factor additive model $\eta((t_1, \ldots, t_K); \beta) = \beta_0 + \sum_{k=1}^{K} \eta_k(t_k; \beta_k)$, $(t_1, \ldots, t_K) \in T_1 \times \ldots \times T_K$.

This situation covers the multi-dimensional Mitscherlisch law of diminishing returns $\eta((t_1, \ldots, t_K); \beta) = \beta_0 + \sum_{k=1}^{K} \beta_{k1} e^{\beta_{k2} t_k}$.

Acknowledgement

A substantial part of the present work was done during a stay at *Uppsala University* which was partly supported by the research grant *Schw 531/2-1* of the *Deutsche Forschungsgemeinschaft*.

References

Atkinson, A. C. and A. N. Donev (1992). *Optimum Experimental Designs*. Clarendon, Oxford.

Bandemer, H. et al. (1977). *Theorie und Anwendung der optimalen Versuchsplanung I. Handbuch zur Theorie*. Akademie-Verlag, Berlin.

Box, G. E. P. and H. L. Lucas (1959). Design of experiments in non-linear situations. *Biometrika* **46**, 77–90.

Chernoff, H. (1953). Locally optimal designs for estimating parameters. *Ann. Math. Statist.* **24**, 586–602.

Fedorov, V. V. (1972). *Theory of Optimal Experiments*. Academic Press, New York.

Ford, I., D. M. Titterington and C. P. Kitsos (1989). Recent advances in nonlinear experimental design. *Technometrics* **31**, 49–60.

Ford, I., B. Torsney and C. F. J. Wu (1992). The use of a canonical form in the construction of locally optimal designs for non-linear problems. *J. R. Statist. Soc., Ser. B* **54**, 569–583.

Kurotschka, V. G. (1984). A general approach to optimum design of experiments with qualitative and quantitative factors. In: J. K. Ghosh and J. Roy (Eds.): *Statistics: Applications and New Directions: Proceedings of the Indian Statistical Institute Golden Jubilee International Conference Calcutta 1981*, 353–368.

Läuter, E. (1974). An experimental design method in the case of a nonlinear parametrization (in Russian). *Math. Operationsforsch. Statist.* **5**, 625–636.

Rafajłowicz, E. and W. Myszka (1992). When product type experimental design is optimal? Brief survey and new results. *Metrika* **39**, 321–333.

Rasch, D. (1990). Optimum experimental design in nonlinear regression. *Commun. Statist., Theory Methods* **A19**, 4789–4806.

Rasch, D., V. Guiard and G. Nürnberg (1992). *Statistische Versuchsplanung*. Fischer, Stuttgart.

Schwabe, R. (1994). Optimal designs for additive linear models. Preprint No A-5-94, Freie Universität Berlin, Fachbereich Mathematik.

Schwabe, R. and W. Wierich (1993). *D*-optimal designs of experiments with non-interacting factors. *J. Statist. Plann. Inference* **44** (to appear).

Silvey, S. D. (1980). *Optimal Design*. Chapman and Hall, London.

Wierich, W. (1986). On optimal designs and complete class theorems for experiments with continuous and discrete factors of influence. *J. Statist. Plann. Inference* **15**, 19–27.

D-Optimal Designs for Generalized Linear Models

RANDY R. SITTER and BEN TORSNEY

Department of Mathematics and Statistics, Carleton University, Ottawa,
Ontario K1S 5B6, Canada
and
Department of Statistics, University of Glasgow, Glasgow, G12 8QW, U.K.

Abstract:

This paper develops some simple methods for obtaining D-optimal designs for generalized linear models with multiple design variables. In some important cases the numerical complexity can be reduced to that of the two parameter case regardless of the original dimension. The form and properties of the obtained D-optimal designs are illustrated and discussed through a few interesting examples.

1 Introduction

Optimal designs for linear models have been studied extensively in the statistical literature. The information matrix is independent of the unknown regression parameters, thus optimal designs can often be characterized and obtained with the help of the equivalence theory of Kiefer and Wolfowitz (see Fedorov, 1972; Silvey, 1980). For generalized linear models (McCullagh and Nelder, 1989) less has been done. Ford, Torsney and Wu (1992) consider c-optimal and D-optimal designs for generalized linear models with simple linear effect and one design variable, and discuss the interest in such optimal designs in this context. By using the geometry of the design space and numerical investigation they find that the D-optimal designs have two or three support points for some link functions. Torsney and Musrati (1993) continued this work. Sitter and Wu (1993) use a different approach to obtain nice characterizations of D-, A- and F-optimal (minimize the length of a Fieller

*Kitsos, C.P., and Müller, W.G., Eds., *Proceedings of MODA4*, Physica Verlag, Heidelberg, 1995

interval) designs for binary response experiments with one design variable and explain some of the numerical results in Ford, Torsney, and Wu (1992).

In this paper we study the more general situation of multiple design variables. In the context of linear models, the addition of more design variables does not significantly change the complexity of the problem or the nature of its solution, however for non-linear models this is not so. Sitter and Torsney (1992) discuss the extension to two design variables for some typical binary response models. They show that in the case of a single design variable there exists a D-optimal design for an unbounded design space, however in the case of two design variables the optimality criterion can be made arbitrarily large by choice of design. A similar situation arises in linear models even for one design variable. Typically, one considers a bounded design space, and chooses the optimal design within this space.

In Section 2 we outline the canonical form of the design problem for generalized linear models which was introduced by Ford, Torsney, and Wu (1992). In Section 3 we describe the geometry of the resulting induced design space starting with the two and three parameter cases and then extending to the general k parameter situation. Once this geometry is understood, methods for obtaining D-optimal designs can be developed for different situations. In Section 4 we illustrate this by developing a general technique for constructing D-optimal designs in the k parameter context for one useful class of bounded design spaces, which does not increase the numerical complexity from the two parameter case. The obtained D-optimal designs have a certain amount of symmetry and are often related to fractional factorials and orthogonal arrays. This may be quite pleasing from a practical sense. We discuss other important classes of bounded design spaces and how the geometry can be used to simplify the search for D-optimal designs. We illustrate the methods in Section 5 as we consider some specific examples. We conclude with a short discussion.

2   A Canonical Form for Generalized Linear Models

Commonly in experimental design with non-linear models a scalar response y has distribution belonging to the exponential family $p(y|\psi)$ with

$$\mathrm{E}(y|\mathbf{x}) = \psi = \psi(\boldsymbol{\theta}^T \mathbf{s}),$$

where $\boldsymbol{\theta}$ is a vector of k unknown parameters, $\mathbf{x}$ is a vector of $k-1$ explanatory variables, $\mathbf{s}^T = (1, \mathbf{x}^T)$ and ψ is a possibly non-linear function. The Fisher information matrix for $\boldsymbol{\theta}$ for a single observation at a given design point $\mathbf{x}$ is given by

$$\mathrm{I}(\boldsymbol{\theta}, \mathbf{x}) = w(\boldsymbol{\theta}^T \mathbf{s}) \mathbf{s} \mathbf{s}^T,$$

where $w(\mu) = (\mathrm{d}\psi/\mathrm{d}\mu)^2/\mathrm{var}(y|\mathbf{x})$, $\mu = \boldsymbol{\theta}^T \mathbf{s}$, and $w(\cdot)$ is assumed measurable. Ford, Torsney and Wu (1992) consider a similar setup and derive a useful canonical form for the design problem. We briefly describe their argument below.

Let $\mathcal{X}$ be the design space from which the design points $\mathbf{x}$ are chosen. Let ξ be a measure denoting the design on $\mathcal{X}$. Thus

$$M(\xi, \boldsymbol{\theta}) = \int w(\boldsymbol{\theta}^T \mathbf{s}) \mathbf{s} \mathbf{s}^T \xi(\mathrm{d}\mathbf{x})$$

is the Fisher information matrix for this design. Most design criteria considered in this context are functions of M. If the criterion is invariant under a linear transformation $\mathbf{s} \to \mathbf{t} = B\mathbf{s}$, where B maps $\mathbf{x}$ to $\mathbf{z}$ implying an induced design space $\mathcal{Z}$, a canonical form of the design problem which can be solved independently of θ arises.

In this article, we will restrict consideration to the design criterion $\det(M)$ or equivalently $\phi(M) = \log\det(M)$. We choose the design ξ on the space $\mathcal{X}$ to maximize ϕ. Since this criterion is invariant under linear transformations of the form above, this problem is equivalent to maximizing

$$\phi(M_\eta) = \log\det(M_\eta)$$

within the design space $\mathcal{Z}$, where

$$M_\eta = \int w(z_1)\mathbf{t}\mathbf{t}^T \eta(\mathrm{d}\mathbf{z}),$$

$$\mathbf{t} = (1, z_1, ..., z_{k-1})^T$$

and the first and second rows of B are chosen to be $(1, 0, ..., 0)^T$ and θ^T, respectively, so that the first component of $\mathbf{z}$ is $z_1 = \theta^T\mathbf{s}$. This restriction on the first two rows of B is useful and will be discussed in the next section. Through this transformation the dependence of the optimal design on θ is replaced with a design space, $\mathcal{Z}$, that varies with θ.

A geometrical approach to the construction of D-optimal designs is sometimes useful. The further induced design space

$$G = G(\mathcal{Z}) = \{g_z \in \mathbf{R}^k : g_z = w(z_1)^{1/2}\mathbf{t}, \mathbf{t} = (1, \mathbf{z}^T)^T, \mathbf{z} \in \mathcal{Z}\}$$

plays an important role in this. The support of D-optimal designs depends heavily on the geometry of this space (see Silvey, 1980 ch. 5). We will use the geometry of G to obtain D-optimal designs in this general context.

3 The Geometry of G

To understand the geometry of G, it is best to start with the two and three parameter cases. By viewing these, the extension to higher dimensions becomes clear.

Ford, Torsney and Wu (1992) investigate the two parameter case and find the optimal designs for a number of common models. For two parameters, $\mathcal{X}$ is a line segment and

$$\psi = \psi(\alpha + \beta x), \quad \theta = (\alpha, \beta)^T, \quad \text{and} \quad \mathbf{s} = (1, x)^T.$$

Choosing B appropriately sets $\mathbf{t} = (1, z)^T$ and induces the design space $\mathcal{Z}$ also a line segment. Thus

$$G = G(\mathcal{Z}) = \{g_z = (g_1, g_2)^T : g_1 = w(z)^{1/2}, g_2 = zw(z)^{1/2}, z \in \mathcal{Z}\},$$

with $G \subset \mathbf{R}^2$. For a particular w, G will be a continuous section of a trajectory in $\mathbf{R}^2$. In some cases, such as many models for binary data, G will be a closed convex curve in $\mathbf{R}^2$ as z ranges over $\mathcal{Z} = (-\infty, \infty)$. In other cases one must restrict to $\mathcal{Z} = [a, b]$ to ensure that the design space is bounded. Even in cases where G is a closed convex curve one may

90

still wish to consider optimal designs over $\mathcal{Z} = [a, b]$ for some a and b (see Torsney and Musrati, 1993 and Ford, Torsney and Wu, 1992). Three common forms for $w(\cdot)$ considered in Ford, Torsney and Wu (1992) are

(1) $w_1(z) = z^c$, $\quad -\infty < c < \infty \qquad \mathcal{Z} \subset \mathbf{R}^+$

(2) $w_2(z) = \exp(z) \qquad\qquad\qquad\quad \mathcal{Z} \subset \mathbf{R}$

(3) $w_3(z) = f(z)^2/F(z)\{1 - F(z)\} \quad \mathcal{Z} \subset \mathbf{R}$,

where $F(z)$, the probability of response at z, is non-decreasing with $F(-\infty) = 0$ and $F(\infty) = 1$, and $f(z)$ is its corresponding density. Ford, Torsney and Wu (1992, pp. 572) show how (1), (2) and (3) arise in the modeling of Binomial, Poisson, Gamma and Inverse Gaussian data.

Fig. 1 sketches the trajectories of various bounded induced design spaces of type G_1, G_2 and G_3, where G_1, G_2 and G_3 refer to (1), (2) and (3), above. Fig. 1(a)-1(e) are of type G_1 for various values of c. Fig. 1(f) is of type G_2 and Fig. 1(g)-1(i) are special cases of G_3 as indicated, with $-\infty < z < \infty$. Ford, Torsney and Wu (1992) discuss the geometry of these curves in detail. We reproduce them here because, as will be illustrated, the geometry of the two parameter case remains important as higher dimensions are considered.

With three parameters, $\mathcal{X}$ is a plane and

$$\psi = \psi(\alpha + \beta_1 x_1 + \beta_2 x_2), \quad \boldsymbol{\theta} = (\alpha, \beta_1, \beta_2)^T, \quad \text{and} \quad \mathbf{s} = (1, x_1, x_2)^T.$$

Choosing B appropriately sets $\mathbf{t} = (1, z_1, z_2)^T$ and induces a design space $\mathcal{Z}$, a curved vertically oriented plane, where

$$G = G(\mathcal{Z}) = \{g_z = (g_1, g_2, g_3)^T : g_1 = w(z_1)^{1/2}, g_2 = z_1 w(z_1)^{1/2}, g_3 = z_2 w(z_1)^{1/2}, \mathbf{z} \in \mathcal{Z}\},$$

with $G \subset \mathbf{R}^3$. For clarity, suppose that g_1 and g_2 label the axes in the horizontal plane and that g_3 labels the vertical axis. Clearly, G will be unbounded if z_2 is unbounded. Due to the invariance of the D criterion to linear transformations, we can, without loss of generality, consider $a_0 \le z_1 \le b_0$ and $-1 \le z_2 \le 1$ by appropriately choosing B. That is, for any bounded $\mathcal{X}$ we can choose B such that the resulting $\mathcal{Z}$ is contained in, and touches all four sides of, the rectangle $\mathcal{Z}_0 = [a_0, b_0] \times [-1, 1]$ for some a_0 and b_0. This will be illustrated through examples in Section 5. Let

$$G_0 = G(\mathcal{Z}_0) = \{g_z = (g_1, g_2, g_3)^T : g_1 = w(z_1)^{1/2}, g_2 = z_1 w(z_1)^{1/2}, g_3 = z_2 w(z_1)^{1/2}, \mathbf{z} \in \mathcal{Z}_0\},$$

and consider $G_0 \supset G$. We can see that for fixed $z_1 = c$, G_0 is a vertical line segment. Also, for $z_2 = d$, a constant, G_0 becomes a continuous trajectory, analogous to the two parameter case. Thus G_0 is a vertical two dimensional object in $\mathbf{R}^3$ and G is a closed region just contained in G_0. To illustrate, Fig. 2 gives the resulting G_0's for the same $w(\cdot)$ functions as in Fig. 1 with the addition of $-1 \le z_2 \le 1$. Though it may not be perfectly clear in Fig. 2, all of these are symmetric in the vertical direction about the horizontal plane defined by the axes.

With the geometry of G_0 (and G) in the three parameter situation and the relationship and fundamental difference between z_1, the argument of $w(\cdot)$, and z_2 in the three parameter case apparent, we are ready to extend to higher dimensions. When $k \ge 3$

$$G = G(\mathcal{Z}) = \{g_z = (g_1, ..., g_k)^T : g_1 = w(z_1)^{1/2}, g_j = z_j g_1, \text{ for } j = 1, ..., k-1, \mathbf{z} \in \mathcal{Z}\},$$

with $G \subset \mathbf{R}^k$. Here $z_2, ..., z_{k-1}$ are all variables of the same nature and we need only consider $-1 \le z_j \le 1$ for $j = 2, ..., k-1$ with $a_0 \le z_1 \le b_0$. The design space G_0 induced

91

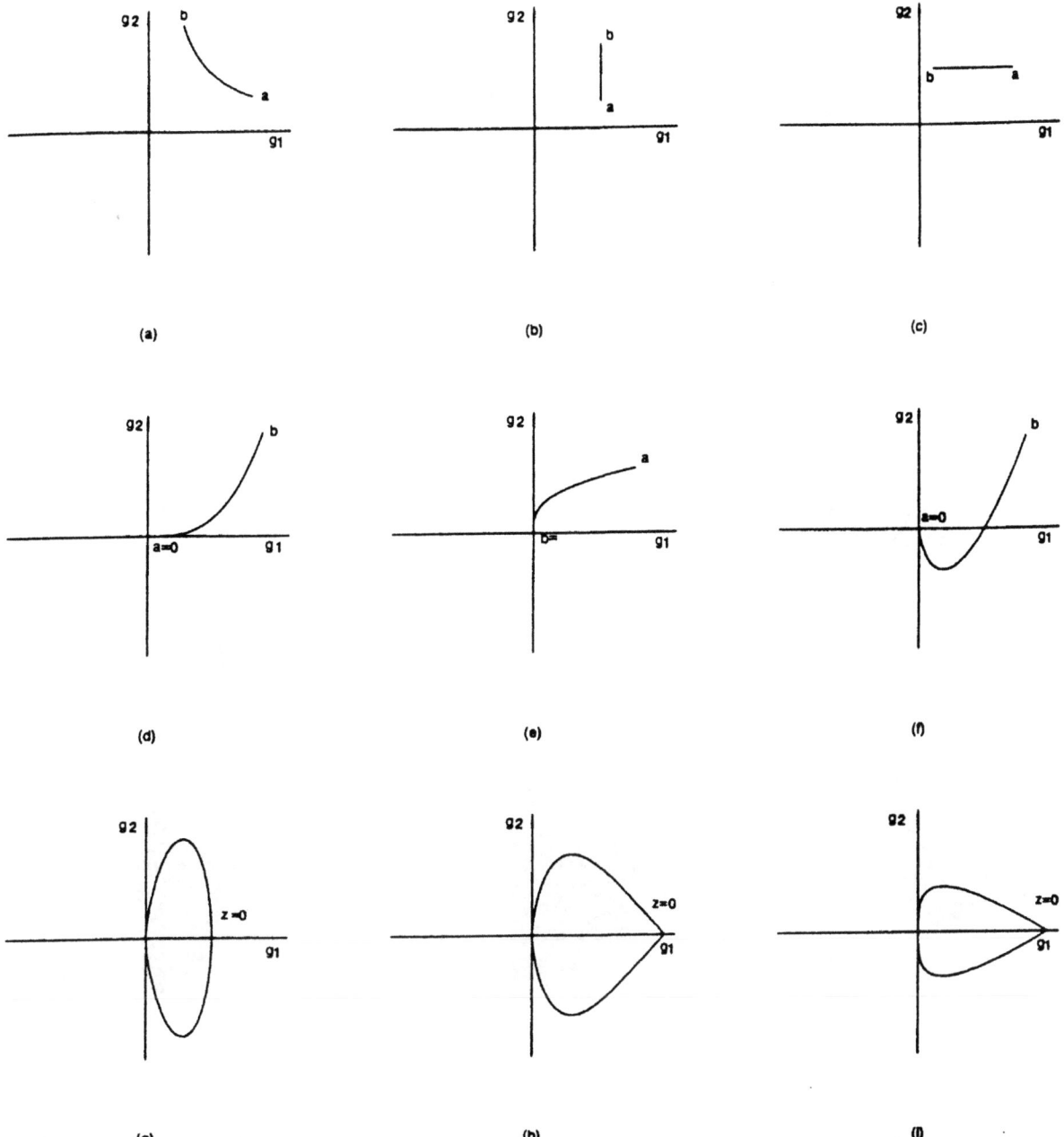

Fig. 1. $G(\cdot)$: (a) $w_1(z)$, $-2 < c < 0$; (b) $w_1(z)$, $c = 0$; (c) $w_1(z)$, $c = -2$; (d) $w_1(z)$, $c > 0$; (e) $w_1(z)$, $c < -2$; (f) $w_2(z)$; (g) $w_3(z)$, logistic; (h) $w_3(z)$, double exponential; (i) $w_3(z)$, double reciprocal.

92

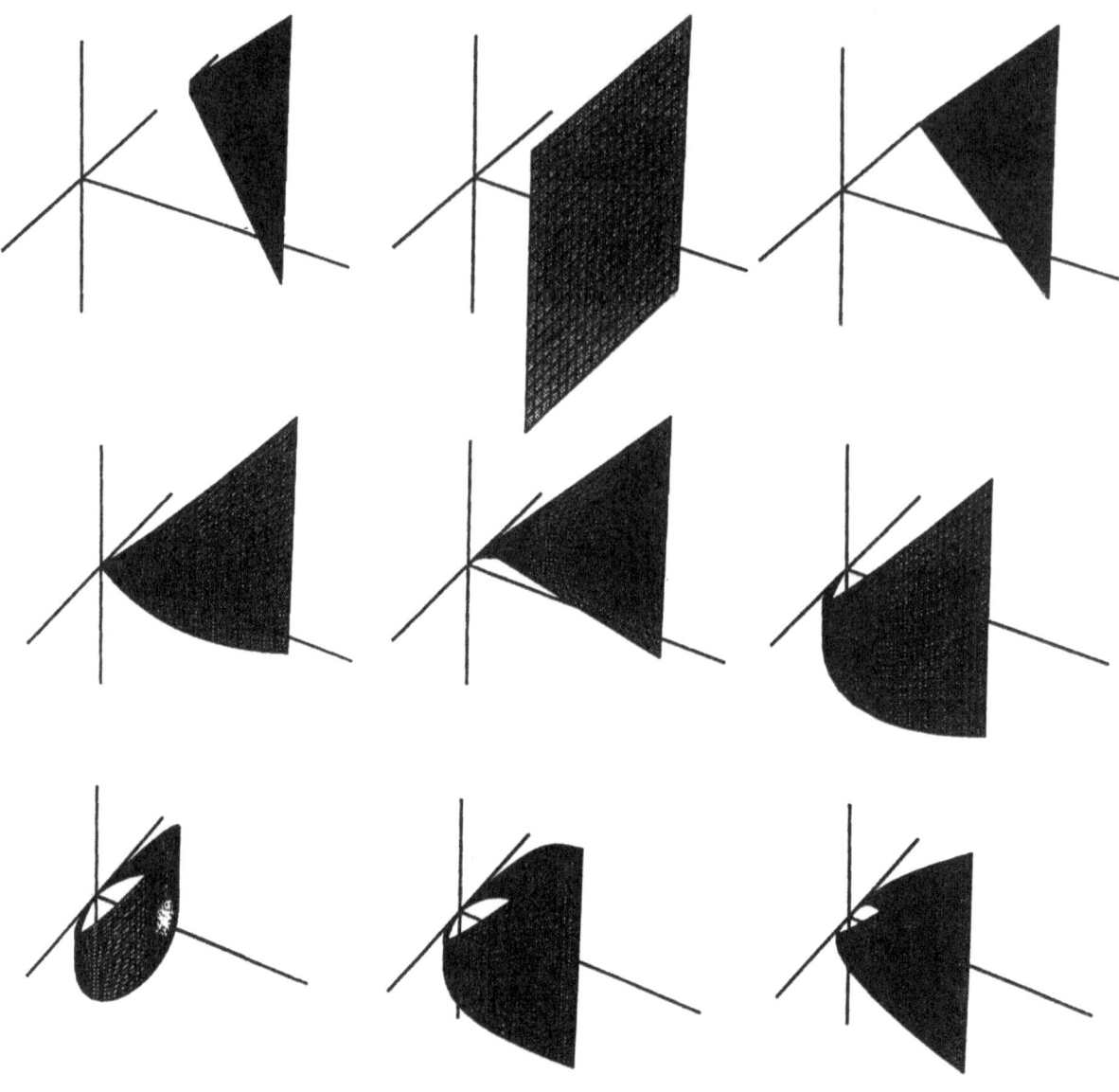

Fig. 2. Three parameter case. $G(\cdot)$: (a) $w_1(z)$, $-2 < c < 0$; (b) $w_1(z)$, $c = 0$; (c) $w_1(z)$, $c = -2$; (d) $w_1(z)$, $c > 0$; (e) $w_1(z)$, $c < -2$; (f) $w_2(z)$; (g) $w_3(z)$, logistic; (h) $w_3(z)$, double exponential; (i) $w_3(z)$, double reciprocal.

by this rectangle is then a $(k-1)$ dimensional hyper-planar object perpendicular to the (g_1, g_2) plane which tracks the trajectory defined by (g_1, g_2) over the range of z_1, and G is a closed region just contained in G_0. Thus, it is clear that the non-linearity of the problem enters only through $z_1 = \theta^T \mathbf{s}$. Whether z_1 must be bounded or may be allowed to be unbounded is a question which does not differ conceptually from the two-dimensional illustration. Once the optimal design on $\mathcal{Z}$ is obtained, one can transform the $\mathbf{z}$ support points back to $\mathbf{x}$ using $(1, \mathbf{x}^T)^T = \mathbf{s} = B^{-1}\mathbf{t}$, where $\mathbf{t}^T = (1, \mathbf{z}^T)$.

At this point, it is desirable to discuss the choice of bounds on the design space, $\mathcal{X}$. In many applications, there exist natural constraints on the original design space. Though the basic principles of optimal design remain the same for any bounded design space $\mathcal{X}$, some will be easier to handle theoretically than others. One important class of regions which is able to handle many practical situations and yet for which optimal designs are less difficult to obtain consists of regions which bound exactly $k-1$ independent linear combinations of $x_1, ..., x_{k-1}$. The simplest example is a rectangle in the $\mathcal{X}$ space. Since the matrix B is only restricted on the choice of rows 1 and 2, the remaining $k-2$ rows can be chosen so that $k-2$ of the bounds on $\mathcal{X}$ translate to $-1 \le z_j \le 1$ for $j = 2, ..., k-1$. The remaining bounded linear combination in the $\mathcal{X}$ space will imply a bound on an additional linear combination of $z_1, ..., z_{k-1}$. Thus the class of bounded design spaces, $\mathcal{X}$, under consideration translates into a class of bounded induced design spaces $\mathcal{Z}$, where $-1 \le z_j \le 1$ for $j = 2, ..., k-1$ and $a \le \sum_{j=1}^{k-1} c_j z_j \le b$ where a, b and $\{c_j\}_{j=1}^{k-1}$ are experimenter chosen constants. If $\mathcal{Z}$ is of this type we say $\mathcal{Z} \in \mathcal{C}$.

It is clear that different $\mathcal{Z} \in \mathcal{C}$ can be classified as to the form of the constraint $a \le \sum_{j=1}^{k-1} c_j z_j \le b$. If $c_2 = c_3 = ... = c_{k-1} = 0$ and $c_1 = 1$, $\mathcal{Z}$ is a rectangular region. This differs fundamentally, from a theoretical perspective, from all other cases in $\mathcal{C}$, since the final bound is on the argument of $w(\cdot)$, and thus $G = G_0$ and is symmetric across the plane defined by (g_1, g_2). It also represents an important case from an applied perspective. Within the class $\mathcal{C}$ there may exist many choices which satisfy the experimenter's constraints. It turns out that rectangular $\mathcal{Z}$ is often a natural choice. It differs from all others only in the fact that one of the bounds is placed on $z_1 = \theta^T \mathbf{s}$. It is often very desirable to place such a bound due to the relationship between z_1 and the $E(y|\mathbf{x})$. When the link function is monotone, which is often the case, a bound on z_1 implies a bound on $E(y|\mathbf{x})$ which will control the range of the data. This can be very important. For example, when dealing with binary data, i.e. Fig. 2 (g) (h) (i), bounding $z_1 = \theta^T \mathbf{s}$ is equivalent to bounding the probability of response. The experimenter would usually like the design to have moderate response probabilities so as to avoid the problems in existence of estimates when there is no response or full response. In the following section, we will develop a simple method for obtaining D-optimal designs when $\mathcal{Z}$ is rectangular, and will discuss the more complicated cases, $\mathcal{Z} \in \mathcal{C}$, and general $\mathcal{Z}$.

4 D-Optimal Designs

Caratheodory's theorem implies there exists a D-optimal design with at least k and at most $k(k+1)/2$ support points (see Silvey, 1980 p. 72). Thus we need only consider designs with a finite number, m, of support points and the matrix M_η can be written

$$M_\eta(\lambda; \mathbf{z}_1, ..., \mathbf{z}_m) = \sum_{i=1}^{m} \lambda_i g_{\mathbf{z}_i} g_{\mathbf{z}_i}^T,$$

where $\lambda = (\lambda_1, ..., \lambda_m)^T$, $0 \le \lambda_i \le 1$ is the mass at the ith support point, $\sum_i \lambda_i = 1$, and $\mathbf{z}_i = (z_{1i}, ..., z_{(k-1)i})^T$ is the ith support point.

With $k = 2$, Caratheodory's theorem implies that the D-optimal design must have two or three support points. Therefore, Ford, Torsney and Wu (1992) were able to use geometrical, analytical and numerical methods to obtain the best two point design and then check to see if it is globally optimal. To check this they use the following necessary and sufficient condition (Kiefer and Wolfowitz, 1960) stated here for general k: we must have

$$h(\mathbf{z}) = g_{\mathbf{z}}^T M_\eta^{-1}(\lambda; \mathbf{z}_1, ..., \mathbf{z}_m) g_{\mathbf{z}} \begin{cases} = & k & \text{for } \mathbf{z} = \mathbf{z}_1, ..., \mathbf{z}_m \\ \le & k & \text{for } \mathbf{z} \in \mathcal{Z} \end{cases} \tag{1}$$

for the design with mass λ_i at support point $\mathbf{z}_i$, $i = 1, ..., m$, to be D-optimal. Thus, in the two parameter case, if this condition does not hold with $k = 2$ for the optimal two point design, the D-optimal design must have three points. When $k \ge 3$ the situation is more complex and is not considered by Ford, Torsney and Wu (1992). In the sequel, we are able to give a general scheme for obtaining D-optimal designs for bounded $\mathcal{X}$ which imply a bounded rectangular $\mathcal{Z}$ that does not increase the amount of numerical work over the two parameter case. However, the obtained design may not be unique and, of the available D-optimal designs, it may not be the one with smallest support. In some cases we are able to show that the obtained D-optimal design has smallest possible support and show that it may have more support points than the number of parameters.

First, the support points of a D-optimal design are the points of contact between G and $SE(G)$, where $SE(G)$ is the smallest ellipsoid centered at the origin which contains G (see Sibson, 1972, Silvey and Titterington, 1973, and Silvey, 1980 ch. 5). From the geometry of G_0 described in the previous section it is clear that for G ($= G_0$ in this case) generated by a rectangular $\mathcal{Z}$ this intersection can only occur at points where $z_j = \pm 1$ for $j = 2, ..., k - 1$. To see this, view Fig. 2. Since the bodies are vertical two dimensional objects, an ellipsoid centered at the origin which contains them could only contact on the upper or lower ridges. Thus we need only consider designs of this form.

4.1 Rectangular $\mathcal{Z}$

Suppose we have an m-point design $\eta = \{(\lambda_i, \mathbf{z}_i)\}$ of the above type. That is $\mathbf{z}_i$ is a $(k-1)$-vector with components $2, ..., k-1$ taking values ± 1. Let $\eta^* = \{(\lambda_i/R, \mathbf{z}_{i1}), ..., (\lambda_i/R, \mathbf{z}_{iR})\}$, where $\mathbf{z}_{i1}, ..., \mathbf{z}_{iR}$ are each $(k-1)$-vectors with components $2, ..., k-1$ taking values ± 1 chosen so that $\sum_{l=1}^R z_{ijl} = 0$ and $\sum_{l=1}^R z_{ijl} z_{ij'l} = 0$ for $2 \le j \ne j' \le k - 1$. Some possible choices are: (i) all possible combinations; or (ii) each $(k - 1)$-vector with components $2, ..., k - 1$ taking values ± 1 chosen from the rows of an $R \times R$ Hadamard matrix with the column of $+$'s removed (fractional factorial, orthogonal array). This implies that all of the off-diagonal elements of $M_{\eta^*}(\mathbf{z})$ except the 1-2 and 2-1 component are zero and the remaining elements are the same as in $M_\eta(\mathbf{z})$. From this it is not difficult to show through simple linear algebra that

$$\det(M_\eta) \le \det(M_{\eta^*}) = S_0^{k-2}[S_0 S_2 - S_1^2], \tag{2}$$

where $S_0 = \sum_{i=1}^m \lambda_i w(z_{i1})$, $S_1 = \sum_{i=1}^m \lambda_i z_{i1} w(z_{i1})$ and $S_2 = \sum_{i=1}^m \lambda_i z_{i1}^2 w(z_{i1})$, with strict inequality if any of the off-diagonal elements except the 1-2 and 2-1 components of M_η are non-zero. Thus we can restrict attention to designs of this type and maximize (2) over choices of λ_i and z_{1i}. The resulting design would have $m \times R$ support points if the number of support points of the design which maximizes (2) has m support points and would be a D-optimal design.

Since, irrespective of k, $\det(M_{\eta^*})$ can be viewed as a function of

$$M(z_1) = \left[\begin{array}{cc} S_0 & S_1 \\ S_1 & S_2 \end{array} \right],$$

we can appeal to Caratheodory's theorem at this point to infer that $2 \leq m \leq 3$. The problem is now reduced in complexity to that of the two parameter problem. The only difference is the S_0^{k-2} multiplier in (2).

It is clear that, even using this approach, there may be many non-equivalent D-optimal designs with the same number of support points. For example, there are five non-equivalent 16×16 Hadamard matrices (see Hall, 1961). Each of these when used in this scheme would yield a different D-optimal design. It is also clear that the resulting design may not be the D-optimal design with smallest support. For example, if $k = 3$ there exists a D-optimal design with at least 3 and at most 6 design points. Using the above approach a 4 or 6 point D-optimal design will be obtained. If a four point design is obtained it does not mean that there does not exist a three point design which has the same value of the determinant, it only implies that there exists no better three point design and any three point D-optimal design, η, must have the property that all of the off-diagonal elements of $M_\eta(\mathbf{z})$ except the 1-2 and 2-1 elements are zero. Otherwise an argument similar to the one preceding (2) will imply that there exists a six point design which is better. In Section 5.1, we will look carefully at some of the cases implied by w_1, w_2 and w_3 of Section 2.

4.2 General $\mathcal{Z}$

The method in Section 4.1 depends heavily on the symmetry of G. For general $\mathcal{Z}$, since $G \subset G_0$ and is thus also a vertical hyper-planer object, $SE(G)$ can only intersect G on its boundaries and thus the D-optimal design must have support points on the boundary of $\mathcal{Z}$. For example, if $\mathcal{Z} \in C$ we still have $z_j = \pm 1$ for $j = 2, ..., k-1$ as in Section 4.1, however, for some fixed design point $\mathbf{z} = (z_1, ..., z_j, ..., z_{k-1})^T \in \mathcal{Z}$, the point $\mathbf{z}^* = (z_1, ..., -z_j, ..., z_{k-1})^T$ may not be in $\mathcal{Z}$. This complicates the optimization greatly. In some cases, the form of $w(\cdot)$ and the geometry of G and G_0 may suggest where G and $SE(G)$ intersect and the solution can thus be obtained.

Another possible simple method which may work that uses the developments in Section 4.1 is as follows. Form G_0 from $\mathcal{Z}_0$ as previously described and use the method of Section 4.1 to obtain a D-optimal design on $\mathcal{Z}_0$. If the solution lies in $\mathcal{Z}$, the obtained design will be D-optimal in $\mathcal{Z}$. In Section 5.2, we give some examples where the above methods are used for $\mathcal{X}$ which imply $\mathcal{Z} \in C$ and general $\mathcal{Z}$.

5 Some Examples

5.1 Rectangular $\mathcal{Z}$ with $a_0 \leq z_1 \leq b_0$

We will use the same definition of convex (concave) to the origin as that given on page 574 of Ford, Torsney and Wu (1992). As in the two parameter case considered there, G_1 is convex to the origin for $-2 \leq c \leq 0$. All other cases considered are concave. Below we discuss the D-optimal designs for various cases numbered 1-5. Ford, Torsney and Wu

(1992) considered similar cases for $k = 2$ and their results remain as special cases below. We use example (1) below to illustrate some of the issues concerning minimal support and uniqueness of the obtained D-optimal designs.

(1) $G = G_1$: $-2 < c < 0$, $a_0 > 0$, $b_0 < \infty$; $c = 0$, $a_0 \geq 0$, $b_0 < \infty$; $c = -2$, $a_0 > 0$, $b_0 \leq \infty$.

It is clear from the convexity to the origin in Fig. 1(a,b,c) and 2(a,b,c) and the obvious extension to general k, that g_z for $z = (a_0, \pm 1, ..., \pm 1)^T$ and $(b_0, \pm 1, ..., \pm 1)^T$ are the only possible contact points between $SE(G)$ and G. Thus $m = 2$, but unlike in the two parameter case the total mass at $z_1 = a_0$ and $z_1 = b_0$ may not be equal. To determine the total mass at $z_1 = a_0$ and $z_1 = b_0$ we must maximize (2), which in this case reduces to maximizing the polynomial

$$\phi(\lambda) = \lambda(1 - \lambda)[\lambda a_0^c + (1 - \lambda)b_0^c]^{k-2}$$

over choices of $0 < \lambda < 1$. The resulting mass λ and $(1 - \lambda)$ at $z_1 = a_0$ and b_0 must then be distributed evenly over all the points corresponding to the various values of $z_j = \pm 1$ for $j = 2, ..., k - 1$. These z_j settings must be balanced as described in the previous section. The case $k = 2$ yields the solution $\lambda = 1/2$ independent of a_0 and b_0, but for $k \geq 3$ this may not be the case. For example, suppose $k = 3$. Then

$$\phi(\lambda) = \lambda^3[b_0^c - a_0^c] + \lambda^2[a_0^c - 2b_0^c] + \lambda b_0^c,$$

which has stationary points

$$\lambda = \frac{2b_0^c - a_0^c \pm \sqrt{a_0^{2c} - a_0^c b_0^c + b_0^{2c}}}{3(b_0^c - a_0^c)}. \tag{3}$$

Without loss of generality let $b_0 = qa_0$ for some $q > 1$. Then (3) becomes

$$\lambda = \frac{2q^c - 1 \pm \sqrt{1 - q^c + q^{2c}}}{3(q^c - 1)}$$

independent of a_0. Here q controls the distance between a_0 and b_0. Table 1 gives the maximizing value of λ for various $q > 1$ and $-2 \leq c \leq 0$. It is not difficult to show that as $q \to 1$, $\phi(\lambda)$ is maximized at $\lambda = 1/2$ and as $q \to \infty$ it is maximized at $\lambda = 2/3$ for any $0 < c \leq -2$. Table 1 confirms this, as well as showing that between these extremes the maximizing value ranges over $1/2 < \lambda < 2/3$. For verification and illustration we will check that the obtained design satisfies (1) for $k = 3$. We will illustrate using the specific case $c = -1$, $a_0 = 1$, and $b_0 = 2$. This implies $q = 2$. From Table 1 we see that the optimal $\lambda = 0.577$. Thus the D-optimal design obtained using the described method would be

$$\{(\lambda, z_1, z_2) : (.2885, 1, 1), (.2885, 1, -1), (.2115, 2, 1), (.2115, 2, -1)\}. \tag{4}$$

Fig. 3 gives a surface plot of the equation in (1) over $z \in \mathcal{Z}$ for this design, where $\mathcal{Z} = \{(z_1, z_2) : 1 \leq z_1 \leq 2, -1 \leq z_2 \leq 1\}$. It is clear from Fig. 3 that condition (1) is satisfied and this design is D-optimal. Note also that in this case it is not difficult to prove that there is no other weighting on the four points which attains the minimum determinant value. In particular there exists no three point D-optimal design. Thus the obtained four point D-optimal design has smallest possible support.

Table 1

Optimal λ values for case 1

c	≈ 1	1.05	2	3	5	10	20	100	∞
0.0	0.500	0.500	0.500	0.500	0.500	0.500	0.500	0.500	0.500
-0.5	0.500	0.503	0.542	0.564	0.587	0.611	0.628	0.650	0.667
-1.0	0.500	0.506	0.577	0.608	0.632	0.650	0.658	0.665	0.667
-2.0	0.500	0.512	0.623	0.648	0.660	0.665	0.666	0.667	0.667

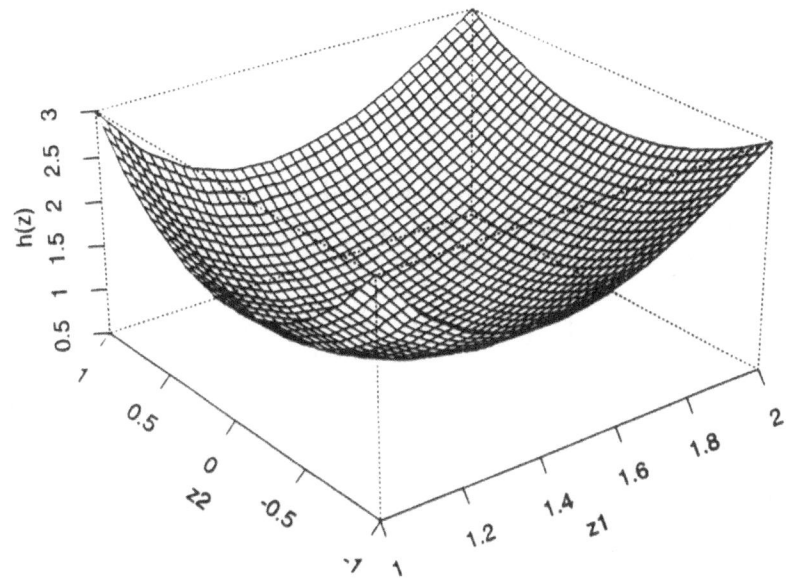

Fig. 3. $h(\mathbf{z})$: $w_1(z)$, $k = 3$, $c = -1$, $a = 1$, $b = 2$. D-optimal design given in (4).

This may not always be the case. For example, suppose $k = 4$ in the above example. The described method would yield an eight point symmetric design

$$\eta_8^* = \left\{ \begin{array}{l} (\lambda, z_1, z_2, z_3): \quad (.16,1,1,1), \quad (.16,1,1,-1), \quad (.16,1,-1,1), \quad (.16,1,-1,-1) \\ \qquad\qquad\quad (.09,2,1,1), \quad (.09,2,1,-1), \quad (.09,2,-1,1), \quad (.09,2,-1,-1) \end{array} \right\}.$$

However, by fixing to these eight possible support points and viewing the restrictions on the weights necessary to imply the appropriate off-diagonal elements of $M_{\eta^*}(\mathbf{z})$ remain zero while the determinant still attains its maximum, we see that there still exists some flexibility in the choice of weights. By such a choice we obtain the following D-optimal design which has positive weight on only six of the eight points:

$$\eta_6^* = \left\{ \begin{array}{l} (\lambda, z_1, z_2, z_3): \quad (.205,1,1,1), \quad (.115,1,1,-1), \quad (.115,1,-1,1) \\ \qquad\qquad\quad (.205,1,-1,-1) \quad (.18,2,1,-1), \quad (.18,2,-1,1) \end{array} \right\}.$$

Similar possibilities arise in the following examples, but the method is similar so we will not present any further results.

(2) $G = G_1$: $c > 0$, $a_0 \geq 0$, $b_0 < \infty$.

98

Fig. 1(d) and 2(d) suggest that G and $SE(G)$ touch at g_z where $\mathbf{z} = (b_0, \pm 1, ..., \pm 1)^T$ and $(z_1, \pm 1, ..., \pm 1)^T$ for one other value of $a_0 \le z_1 = z < b_0$. Thus $m = 2$, and maximizing (2) reduces to maximizing

$$\phi(z, \lambda) = \lambda(1 - \lambda)(b_0 - z)^2 z^c [\lambda z^c + (1 - \lambda) b_0^c]^{k-2}$$

for $0 < \lambda < 1$ and $a_0 \le z < b_0$. For example, when $k = 3$, $a_0 = 0$, $b_0 = 1$ and $c = 1$ we have

$$\phi(z, \lambda) = \lambda(1 - \lambda)(1 - z)^2 z [\lambda z + 1 - \lambda]$$

which is maximized at $\lambda = 0.4$, $z = 0.375$. Thus the D-optimal design obtained using the described method would be

$$\{(\lambda, z_1, z_2) : (.2, .375, 1), (.2, .375, -1), (.3, 1, 1), (.3, 1, -1)\}. \tag{5}$$

If, however, $0.375 < a_0 < 1$ then ϕ is maximized over z at $z = a_0$. For example if $a_0 = 0.425$, $\phi(z, \lambda)$ is maximized at $\lambda = 0.41$, $z = 0.425$. Thus the D-optimal design obtained using the described method would be

$$\{(\lambda, z_1, z_2) : (.205, .425, 1), (.205, .425, -1), (.295, 1, 1), (..295, 1, -1)\}. \tag{6}$$

It is not difficult to show that equation (1) is satisfied over $\mathbf{z} \in \mathcal{Z}$ for designs (5) and (6).

(3) $G = G_1$: $c < -2$, $a_0 > 0$, $b_0 \le \infty$; and $G = G_2$: $a_0 \ge -\infty$, $b_0 < \infty$.

Solutions can be obtained in case (3) by similar arguments to those used in case (2).

(4) $G = G_3$: $F \in \mathcal{F}$, $a_0 = -\infty$, $b_0 = \infty$.

Here $\mathcal{F}$ is the same as $\mathcal{F}_D$ in Ford, Torsney and Wu (1992) and denotes the set of F such that for $-\infty < z < \infty$ G is closed, bounded and convex. This case has been examined in detail for $k = 2$ by Ford, Torsney and Wu (1992), Torsney and Musrati (1993) and Sitter and Wu (1993), and for $k = 3$ by Sitter and Torsney (1992). These results extend to general k directly. For example, if we restrict to symmetric F', and apply arguments similar to those of Sitter and Wu (1993) for the two parameter case one can show that to maximize (2) only designs symmetric about zero in z_1 with at most three points need be considered. Here we make the same additional assumptions on F as Sitter and Wu (1993)'s cases (i) and (ii), which are satisfied by those in Fig. 1(g,h,i) and the probit. So maximizing (2) reduces to maximizing

$$\phi(\lambda, z_1) = [\lambda w(z_1) + (1 - \lambda) w(0)]^{k-1} \lambda z_1^2 w(z_1),$$

over $0 < \lambda < 1$ and $-\infty < z_1 < \infty$, where $\lambda/2$ is the total mass at $\pm z_1$ and $(1 - \lambda)$ is the total mass at zero. By a similar development as that of Sitter and Wu (1993), it can be shown that if

$$u(z_1) = \begin{cases} \frac{w(0)}{k[w(0) - w(z_1)]} & \text{if } \frac{w(z_1)}{w(0)} \le 1 - \frac{1}{k} \\ 1 & \text{otherwise} \end{cases}$$

then the optimal value of z_1, is given by a solution of

$$z_1 u(z_1) + \frac{2w(z_1)}{k w'(z_1)} = 0. \tag{7}$$

If z_1^* is the optimal value of z_1, then the optimal value of λ is $u(z_1^*)$. The resulting D-optimal designs for the $w(z)$'s given in Fig. 1(g,h,i) when $k = 3$ are given in Sitter and Torsney (1992).

To illustrate how this method can be used for arbitrary k, we will consider the case $k = 9$. The optimal values of z_1 and λ, the solutions to (7) above, for the F functions in Fig. 1(g,h) are given in Tables 2 and 3, respectively, where the total mass at $-z_1$, 0, and z_1 has been evenly distributed over the values of $(z_2, ..., z_8)^T$ which were determined using the rows of the 8×8 Hadamard matrix given on page 322 of Wolter (1985). One can see that the method yields a 16 point design in the first case and a 24 point design in the second. This difference in the number of support points arises from the fact that with the double exponential the above procedure yields a positive mass at $z_1 = 0$, whereas the logit does not. This is analogous to the two parameter case where the logit has a two point D-optimal design while the double exponential has a three point D-optimal design. It is interesting to note that the z_1 values for the double exponential in Table 3 are identical to those obtained in the two (Sitter and Wu, 1993) and three (Sitter and Torsney, 1992) parameter cases. In fact, it can be shown that these same three values will be obtained for any k, with different weights. This property holds more generally for any symmetric $f(\cdot) = F'(\cdot)$ which falls under case (ii) of Sitter and Wu (1993), and has a three point solution. Translating back to a particular $\mathcal{X}$ space which bounds $k-2$ linear combinations of the x_i's amounts to choosing B appropriately and applying B^{-1} to the transposed rows of Table 2.

(5) $G = G_3$: $F \in \mathcal{F}$, $a_0 > -\infty$, $b_0 \leq \infty$; $a_0 \geq -\infty$, $b_0 < \infty$.

This case has been examined in detail for $k = 2$ by Ford, Torsney and Wu (1992) (for $k = 3$ see Sitter and Torsney, 1992). These extend to general k using similar arguments as above.

Table 2

Optimal design for the logit with $k = 9$

λ_i	i	z_1	z_2	z_3	z_4	z_5	z_6	z_7	z_8	n_i
						design variables				$n = 400$
1/16	1	.679	+1	+1	+1	+1	+1	+1	+1	25
1/16	2	.679	−1	+1	−1	+1	−1	+1	−1	25
1/16	3	.679	−1	−1	+1	+1	−1	−1	+1	25
1/16	4	.679	+1	−1	−1	+1	+1	−1	−1	25
1/16	5	.679	+1	+1	+1	−1	−1	−1	−1	25
1/16	6	.679	−1	+1	−1	−1	+1	−1	+1	25
1/16	7	.679	−1	−1	+1	−1	+1	+1	−1	25
1/16	8	.679	+1	−1	−1	−1	−1	+1	+1	25
1/16	9	-.679	+1	+1	+1	+1	+1	+1	+1	25
1/16	10	-.679	−1	+1	−1	+1	−1	+1	−1	25
1/16	11	-.679	−1	−1	+1	+1	−1	−1	+1	25
1/16	12	-.679	+1	−1	−1	+1	+1	−1	−1	25
1/16	13	-.679	+1	+1	+1	−1	−1	−1	−1	25
1/16	14	-.679	−1	+1	−1	−1	+1	−1	+1	25
1/16	15	-.679	−1	−1	+1	−1	+1	+1	−1	25
1/16	16	-.679	+1	−1	−1	−1	−1	+1	+1	25

5.2 General $\mathcal{Z}$

Consider a bounded rectangular region in $\mathcal{X}$ with $a_j \leq x_j \leq b_j$ for $j = 1, ..., k-1$. For simplicity, assume $a_j = -1$ and $b_j = 1$ for $j = 2, ..., k-1$, $a_1 = a$, $b_1 = b$, and

Table 3

Optimal design for the double exponential with $k = 9$

λ_i	i	design variables								$n = 400$
		z_1	z_2	z_3	z_4	z_5	z_6	z_7	z_8	n_i
.00783	1	1.593	+1	+1	+1	+1	+1	+1	+1	3
.00783	2	1.593	−1	+1	−1	+1	−1	+1	−1	3
.00783	3	1.593	−1	−1	+1	+1	−1	−1	+1	3
.00783	4	1.593	+1	−1	−1	+1	+1	−1	−1	3
.00783	5	1.593	+1	+1	+1	−1	−1	−1	−1	3
.00783	6	1.593	−1	+1	−1	−1	+1	−1	+1	3
.00783	7	1.593	−1	−1	+1	−1	+1	+1	−1	3
.00783	8	1.593	+1	−1	−1	−1	−1	+1	+1	3
.109	9	0	+1	+1	+1	+1	+1	+1	+1	44
.109	10	0	−1	+1	−1	+1	−1	+1	−1	44
.109	11	0	−1	−1	+1	+1	−1	−1	+1	44
.109	12	0	+1	−1	−1	+1	+1	−1	−1	44
.109	13	0	+1	+1	+1	−1	−1	−1	−1	44
.109	14	0	−1	+1	−1	−1	+1	−1	+1	44
.109	15	0	−1	−1	+1	−1	+1	+1	−1	44
.109	16	0	+1	−1	−1	−1	−1	+1	+1	44
.00783	17	-1.593	+1	+1	+1	+1	+1	+1	+1	3
.00783	18	-1.593	−1	+1	−1	+1	−1	+1	−1	3
.00783	19	-1.593	−1	−1	+1	+1	−1	−1	+1	3
.00783	20	-1.593	+1	−1	−1	+1	+1	−1	−1	3
.00783	21	-1.593	+1	+1	+1	−1	−1	−1	−1	3
.00783	22	-1.593	−1	+1	−1	−1	+1	−1	+1	3
.00783	23	-1.593	−1	−1	+1	−1	+1	+1	−1	3
.00783	24	-1.593	+1	−1	−1	−1	−1	+1	+1	3

$\boldsymbol{\theta}^T = (0, 1, ..., 1)$. Thus $z_1 = \boldsymbol{\theta}^T \mathbf{s} = \sum_{j=1}^{k-1} x_j$. Let $z_j = -x_j$ for $j = 2, ..., k-1$. Then the bounds $a_j \leq x_j \leq b_j$ translate into $a \leq \sum_{j=1}^{k-1} z_j \leq b$ and $-1 \leq z_j \leq 1$ for $j = 2, ..., k-1$.

(1) $G = G_1$: $-1 < c < 0$, $a - k + 2 > 0$, $b < \infty$; $c = 0$, $a - k + 2 \geq 0$, $b < \infty$; $c = -2$, $a - k + 2 > 0$, $b \leq \infty$.

This corresponds to example (1) of Section 5.1; note that the smallest value of z_1 possible with $\mathbf{z} \in \mathcal{Z}$ is $a_0 = a - k + 2$. Viewing Fig. 2 (a, b, c), it is clear that in this case with $k = 3$ the objects will look similar except that the ends will not be vertical. These ends will be angled straight lines when projected onto the (g_1, g_3) plane. It is then obvious from the convexity to the origin in Fig. 2 (a, b, c) for $k = 3$ and the extension to general k that $g_{\mathbf{z}^*}$, where $\mathbf{z}^* = (z_1^*, ..., z_{k-1}^*)^T$ satisfies $z_j^* = \pm 1$ for $j = 2, ..., k-1$ and $\sum_{j=1}^{k-1} z_j^* = a$ or b, are the only possible contact points between SE(G) and G. So the problem reduces to finding the necessary weights at these support points. To determine the weights numerically, the algorithms discussed in Torsney (1983, 1988) and in Torsney and Alahmadi (1992) can be employed.

(2) $G = G_3$: $F \in \mathcal{F}$, $a > -\infty$, $b \leq \infty$; $a \geq -\infty$, $b < \infty$.

This corresponds to example (5) of Section 5.1. Let us assume the same rectangular region in $\mathcal{X}$ as in the previous example. This implies $\mathcal{Z}$ defined by $a \leq z_1 + z_2 \leq b$ and $-1 \leq z_2 \leq 1$ in the case $k = 3$. Let

$$B = \begin{pmatrix} 1 & 0 & 0 \\ 0 & 1 & 1 \\ 0 & -1 & 0 \end{pmatrix}.$$

Restrict to this case and let $a = -b = -2.5$ and $F(t) = e^t/(1 + e^t)$, i.e. the logit model.

Using the second method suggested in Section 4.2, we can consider the rectangular region, $\mathcal{Z}_0$ defined by $-\infty < z_1 < \infty$ and $-1 \leq z_2 \leq 1$ which contains $\mathcal{Z}$. The optimal design for this case obtained using the method described in example 4 of Section 5.1 has four symmetrically placed points $(\lambda_i, z_{1i}, z_{2i})$,

$$(.25, 1.22, 1) \quad (.25, 1.22, -1) \quad (.25, -1.22, 1) \quad (.25, -1.22, -1).$$

It is clear that all four points lie in $\mathcal{Z}$, so the design is D-optimal in this region. If we now transform back to $\mathcal{X}$, by applying B^{-1} to each $\mathbf{t} = (1, \mathbf{z}^T)^T$ generated by the above support points, we see that the D-optimal design is given by the points $(\lambda_i, x_{1i}, x_{2i})$,

$$(.25, 2.22, -1) \quad (.25, 0.22, 1) \quad (.25, -0.22, -1) \quad (.25, -2.22, 1).$$

None of these points lie on the corners of the rectangular $\mathcal{X}$ and the shape of the region defined by joining the four points with lines is not a rectangle in $\mathcal{X}$. Thus we see the effect of the non-linearity on the design.

(3) $G = G_1$: $c > 0$, $a_0 \geq 0$, $b_0 < \infty$.

This corresponds to example (2) of Section 5.1. Let $k = 3$, $c = 1$ and $\boldsymbol{\theta}^T = (1/2, 1/2, 1/2)$ and consider a bounded triangular region, $\mathcal{X}$, with vertices at $(0,1)^T$, $(1,0)^T$ and $(0,0)^T$. Choose

$$B = \begin{pmatrix} 1 & 0 & 0 \\ 1/2 & 1/2 & 1/2 \\ 0 & 1 & -1 \end{pmatrix}.$$

Then $\mathcal{Z}$ becomes a triangle with vertices at $(1, -1)$, $(1, 1)$ and $(1/2, 0)$. So $\mathcal{Z} \supset \mathcal{Z}_0$, where $\mathcal{Z}_0 = [a_0, b_0] \times [-1, 1]$ with $a_0 = 1/2$ and $b_0 = 1$. So $\mathcal{Z}_0$ is of the form discussed in example (2) of Section 5.1 with $0.375 < a_0 < 1$. Viewing Fig. 1(d) and Fig. 2(d), it is clear that $\mathcal{Z}$ will be a triangular region on the vertical curved plane depicted, with one edge corresponding to the long vertical edge in Fig. 2(d) and with the other vertex on the curved plane and in the (g_1, g_2) plane. The solution (6) suggests that in this case the only possible contact points between SE(G) and G will occur at the vertices of $\mathcal{Z}$. The symmetry of $\mathcal{Z}$ further suggests putting equal mass at $(1, 1)$ and $(1, -1)$. Assuming this is so, the problem reduces to maximizing

$$\phi(\lambda) = \lambda^2 (1 - \lambda)/8,$$

where $\lambda/2$ is the mass at $(1, 1)$ and at $(1, -1)$ and $(1 - \lambda)$ is the mass at $(1/2, 0)$. This is maximized at $\lambda = 2/3$ which implies a design

$$\{(\lambda, z_1, z_2) : (1/3, 1, 1), (1/3, 1, -1), (1/3, 1/2, 0)\}.$$

It is not difficult to show that (1) is satisfied for this design and thus it is D-optimal.

6 Discussion

We have developed methods for deriving D-optimal designs for generalized linear models with multiple design variables in some simple bounded design spaces. The results are interesting and illustrate how much more difficult the problem is than in the linear case. Clearly, further work, both theoretical and numerical, is needed on obtaining designs for more complex bounds on $\mathcal{Z}$ and for obtaining designs with smallest possible support. We feel that the examples and derivations in this paper could be important in guiding the development of algorithms for obtaining D-optimal designs in this context.

102

Acknowledgements

The authors would like to thank C.F.J. Wu for some helpful comments, and Elaine Leduc for her work on (2) and Fig. 1 and 2 while supported by an NSERC Summer '92 Undergraduate Research Assistantship.

References

Fedorov, V.V. (1972) *Theory of Optimal Experiments*. Academic Press, New York.

Ford, I., Torsney, B., and Wu, C.F.J. (1992) The use of a canonical form in the construction of locally optimal designs for nonlinear problems. *J. R. Statist. Soc.* B, **54**, 569-583.

Hall, M. Jr. (1961) Hadamard matrices of order 16, *Research Summary*, No. 36-10, Vol. 1, pp. 21-26, Jet Propulsion Laboratory, Pasadena, California.

Kiefer, J. and Wolfowitz, J. (1960) The equivalence of two extremum problems. *Can. J. Math.*, **12**, 363-366.

McCullagh, P. and Nelder, J.A. (1989) *Generalized Linear Models*. 2nd Ed. Chapman and Hall, London.

Sibson, R. (1972) Discussion on results in the theory and construction of D-optimum experimental designs (by H.P. Wynn). *J. R. Statist. Soc.* B, **34**, 174-175.

Silvey, D. (1980) *Optimal Design*. Chapman and Hall, London.

Silvey, D. and Titterington, D.M. (1973) A geometric approach to optimal design theory. *Biometrika*, **60**, 21-32.

Sitter, R.R. and Torsney, B. (1992) Optimal designs for binary response experiments with two design variables. Technical Report Series of the Laboratory for Research in Statistics and Probability at Carleton University, Ottawa, No. 210.

Sitter, R.R. and Wu, C.F.J. (1993) Optimal designs for binary response experiments: Fieller, D and A criteria. *Scand. J. Statist.*, **20**, 329-341.

Torsney, B. (1983) A moment inequality and monotonicity of an algorithm. In *Proc. Int. Sym. on Semi-infinite Programming and Appl.*, (Kortanek, K.O. and Fiacco, A.V. eds) at Univ. of Texas, Austin. Lecture Notes in Economics and Mathematical Systems, 215, 249-260.

Torsney, B. (1988) Computing optimizing distributions with applications in design, estimation and image processing. *Optimal Design and Analysis of Experiments* (Edited by Y. Dodge, V. Fedorov, and H.P. Wynn), Elsevier Science Publishers B.V. (North Holland), 361-370.

Torsney, B. and Alahmadi, A.M. (1992) Further development of algorithms for constructing optimizing distributions. *Model Oriented Data Analysis. Proceedings of the 2nd IIASA Workshop* in St. Kyrik, Bulgaria (V. Fedorov, W.G. Müller, I.N. Vuchkov, eds) Physica-Verlag, 121-129.

Torsney, B. and Musrati, A.K. (1993) On the construction of optimal designs with applications to binary response and to weighted regression models. *Proceedings of the 3rd Model Oriented Data Analysis Workshop* in Petrodvorets, Russia, (W.G. Müller, H.P. Wynn, A.A. Zhigljavsky, Eds) Physica–Verlag, 37-52.

Wolter, K.M. (1985) *Introduction to Variance Estimation*. Springer-Verlag, New York.

Further Results on Optimal Designs for Generalized Tic Polynomials on the Simplex

RALF-DIETER HILGERS

Institute of Medical Documentation and Statistics, University of Cologne,
Joseph-Stelzmannstraße 9, 50924 Cologne, Germany.

Abstract:

In a recent paper Hilgers and Bauer (1994) discussed D-optimal designs for (generalized) tic polynomials in q variables on the simplex. The simplex is an appropriate factor space in mixture experiments, where the response also depends on the total amount of the components. These results are extended in two different aspects. Firstly the change of the optimal design under variation of the optimality criterion and secondly the exclusion of some of the regression functions are studied. In particular the support of the D-optimal extended simplex centroid design is also φ_p-optimal. On the other hand, the deletion of the linear terms in the regression model seems to result in the deletion of the design points belonging to the corresponding regression functions.

1 Introduction

The methodology to describe the quantitative response in experiments where q components are in mixture is based on Scheffé (1958, 63). He proposed that the response only depends on the q components but not on the total amount. So the resulting factor space is the (unit)–q–simplex

$$\mathcal{U}_q = \left\{ \mathbf{x} \in \mathbb{R}^q : 0 \leq x_r \leq 1, \ 1 \leq r \leq q, \ \sum_{r=1}^{q} x_r = 1 \right\}.$$

Scheffé (1963) introduced multinomials of degree n in q variables

$$\eta(\mathbf{x}) = \vartheta_0 + \sum_{\ell=1}^{n} \sum_{1 \leq s_1 < \cdots < s_\ell \leq q} \vartheta_{s_1 \cdots s_\ell} x_{s_1} \cdots x_{s_\ell}$$

*Kitsos, C.P., and Müller, W.G., Eds., *Proceedings of MODA4*, Physica Verlag, Heidelberg, 1995

on $\mathcal{U}_q$. Here the restriction $\sum_{r=1}^{q} x_r = 1$ of the factor space leads to models, without intercept, called *special n-tic polynomials in q variables*. Subject to this model designs supported by the *simplex centroids (up to depth n)*, which consists of all barycenters up to depth n, i.e. the points $(1, 0, \ldots, 0); (1/2, 1/2, 0, \ldots, 0); \ldots; (1/n, \ldots, 1/n, 0, \ldots, 0)$ and their images under all permutations of the indices were suggested. D–optimality of the uniformly weighted design was calculated by Kiefer (1961) for $n = 2$, Uranisi (1964) for $n = 3$, Atwood (1969) for $n = q$.

Piepel (1985) investigated experiments where the response does not only depend on the proportion of the components but also on the total amount of the mixture. For example, in medical experiments, the response to a drug also depends on the total amount of the active substances. Because of toxicity or side effects, however, the total amount to the active substances has to be constrained. Under these conditions, it seems to be preferable to consider the q-dimensional simplex

$$\mathcal{S}_q = \left\{ \mathbf{x} \in \mathbb{R}^q : 0 \leq x_r \leq 1,\ 1 \leq r \leq q,\ \sum_{r=1}^{q} x_r \leq 1 \right\} \tag{1}$$

as an adequate factor space to describe a mixture in q-components. On this factor space Hilgers and Bauer (1994) introduced *generalized tic polynomials in q variables* on the q-dimensional simplex. In the following a slightly modified form of this model is used:

$$\eta(\mathbf{x}) = \vartheta_0 + \sum_{\ell \in \mathcal{J}} \sum_{1 \leq s_1 < \cdots < s_\ell \leq q} \vartheta_{s_1 \cdots s_\ell} x_{s_1} \cdots x_{s_\ell} , \tag{2}$$

where $\mathcal{J} \subseteq \{1, \ldots, q\}$. This means, that for $\ell \in \mathcal{J}$ the set of all tic-type regression functions of order ℓ are included in (2). If (2) consists of all linear ($\ell = 1$) and a subset of the orders $\ell = 2, \ldots, q$, i.e. $\mathcal{J} \subseteq \{1, \ldots, q\}$ and $\mathcal{J} \cap \{1\} \neq \emptyset$ then the D–optimum design can only be supported by the origine and the set of all barycenters of depth 1 up to q. A more concrete form of the D–optimum design results, if $\mathcal{J} = \{1, k, \ldots, q\}$ where $2 \leq k \leq q$. In those cases points of support can only be the origine and the set of the corresponding barycenters of depth k up to q. So the D–optimum design is uniformly weighted. Furthermore if $\mathcal{J} = \{1\}, \{1, 2\}$ or $\{1, 2, 3\}$ the uniformly weighted D–optimum design is supported by the origine and the set of barycenters of depth 1, 1 and 2, 1, 2 and 3 respectively.

In this paper optimal designs for models of the form (2) were investigated on $\mathcal{S}_q$. Section 2 includes preliminaries while in Section 3 φ_p–optimality of (2) will be studied. Finally in Section 4 D–optimum designs of some special forms of (2) without linear terms were considered.

2 Preliminaries

Basic model: In a standard (linear) regression equation $\eta(\mathbf{x}) = \sum_{r=1}^{m} \vartheta_r f_r(\mathbf{x})$ the $f_1, \ldots, f_m$ are continuous, real valued, linear independent regression functions on a compact set $\mathcal{X}$ and $\vartheta_1, \ldots, \vartheta_m$ are m unknown parameters. The outcome $\eta(\mathbf{x})$ of an experiment $\mathbf{x} \in \mathcal{X}$ is considered as a realization of the random variable $\mathcal{Y}$ with mean $\sum_{r=1}^{m} \vartheta_r f_r(\mathbf{x})$ and unknown variance σ^2 independent of $\mathbf{x}$. Here the observations in different experimental points are assumed to be uncorrelated. To estimate the coefficient vector ϑ observations at various values of the experimental point $\mathbf{x}$ are taken. The generalization of the proportions of observations in the pairwise different experimental points to probabilities leads to the (discrete) probability measures δ on $\mathcal{X}$ called designs. The support of a design δ is defined by the set $\text{supp}(\delta) = \{\mathbf{x} \in \mathcal{X} : \delta(\mathbf{x}) > 0\}$. Obviously at least m supporting points are necessary to estimate the m parameters of the model.

Optimality criteria: Designs δ are ordered by values of concave functions of the information matrix $\mathbf{M}(\delta) = \int \mathbf{f}(\mathbf{x})\mathbf{f}^t(\mathbf{x})\delta(\mathbf{x})$. In this context the information matrix is proportional to the covariance matrix of the best linear unbiased estimator of the parameter vector $\vartheta_1, \ldots, \vartheta_m$. Kiefer (1974) introduced the matrix mean called φ_p–criteria. He called a design φ_p-optimal, iff $\varphi_p(\mathbf{M}(\delta)) = [(1/k)\text{tr}(\mathbf{M}^{-p}(\delta))]^{(1/p)}$ is minimized by $\mathbf{M}(\delta)$, respectively δ. Moreover, he found a design to be φ_p–optimal ($p \geq -1$), iff it satisfies $\sup_{\mathbf{x}\in\mathcal{X}} d_p(\mathbf{x}; \delta^*) = \text{tr}(\mathbf{M}^{-p}(\delta^*))$ for all $\mathbf{x}$, where $d_p(\mathbf{x}; \delta) = \sup_{\mathbf{x}\in\mathcal{X}} \mathbf{f}^t(\mathbf{x})\mathbf{M}^{-(p+1)}(\delta)\mathbf{f}(\mathbf{x})$. Since $\int d_p(\mathbf{x}; \delta^*)d\delta^*(\mathbf{x}) = \text{tr}(\mathbf{M}^{-p}(\delta^*))$, an optimum design can have points of support only where $d_p(\mathbf{x}; \delta^*) = \text{tr}(\mathbf{M}^{-p}(\delta^*))$. Particularly the setting above leads to A–optimality for $p = 1$, to E–optimality for $p = \infty$, and to D–optimality with the well known equivalence to G–optimality for $p = 0$.

3 φ_p–optimality for Generalized Tic Polynomials

In the following φ_p–optimality for generalized tic polynomials (2) in q variables on the q dimensional simplex will be considered.

THROUGHOUT THIS SECTION IT IS ASSUMED, THAT ALL LINEAR TERMS ARE INCLUDED IN THE MODEL,

$$\text{i.e. } 1 \in \mathcal{J}.$$

For φ_p–optimality, the argumentation of Atwood (1969) and Hilgers and Bauer (1994) concerning D–optimality will to be adopted. Firstly the following corollary generalizes the D–optimality result lemma 2.1 of Hilgers and Bauer (1994) to φ_p–optimality:

Corollar 3.1 *A φ_p–optimal design for the regression model (2) with $1 \in \mathcal{J}$ and $\mathcal{J} \subseteq \{1, \ldots, q\}$ on $\mathcal{S}_q$ can only be supported on $\{\mathbf{0}\} \cup \mathcal{U}_q$.*

Proof: The proof follows the lines of the argumentation of lemma 2.1 in Hilgers and Bauer (1994) and theorem 2.1 of Atwood (1969). Atwood analysed the possible extrema of the generalized variance $d(\mathbf{x}; \delta) = \sup_{\mathbf{x}\in\mathcal{X}} \mathbf{f}^t(\mathbf{x})\mathbf{M}^{-1}(\delta)\mathbf{f}(\mathbf{x})$ on $\mathcal{S}_q$. However, here the extrema of $d_p(\mathbf{x}; \delta) = \sup_{\mathbf{x}\in\mathcal{X}} \mathbf{f}^t(\mathbf{x})\mathbf{M}^{-(p+1)}(\delta)\mathbf{f}(\mathbf{x})$ have to be analysed. Obviously the extrema are not affected by choosing $\mathbf{M}^{-(p+1)}$ instead of $\mathbf{M}^{-1}$ in the generalized variance function. ∎

Now possible supporting points of a φ_p–optimal design on $\mathcal{U}_q$ have to be identified. Again the argumentation follows the lines of theorem 2.1 of Atwood (1969) and corollary 2.2 of Hilgers and Bauer (1994). The modified formulation of corollary 2.2 for φ_p–optimal designs leads to:

Corollar 3.2 *Consider the regression model (2) with $\mathcal{J} \cap \{1\} \neq \emptyset$ and $\mathcal{J} \subseteq \{1, \ldots, q\}$, on the factor space $\mathcal{X} \subset \mathbb{R}^q$ which is symmetric under the interchange of some two coordinates. Let I be a line segment in $\mathcal{X}$, satisfying $\sum_{r=1}^q x_r = \alpha$, which is invariant under the interchange of those same two coordinates, and on which all coordinates but those two are constant.*

a) *If the regression functions $f(\mathbf{x}) = x_{s_1} \cdots x_{s_\ell}, 1 \leq s_1 < \cdots < s_\ell \leq q$, are zero on I for all $\ell \in \mathcal{J}$, then a φ_p-optimal design can be supported on I only at the endpoints.*

b) *If at least one regression function $f(\mathbf{x}) = x_{s_1} \cdots x_{s_\ell}, \ell \in \mathcal{J}$, does not vanish on I, then a φ_p-optimal design can be supported on I only at the midpoint and at the endpoints.*

Remark: If the regression equation is of the form $\eta(\mathbf{x}) = \vartheta_0 + \vartheta_1 x_1 + \vartheta_2 x_2 + \vartheta_3 x_3 + \vartheta_4 x_1 x_2$, i.e. the model is a linear function of one (or more) component, then the barycenters corresponding to the "missing" regression functions of higher order are ruled out.

Consequently the generalized tic polynomials (2) with $\mathcal{J} \subseteq \{1, \ldots, q\}$ and $\mathcal{J} \cap \{1\} \neq \emptyset$ defined on $\mathcal{S}_q$ can only be the supported in origine and a subset of all barycenters of depth 1 up to q.

Lemma 2 *(cf. Hilgers and Bauer, 1994)*
Let the regression model be of the form (2) on $\mathcal{S}_q$.

a) *For $\mathcal{J} \subseteq \{1, \ldots, q\}$ and $\mathcal{J} \cap \{1\} \neq \emptyset$ a φ_p-optimal design can only be supported by the origine, the set of all barycenters of depth 1 up to q.*

b) *In particular, for $\mathcal{J} = \{1, k, \ldots, q\}$, $2 \leq k \leq q$, a φ_p-optimal design can only be supported by the origine, the set of all vertices (barycenters of depth 1) and the barycenters of depth k up to q.*

Proof: The proof follows the lines of the argumentation which leads to theorem 2.3 part c) in Hilgers and Bauer (1994). For part b) above one has to notice, that on line segments on $\mathcal{U}_q$ where the regression model is a linear function in two components, the generalized variance is only a quadratic function in one component. ∎

Remarks:

a) It is worth to be noted, that the linear functions have to be included in the regression setup and that the application of corollary 2.2 assumes the symmetry of the regression setup.

b) The results of Hilgers and Bauer (1994) concerning minimum support designs in some cases allow the calculation of the weights for this specific φ_p-optimality criteria. Especially for A-optimality the criteria derived in Pukelsheim and Torsney (1991) are applicable. The weights of the A- and E-optimal design of the model $\eta(\mathbf{x}) = \vartheta_0 + \sum_{r=1}^{q} \vartheta_r x_r$ were given in Hilgers and Bauer (1994). Other concrete forms of the φ_p-optimal designs can be found by numerical computation.
φ_p-optimal designs for the special n-tic polynomials on $\mathcal{U}_q$ have been considered by various authors, cf. Kiefer (1975), Galil and Kiefer (1977), Kiefer (1978), Lorenz (1991). However, generally the results cannot be transferred to the simplex, because the φ_p-optimality criteria lacks of transformation invariance. As an example one might compare the results of the linear model $\eta(\mathbf{x}) = \sum_{r=1}^{q} \vartheta_r x_r$ on $\mathcal{U}_q$ with the corresponding model $\eta(\mathbf{x}) = \vartheta_0 + \sum_{r=1}^{q} \vartheta_r x_r$ on $\mathcal{S}_q$. On $\mathcal{U}_q$ the uniformly weighted design supported on all vertices is universally optimal, for the model $\eta(\mathbf{x}) = \sum_{r=1}^{q} \vartheta_r x_r$, whereas the uniformly weighted design supported on all vertices and the origine is only D-optimal for the model $\eta(\mathbf{x}) = \vartheta_0 + \sum_{r=1}^{q} \vartheta_r x_r$ on $\mathcal{S}_q$, cf. Hilgers (1991).

c) The above results can be formulated analogously for the corresponding polynomials without constant term, cf. Scheffé (1963), on S_q and on $\mathcal{U}_q$ as well. However, beause of the absence of the intercept in these models, the origine as supporting point is ruled out in the corresponding designs.

4 Examples for Generalized Tic Polynomials without Linear Terms

In the previous section results were derived for regression models including all linear terms, i.e. the regression functions $f(\mathbf{x}) = x_r, 1 \le r \le q$. The inclusion of all linear functions in the model is essential for the proofs of restricting the possible points of support. The argumentation does not hold if the set of regression functions is given by $x_1, x_3, x_1 x_2, x_2 x_3, x_1 x_3$ and the line segment under consideration is of the form $I = \{\mathbf{x} \in \mathcal{X} : \mathbf{x} = x_2 \mathbf{e}_2\}$. On this line the generalized variance is constant.

In this section D–optimal designs for three models will be considered. The common property of these models is the absence of some of the linear regression terms.

The first model is embedded in (2) by the choice $\mathcal{J} = \{q\}$. In applications this model has the following interpretation. In the case of not being able to make experiments with mono-substances and lower order mixtures, it is best to compare placebo with the equimixing dose. As one might expect from theorem 2.3 c) of Hilgers and Bauer (1994) the following result can be stated:

Lemma 3 *Let the regression be of the form*

$$\eta(\mathbf{x}) = \vartheta_0 + \vartheta_1 x_1 \cdots x_q$$

defined on S_q. The unique D–optimal design to estimate the coefficients by their BLUE *is given by the uniformly weighted design with support $\{0; (\sum_{r=1}^q \mathbf{e}_r)/q\}$.*

Proof: Obviously an orthogonal regression function set is given by $\{1 - q^q(x_1 \cdots x_q); q^q(x_1 \cdots x_q)\}$. Using arithmetic - geometric mean inequality it can be shown that the generalized variance is bounded by

$$\begin{aligned} \mathbf{g}^t(\mathbf{x})\mathbf{M}^{-1}\mathbf{g}(\mathbf{x}) &= 2[(1 - q^q x_1 \cdots x_q)^2 + (q^q x_1 \cdots x_q)^2] \\ &= 2[1 + 2q^q x_1 \cdots x_q(q^q x_1 \cdots x_q - 1)] \\ &\le 2 . \end{aligned}$$

■

Let $\mathcal{J} = \{2\}$ in (2), the model includes only all two-way interactions.

Lemma 4 *The unique D–optimal design to estimate the coefficients of the model*

$$\eta(\mathbf{x}) = \vartheta_0 + \sum_{1 \le r < s \le q} \vartheta_{rs} x_r x_s$$

defined on S_q by their BLUE *is given by the uniformly weighted design with support $\{0; (\mathbf{e}_r + \mathbf{e}_s)/2, 1 \le r < s \le q\}$.*

Proof: Here the orthogonal regression functions are given by $\{4x_r x_s, 1 \leq r < s \leq q;\ 1 - 4\sum_{1 \leq r < s \leq q} x_r x_s\}$. Notice that here the number of $k = 1 + q(q-1)/2$ supporting points equals the number of regression functions. So D-optimality can be stated by:

$$
\begin{aligned}
\frac{1}{k}\mathbf{g}^t(\mathbf{x})\mathbf{M}^{-1}\mathbf{g}(\mathbf{x}) &= \Big(1 - 4\sum_{1 \leq r < s \leq q} x_r x_s\Big)^2 + 16\sum_{1 \leq r < s \leq q}(x_r x_s)^2 \\
&= 1 - 4\sum_{1 \leq r \neq s \leq q} x_r x_s + 4\Big(\sum_{1 \leq r \neq s \leq q} x_r x_s\Big)^2 + 8\sum_{1 \leq r \neq s \leq q}(x_r x_s)^2 \\
&= 1 - 4\sum_{1 \leq r \neq s \leq q} x_r x_s\Big(1 - 2x_r x_s - \sum_{1 \leq r \neq s \leq q} x_r x_s\Big) \\
&= 1 - 4\sum_{1 \leq r \neq s \leq q} x_r x_s\Big(1 - 2x_r x_s - (\sum_{r=1}^{q} x_r)^2 + \sum_{r=1}^{q} x_r^2\Big) \\
&= 1 - 4\sum_{1 \leq r \neq s \leq q} x_r x_s\Big(1 - (\sum_{r=1}^{q} x_r)^2 + (x_r - x_s)^2 + \sum_{\substack{u \neq r,s \\ u=1}}^{q} x_u^2\Big) \\
&\leq 1 .
\end{aligned}
$$

∎

The last example is an incomplete two tic polynomial deviating from the general form of the model (2). This may be applied in practice, when one substance can be used only in combination with a second substance.

Lemma 5 *Let the regression be of the form*

$$\eta(\mathbf{x}) = \vartheta_0 + \vartheta_1 x_1 + \vartheta_2 x_1 x_2$$

defined on S_q. The unique D-optimal design to estimate the coefficients by their BLUE is given by the uniformly weighted design with support $\{0;\ \mathbf{e}_1;\ (\mathbf{e}_1 + \mathbf{e}_2)/2\}$.

Proof: Again the orthogonal regression function set with respect to the postulated design $\{4x_1 x_2;\ x_1 - 2x_1 x_2;\ 1 - x_1 - 2x_1 x_2\}$ is used to facilitate the calculation. The equivalence between D- and G-optimality leads to the following inequality

$$
\begin{aligned}
\mathbf{g}^t(\mathbf{x})\mathbf{M}^{-1}\mathbf{g}(\mathbf{x}) &= 3\left[(4x_1 x_2)^2 + (x_1 - 2x_1 x_2)^2 + (1 - x_1 - 2x_1 x_2)^2\right] \\
&= \left[1 + 2x_1^2 - 2x_1 - 4x_1 x_2 + 24x_1^2 x_2^2\right] \\
&= \left[1 + 2x_1(x_1 + x_2 - 1) + 24x_1^2 x_2^2(-(1/4) + x_1 x_2)\right] \leq 3 .
\end{aligned}
$$

∎

5 Further Comments

The main result of section 2 is that the support of the D-optimal design works well under other optimality criteria, cf. Hilgers and Bauer (1994). However, in contrary to the D-optimality most of the other optimality criteria lack of transformation invariance which is one of the favourable properties of the D-optimality for practical application. In particular in the context of mixture experiments the factor space often has to be transferred by the introduction of pseudo components, cf. Kurotori (1966), Piepel (1983) and Crosier (1986).

In section 3 the common property of the optimal designs seems to be, that the effect of ruling out linear terms in the model leads to the deletion of the corresponding supporting points. With the results of section 2 and Hilgers and Bauer (1994) this seems to work in a more general case too.

However, the setting of such not fully saturated model must be critizised from the practical point of view. Linear transformations of the variables yield to the inclusion of all terms up to a fixed order. Nevertheless, there seems to be a more general result for the relation between the absence of some of the regression functions and the corresponding supporting points.

Acknowledgement

This paper partly includes results of the author's PhD thesis at the University of Dortmund. The author is grateful to P. Bauer for useful discussions.

References

Atwood, C.L.: Optimal and efficient designs of experiments. *Ann. Statist.*, **40** (1969), 1570-1602.

Crosier, R. B.: The geometry of constrained mixture experiments. *Technometrics*, **28** (1986), 95-102.

Galil, Z. and Kiefer, J.: Comparison of simplex designs for quadratic mixture models. *Technometrics*, **19** (1977), 445-453.

Hilgers, R.-D.: Optimale Versuchsplanung in Mischungs-Mengen-Experimenten. Ph.D. Thesis, University of Dortmund, (1991).

Hilgers, R.-D. and Bauer P.: Optimal designs for mixture amount experiments. *submitted to J. Statist. Plann. Inf.* (1994).

Kiefer, J.: Optimum designs in regression problems II. *Ann. Math. Statist.*, **32** (1961), 298-325.

Kiefer, J.: General equivalence theory for optimum designs (approximate theory). *Ann. Statist.*, **2** (1974), 849-879.

Kiefer, J.: Optimal design: Variation in structure and performance under change of criterion. *Biometrika*, **62** (1975), 277-288.

Kiefer, J.: Asymptotic approach to families of design problems. *Commun. Statist. A*, **7** (1978), 1347-1362.

Kurotori, I. S.: Experiments with mixtures of components having lower bounds. *Industrial Quality Control*, **22** (1966), 592-596.

Lorenz, W.: Optimale Versuchsplanung für Mischungsexperimente bei polynomialer Regression. Diplomarbeit an der Freien Universität Berlin, 1991

Piepel, G. F.: Defining consistent constrained regions in mixture experiments. *Technometrics*, **25** (1983), 97-101.

Piepel, G. F. and Cornell, J. A.: Response depends on the total amount. *Technometrics*, **27** (1985), 219-227.

Pukelsheim, F. and Torsney B.: Optimal weights for experimental designs on linearly independent support points. *Ann. Statist.*, **19** (1991), 1614-1625.

Scheffé, H.: Experiments with mixtures. *J. Roy. Statist. Soc. B*, **20** (1958), 344-360.

Scheffé, H.: The simplex centroid design for experiments with mixtures. *J. Roy. Statist. Soc. B*, **25** (1963), 235-263.

Uranisi, H.: Optimum design for the special cubic regression on the q-simplex. *Mathematical Reports* (General Ed. Dept., Kuyushu University), **1** (1964), 7-12.

Relations between Spring and Chemical Balance Weighing Designs with the Diagonal Covariance Matrix of Errors

BRONISŁAW CERANKA and KRYSTYNA KATULSKA

Department of Mathematical and Statistical Methods, Agricultural University, 60-637 Poznań, Poland
and
Faculty of Mathematics and Computer Science, Adam Mickiewicz University, 60–769 Poznań, Poland.

Abstract:

The paper deals with the problem of estimating the individual weights of objects with minimum variances by using a weighing design with the diagonal covariance matrix of errors in the model. The necessary and sufficient conditions for optimum biased spring balance weighing designs with the diagonal covariance matrix of errors and for optimum chemical balance weighing designs with the diagonal covariance matrix of errors are given and the relations between these designs are investigated. Also new optimum weighing designs are found.

1 Introduction

The results of n weighing operations determining the individual weights of p light objects fit into the linear model

$$y = X\omega + e, \tag{1}$$

where y is the $n \times 1$ observed vector of the recorded weights, ω is the $p \times 1$ vector of the unknown weights (parameters) of the objects and $X = (x_{ij})$, $i = 1,\ldots,n$, $j = 1,\ldots,p$ is a matrix of known elements. A typical element x_{ij} is given by

$x_{ij} = +1$ if the jth object is placed on the right pan during the ith weighing operation,

$\quad = -1$ if the jth object is placed on the left pan during the ith weighing operation,

$\quad = 0$ if the jth object is not utilized in either pan during the ith weighing operation

in the case of a chemical balance;

---

*Kitsos, C.P., and Müller, W.G., Eds., *Proceedings of MODA4*, Physica Verlag, Heidelberg, 1995

112

$x_{ij} = +1$ if the jth object is placed on the pan during the ith weighing operation,

$\quad = 0$ if the jth object is not utilized during the ith weighing operation in a spring balance.

The vector **e** is an $n \times 1$ random vector of errors, with $E(\mathbf{e}) = \mathbf{0}_n$ and $E(\mathbf{ee}') = \sigma^2 \mathbf{G}$, where $\mathbf{0}_n$ denotes the $n \times 1$ vector with zero elements everywhere, "E" stands for the expectation, $\mathbf{e}'$ denotes the transpose of **e** and **G** is the known $n \times n$ diagonal positive definite matrix. The matrix **X** is called the weighing design matrix and we refer to it as the matrix of a weighing design with the covariance matrix $\sigma^2 \mathbf{G}$.

The normal equations estimating ω are of the form

$$\mathbf{X}'\mathbf{G}^{-1}\mathbf{X}\hat{\omega} = \mathbf{X}'\mathbf{G}^{-1}\mathbf{y}, \tag{2}$$

where (the solution) $\hat{\omega}$ is the column vector of the weights estimated by the least squares method.

The parameter vector ω is estimable if and only if rank $(\mathbf{X}) = p$ in which case

$$\hat{\omega} = (\mathbf{X}'\mathbf{G}^{-1}\mathbf{X})^{-1}\mathbf{X}'\mathbf{G}^{-1}\mathbf{y} \tag{3}$$

and the covariance matrix of $\hat{\omega}$ is

$$Var(\hat{\omega}) = \sigma^2(\mathbf{X}'\mathbf{G}^{-1}\mathbf{X})^{-1}. \tag{4}$$

Katulska (1989) showed that the minimum attainable variance for each of the estimated weights for a chemical balance weighing design with the covariance matrix $\sigma^2 \mathbf{G}$ is $\frac{\sigma^2}{(tr\mathbf{G}^{-1})}$, i.e.

$$Var(\hat{\omega}_j) \geq \frac{\sigma^2}{tr\mathbf{G}^{-1}}, \quad j = 1, \ldots, p. \tag{5}$$

For the particular case of $\mathbf{G} = \mathbf{I}_n$, where $\mathbf{I}_n$ is a unit matrix of order n we have the result of Hotelling (1944).

In the special case of a spring balance weighing design when bias is present, it can be assumed to be one object and its value is estimated by taking the column of the elements 1 in **X** corresponding to the bias, i.e.

$$\mathbf{X} = \left[\mathbf{1}_n \vdots X_1\right], \tag{6}$$

where $\mathbf{1}_n$ is the $n \times 1$ column vector of ones.

Ceranka and Katulska (1990, 1992a) showed that the minimum attainable variance for each of the estimated weights for a biased spring balance weighing design with the diagonal covariance matrix of errors is $4\sigma^2/(tr\mathbf{G}^{-1})$, i.e.

$$Var(\hat{\omega}_j) \geq \frac{4\sigma^2}{tr\mathbf{G}^{-1}}, \quad j = 2, \ldots, p. \tag{7}$$

For the particular case of $\mathbf{G} = \mathbf{I}_n$ we have the result of Moriguti (1954).

2 Optimum Weighing Designs

Definition 2.1. A nonsingular chemical balance weighing design **X** with the covariance matrix $\sigma^2 \mathbf{G}$ is said to be optimum for the estimated individual weights if

$$Var(\hat{\omega}_j) = \frac{\sigma^2}{tr\mathbf{G}^{-1}}, \quad j = 1, \ldots, p.$$

Definition 2.2. A nonsingular biased spring balance weighing design $\mathbf{X}$ with the co-variance matrix $\sigma^2\mathbf{G}$ is said to be optimum for the estimated individual weights if

$$Var(\hat{\omega}_j) = \frac{4\sigma^2}{tr\mathbf{G}^{-1}}, \quad j = 2, \ldots, p.$$

The following theorems give necessary and sufficient conditions under which a chemical balance weighing design with the covariance matrix $\sigma^2\mathbf{G}$ or a biased spring balance weighing design with the same covariance matrix $\sigma^2\mathbf{G}$ is optimum for the estimated individual weights.

Theorem 2.1. A nonsingular chemical balance weighing design $\mathbf{X}$ with the covariance matrix $\sigma^2\mathbf{G}$ is optimum for the estimated individual weights if and only if

$$\mathbf{X}'\mathbf{G}^{-1}\mathbf{X} = (tr\mathbf{G}^{-1})\mathbf{I}_p. \tag{7}$$

This result was originally proved by Katulska (1989).

For the case of a biased spring balance weighing design we have

Theorem 2.2. A nonsingular biased spring balance weighing design $\mathbf{X} = \left[\mathbf{1}_n \vdots X_1\right]$ with the covariance matrix $\sigma^2\mathbf{G}$ is optimum for the estimated individual weights if and only if

$$\mathbf{X}_1'\mathbf{G}^{-1}\mathbf{X}_1 = \frac{tr\mathbf{G}^{-1}}{4}(\mathbf{I}_{p-1} + \mathbf{1}_{p-1}\mathbf{1}_{p-1}'). \tag{8}$$

This theorem was proved by Ceranka and Katulska (1990).

We now investigate relations between optimum chemical balance weighing designs and optimum biased spring balance weighing designs.

Theorem 2.3. A biased spring balance weighing design $\mathbf{X} = \left[\mathbf{1}_n \vdots X_1\right]$ with the co-variance matrix $\sigma^2\mathbf{G}$ is optimum for the estimated individual weights if and only if the chemical balance weighing design $\mathbf{X}^* = 2\mathbf{X} - \mathbf{1}_n\mathbf{1}_p'$ with the same covariance matrix $\sigma^2\mathbf{G}$ is optimum for the estimated individual weights.

Proof. Let $\mathbf{X} = \left[\mathbf{1}_n \vdots X_1\right]$ be the matrix of an optimum biased spring balance weighing design with the covariance matrix $\sigma^2\mathbf{G}$ for the estimated individual weights. Then $\mathbf{X}^* = \left[\mathbf{1}_n \vdots 2\mathbf{X}_1 - \mathbf{1}_n\mathbf{1}_{p-1}'\right]$ and

$$\mathbf{X}^{*\prime}\mathbf{G}^{-1}\mathbf{X}^* = \begin{bmatrix} \mathbf{1}_n'\mathbf{G}^{-1}\mathbf{1}_n & 2\mathbf{1}_n'\mathbf{G}^{-1}\mathbf{X}_1 - \mathbf{1}_n'\mathbf{G}^{-1}\mathbf{1}_n\mathbf{1}_{p-1}' \\ 2\mathbf{X}_1'\mathbf{G}^{-1}\mathbf{1}_n - \mathbf{1}_{p-1}\mathbf{1}_n'\mathbf{G}^{-1}\mathbf{1}_n & (2\mathbf{X}_1' - \mathbf{1}_{p-1}\mathbf{1}_n')\mathbf{G}^{-1}(2\mathbf{X}_1 - \mathbf{1}_n\mathbf{1}_{p-1}') \end{bmatrix}. \tag{9}$$

From the assumption it follows that the condition (2.2) is satisfied. This implies

$$\mathbf{X}^{*\prime}\mathbf{G}^{-1}\mathbf{X}^* = \begin{bmatrix} \mathbf{1}_n'\mathbf{G}^{-1}\mathbf{1}_n & \mathbf{0}_{p-1}' \\ \mathbf{0}_{p-1} & (\mathbf{1}_n'\mathbf{G}^{-1}\mathbf{1}_n)\mathbf{I}_{p-1} \end{bmatrix} = (tr\mathbf{G}^{-1})\mathbf{I}_p.$$

Hence, $\mathbf{X}^*$ is the matrix of an optimum chemical balance weighing design with the covariance matrix $\sigma^2\mathbf{G}$ for the estimated individual weights.

Now, let $\mathbf{X} = \left[\mathbf{1}_n \vdots X_1\right]$ denote the matrix of a biased spring balance weighing design with the covariance matrix $\sigma^2\mathbf{G}$ and $\mathbf{X}^* = 2\mathbf{X} - \mathbf{1}_n\mathbf{1}_p'$ is the matrix of an optimum chemical balance weighing design with the covariance matrix $\sigma^2\mathbf{G}$ for the estimated individual weights. From (2.3) and (2.1) it follows immediately that

$$2\mathbf{1}_n'\mathbf{G}^{-1}\mathbf{X}_1 - \mathbf{1}_n'\mathbf{G}^{-1}\mathbf{1}_n\mathbf{1}_{p-1}' = \mathbf{0}_{p-1}'$$

and

$$4\mathbf{X}_1'\mathbf{G}^{-1}\mathbf{X}_1 - 2\mathbf{1}_{p-1}\mathbf{1}_n'\mathbf{G}^{-1}\mathbf{X}_1 - 2\mathbf{X}_1'\mathbf{G}^{-1}\mathbf{1}_n\mathbf{1}_{p-1}' + \mathbf{1}_{p-1}\mathbf{1}_n'\mathbf{G}^{-1}\mathbf{1}_n\mathbf{1}_{p-1}' = (\mathbf{1}_n'\mathbf{G}^{-1}\mathbf{1}_n)\mathbf{I}_{p-1}.$$

Hence

$$\mathbf{X}_1'\mathbf{G}^{-1}\mathbf{X}_1 = \frac{tr\mathbf{G}^{-1}}{4}(\mathbf{I}_{p-1} + \mathbf{1}_{p-1}\mathbf{1}_{p-1}').$$

It denotes that $\mathbf{X} = [\mathbf{1}_n \vdots \mathbf{X}_1]$ is the matrix of an optimum biased spring balance weighing design with the covariance matrix $\sigma^2\mathbf{G}$ for the estimated individual weights. So, the theorem is proved.

For the particular case of $\mathbf{G} = \mathbf{I}_n$ Theorem 2.3 was proved by Ceranka and Katulska (1992b).

In the next section we apply Theorem 2.3 to obtain new optimum weighing designs with the covariance matrix $\sigma^2\mathbf{G}$ for the estimated individual weights.

3   Some Optimum Weighing Designs

It is obvious that we have many interesting possibilities of patterns of matrix $\mathbf{G}$. The constructions of optimum biased spring balance weighing designs with the covariance matrix $\sigma^2\mathbf{G}$ for each of forms of $\mathbf{G}$ must be investigated seperately.

Ceranka and Katulska (1990) constructed optimum biased spring balance weighing designs with the covariance matrix $\sigma^2\mathbf{G}$ for the estimated individual weights, when the matrix $\mathbf{G}$ was of the form

$$\mathbf{G} = \begin{cases} \begin{bmatrix} a\mathbf{I}_c & \mathbf{0}_c\mathbf{0}_d' & \mathbf{0}_c\mathbf{0}_{n-c-d}' \\ \mathbf{0}_d\mathbf{0}_c' & a_1\mathbf{I}_d & \mathbf{0}_d\mathbf{0}_{n-c-d}' \\ \mathbf{0}_{n-c-d}\mathbf{0}_c' & \mathbf{0}_{n-c-d}\mathbf{0}_d' & \mathbf{I}_{n-c-d,} \end{bmatrix}, & \begin{array}{l} a>0,\ a_1>0, \\ c>0,\ d>0, \end{array} \\[20pt] \begin{bmatrix} a_1\mathbf{I}_d & \mathbf{0}_d\mathbf{0}_{n-d}' \\ \mathbf{0}_{n-d}\mathbf{0}_d' & \mathbf{I}_{n-d} \end{bmatrix}, & \begin{array}{l} a_1>0, \\ c=0,\ d>0, \end{array} \\[20pt] \begin{bmatrix} a\mathbf{I}_c & \mathbf{0}_c\mathbf{0}_{n-c}' \\ \mathbf{0}_{n-c}\mathbf{0}_c' & \mathbf{I}_{n-c} \end{bmatrix}, & \begin{array}{l} a>0, \\ c>0,\ d=0. \end{array} \end{cases} \tag{10}$$

Constructions of optimum biased spring balance weighing designs with the covariance matrix $\sigma^2\mathbf{G}$ for the estimated individual weights for the particular case of $\mathbf{G}$ when $c=1$ and $d=0$ was considered by Ceranka and Katulska (1992a).

Suppose further that the matrix $\mathbf{X}$ is partitioned in the same way as matrix $\mathbf{G}$ given by (3.1), i.e.

$$
\mathbf{X} = \left\{
\begin{array}{ll}
\begin{bmatrix} 1_c & 0_c0'_v \\ 1_d & 1_d1'_v \\ 1_b & \mathbf{N}' \end{bmatrix}, & c > 0, \quad d > 0, \\[6pt]
\begin{bmatrix} 1_d & 1_d1'_v \\ 1_b & \mathbf{N}' \end{bmatrix}, & c = 0, \quad d > 0, \\[6pt]
\begin{bmatrix} 1_c & 0_c0'_v \\ 1_b & \mathbf{N}' \end{bmatrix}, & c > 0 \quad d = 0,
\end{array}
\right.
\tag{11}
$$

respectively, where $p = v + 1$, $n = b + c + d$ and $\mathbf{N}$ denotes the binary incidence matrix of a balanced incomplete block (BIB) design with parameters v, b, r, k, λ. Ceranka and Katulska (1990) investigated when the design matrices given by (3.2) are the matrices of optimum spring balance weighing designs with the covariance matrix $\sigma^2\mathbf{G}$, where $\mathbf{G}$ is given by (3.1), for the estimated individual weights. They proved the following theorem.

Theorem 3.1. A biased spring balance weighing design $\mathbf{X}$ given by (3.2) with the covariance matrix $\sigma^2\mathbf{G}$, where $\mathbf{G}$ is given by (3.1) is optimum for the estimated individual weights if and only if

$$
\frac{c}{a} + \frac{d}{a_1} + b = 4(r - \lambda)
$$

and

$$
r = 2\lambda + \frac{d}{a_1}.
$$

Now, from Theorem 2.3 and Theorem 3.1 we have
Theorem 3.2. A chemical balance weighing design

$$
\mathbf{X}^* = \left\{
\begin{array}{ll}
\begin{bmatrix} 1_c & -1_c1'_v \\ 1_d & 1_d1'_v \\ 1_b & 2\mathbf{N}' - 1_b1'_v \end{bmatrix}, & a > 0, d > 0, \\[6pt]
\begin{bmatrix} 1_d & 1_d1'_v \\ 1_b & 2\mathbf{N}' - 1_b1'_v \end{bmatrix}, & c = 0, d > 0, \\[6pt]
\begin{bmatrix} 1_c & -1_c1'_v \\ 1_b & 2\mathbf{N}' - 1_b1'_v \end{bmatrix}, & c > 0, d = 0,
\end{array}
\right.
\tag{12}
$$

with the covariance matrix $\sigma^2\mathbf{G}$, where $\mathbf{G}$ is given by (3.1) is optimum for the estimated individual weights if and only if

$$
\frac{c}{a} + \frac{d}{a_1} + b = 4(r - \lambda)
$$

and

$$
r = 2\lambda + \frac{d}{a_1}.
$$

The parameter combinations of the existing BIB designs leading to optimum biased spring balance weighing designs X given by (3.2) with the covariance matrix $\sigma^2\mathbf{G}$, where $\mathbf{G}$

116

is given by (3.1), for the estimated individual weights are given by Ceranka and Katulska (1990).

Now, we will consider the construction of an optimum biased spring balance weighing design for the estimated individual weights from the optimum chemical balance weighing design. The main idea of this construction is contained in the following

Theorem 3.3. If $\mathbf{X}_i^*$, $i = 1, \ldots, t$ is the $n_i \times p$ matrix of an optimum chemical balance weighing design with the covariance matrix $\sigma^2 \mathbf{G}_i$ for the estimated individual weights then $\mathbf{X}^* = \left[\mathbf{X}_1^{*\prime} \vdots \ldots \vdots \mathbf{X}_t^{*\prime} \right]'$ is the $\sum_{i=1}^{t} n_i \times p$ matrix of an optimum chemical balance weighing design with the covariance matrix $\sigma^2 \mathbf{G}$ where $\mathbf{G} = diag\{\mathbf{G}_1, \ldots, \mathbf{G}_t, \}$, for the estimated individual weights.

This theorem was proved by Katulska (1989). From the above theorem and Theorem 2.3 we have

Theorem 3.4. If $\mathbf{X}_i^* = \left[\mathbf{1}_{n_i} \vdots \mathbf{X}_{1_i}^* \right]$, $i = 1, \ldots, t$, is the $n_i \times p$ matrix of an optimum chemical balance weighing design with the covariance matrix $\sigma^2 \mathbf{G}_i$ for the estimated individual weights then $\mathbf{X} = \left[\mathbf{X}_1' \vdots \ldots \mathbf{X}_t' \right]'$, where $\mathbf{X}_i = \frac{1}{2}(\mathbf{X}_i^* + \mathbf{1}_{n_i} \mathbf{1}_p')$ is the $\sum_{i=1}^{t} n_i \times p$ matrix of an optimum biased spring balance weighing design with the covariance matrix $\sigma^2 \mathbf{G}$, where $\mathbf{G} = diag\{\mathbf{G}_1, \ldots, \mathbf{G}_t\}$, for the estimated individual weights.

References

Ceranka B., Katulska K. (1990). Constructions of optimum biased spring balance weighing designs with the diagonal covariance matrix of errors. Computational Statistics and Data Analysis 10, 121-131.

Ceranka B., Katulska K. (1992a). Optimum biased spring balance weighing designs with non-homogeneity of the variances of errors. Journal of Statistical Planning and Inference 30, 185-193.

Ceranka B., Katulska K. (1992b). Relations between optimum biased spring balance weighing designs and optimum chemical balance weighing designs. Probastat'91, Proceedings of the 10th Conference on Probability and Mathematical Statistics, Bratislava, Czechoslovakia, August, 26-30, 1991, 95-101.

Hotelling H. (1944). Some improvements in weighing and other experimental techniques. Ann. Math. Statist., 15, 297-306.

Katulska K. (1989). Optimum chemical balance weighing designs with non-homogeneity of the variances of errors. J. Japan Statist. Soc., 19, 95-101.

Moriguti S. (1954). Optimality of orthogonal designs. Rep. Statist. Appl. Res., 3, 1-24.

Optimal Design for Experiments with Potentially Failing Trials

PETER HACKL

Department of Statistics, University of Economics and Business Administration, Augasse 2-6, 1090 Vienna, Austria.

Abstract:

We discuss the problem of optimal allocation of the design points of an experiment for the case where the trials may fail with non-zero probability. Numerical results for D-optimal designs are given for estimating the coefficients of a polynomial regression. For small sample sizes these designs may deviate substantially from the corresponding designs in the case of certain response. They can be less efficient, but are less affected by failing trials.

1 Introduction

Application of the standard theory of optimal design is based on the assumption that all trials of an experiment result in corresponding observations of the response variable. However, we are never sure that the responses of all trials will really be available when the experiment is actually performed. This leads us to consider designs for potentially failing trials, i.e., designs where the probability for getting a response is less than one for some or all sites in the design space.

The problem of potentially failing trials has been treated by Hedayat and Majumdar (1983) and later by Das and Sinha (1994) in the context of redesigning experiments due to shortened resources: Which observation should be dropped given the probabilities that the trials will be failing at the sites of the planned experiment. Our paper discusses a more general question: What is the optimal experiment in the case of possibly failing trials.

The paper is organized as follows. In Section 2 we state the problem and present criteria that allows us to assess the optimality of candidate designs. In Section 3 we show on the basis of a quadratic polynomial how D-optimal designs for estimating the model coefficients change due to failing trials.

---

*Kitsos, C.P., and Müller, W.G., Eds., *Proceedings of MODA4*, Physica Verlag, Heidelberg, 1995

2 Statement of the Problem

Let us assume that the response variable Y is determined by a regressor variable x according to the model

$$Y = \beta_1 + \beta_2 f_2 + \ldots + \beta_m f_m + u = f^T \beta + u \qquad (1)$$

where the components of $f = [1, f_2, \ldots, f_m]^T$ are functions of x: $f_i = f_i(x)$, and β is an m-vector, and the error term u has expectation zero and variance σ^2 for all x; the error terms are assumed to be pairwise independent.

Our aim is to estimate the coefficient vector β. For that reason we plan to take a sample of Y's at a set of sites. The set $\mathcal{X}$ of all sites of interest is the design space for our experiment. A design is represented by a measure ξ over $\mathcal{X}$; given that $\mathcal{X}$ comprises k points, the design is written as

$$\xi = \left\{ \begin{array}{ccc} x_1 & \cdots & x_k \\ n_1 & \cdots & n_k \end{array} \right\}$$

or shorter as $\xi = \{x_1, \ldots, x_k; n_1, \ldots, n_k\}$, where x_i, $i = 1, \ldots, k$, are distinct elements from $\mathcal{X}$ and the integers n_i, $i = 1, \ldots, k$, indicate the number of observations taken at x_i, and $\sum_{i=1}^{k} n_i = r$.

For such an r-trial design, the information matrix for the parameter vector β is $F^T F = \sum_{i=1}^{k} f_i f_i^T n_i$; the matrix F consists of k rows f_i^T. The information matrix per observation is

$$M(\xi) = \frac{1}{r} \sum_{i=1}^{k} f_i f_i^T n_i = \frac{1}{r} F^T F = \frac{\sigma^2}{r} [D_{\hat{\beta}}(\xi)]^{-1} \qquad (2)$$

where $D_{\hat{\beta}}(\xi)$ is the corresponding covariance matrix of the LS-estimator $\hat{\beta}$. The covariance matrix is, given a quadratic loss function, a measure of the expected loss or risk in estimating β by $\hat{\beta}$. Consequently, the information matrix M and the covariance matrix D are important devices for the choice of an experimental design. An optimal design is a design ξ^* that minimizes a suitable measure of imprecision or risk: The most popular criterion is the D–optimality in which $\Psi[M(\xi)] = \log |M(\xi)|^{-1} = -\log |M(\xi)|$ is minimized. This implies the minimization of the determinant $|D_{\hat{\beta}}(\xi)|$ of the expected loss.

In the case of potentially failing trials, i.e., when the probability is less than one that a planned observations is actually observed, the covariance matrix or the expected loss has to take this fact into account: Let $1 - p_i$ be the probability that the trial at site x_i fails ($i = 1, \ldots, k$); and let us assume that the events that trials are failing are independent. As a consequence, the number ρ of observations that are available after the execution of an r-trial experiment for estimating β fulfills the inequality $\rho \leq r$. As the LS-estimator $\hat{\beta}$ requires that ρ is at least m, the event that $\hat{\beta}$ cannot be calculated on the basis of the realized sample has a non-zero probability.

For an r-trial design ξ_u for an experiment with potentially failing trials, the expected loss in estimating the m-vector β of coefficients is the weighted sum of covariance matrices that correspond to the various sets of possibly available observations:

$$D_{\hat{\beta}}(\xi_u) = \sum_{\rho=1}^{r} \frac{1}{\rho} \sum_{j \in J_\rho} p_{\rho,j}^{(r)} \left[M(\xi_{\rho,j}^{(r)}) \right]^{-1} ; \qquad (3)$$

here, for each ρ the set J_ρ corresponds to the set of all possible subsets of size ρ from the sites $\mathcal{X}(\xi_u)$ of ξ_u, i.e., j counts the ρ-trial designs $\xi_{\rho,j}^{(r)}$; the weight for the design $\xi_{\rho,j}^{(r)}$ is the probability

$$p_{\rho,j}^{(r)} = \prod_{i \in I_{\rho,j}^{(a)}(\xi_u)} p_i \prod_{i \in I_{\rho,j}^{(m)}(\xi_u)} (1 - p_i) \tag{4}$$

where the index sets $I_{\rho,j}^{(a)}(\xi_u)$ and $I_{\rho,j}^{(m)}(\xi_u)$ correspond to the subsets of sites where the responses are available and missing, respectively; the information matrices per observation are

$$M(\xi_{\rho,j}^{(r)}) = \frac{1}{\rho} \sum_{i \in I_{\rho,j}^{(a)}(\xi_u)} f_i f_i^T \tag{5}$$

The determinant $|D_{\hat{\beta}}(\xi)|$ of the expected loss is unbounded if the information matrix $M(\xi)$ for the design ξ is singular; of course, such a design is no candidate for being chosen as an optimal design. Given a realistic structure of the p_i, $i = 1, \ldots, k$, we have to expect that singular information matrices are found among those in (3) for any design ξ_u. This means that $D_{\hat{\beta}}(\xi_u)$ is with high probability unbounded. We have therefore to look for a appropriate policy with respect to designs with singular information matrix. Our suggestion is to add a suitably chosen penalty term which is a multiple of the probability that a design results in a singular information matrix:

$$|D_{\hat{\beta}}(\xi_u)| = \left| \sum_{\rho=1}^{r} \frac{1}{\rho} \sum_{j \in J_\rho^r} p_{\rho,j}^{(r)} \left[M(\xi_{\rho,j}^{(r)}) \right]^{-1} + \lambda P_{r,\rho}^s I_m \right| ; \tag{6}$$

here, J_ρ^r and J_ρ^s correspond to the set of designs with regular and singular information matrix, respectively; λ is the penalty and $P_{r,\rho}^s$ is the probability that a design results in a singular information matrix. The effect of the choice of λ is analyzed in the following section.

An optimal design ξ for estimating β minimizes the determinant $|D_{\hat{\beta}}(\xi)|$ of the covariance matrix of the estimate $\hat{\beta}$. In the case of certain response the basic idea of algorithms for obtaining D-optimal designs is the relation

$$|M_{r+1}| = |M_r| \left[\frac{r}{r+1} + \frac{d(x, \xi_r)}{r} \right]$$

where

$$d(x, \xi_r) = \frac{r \operatorname{Var}\{\hat{y}_x\}}{\sigma^2} = f^T(x) M_r^{-1} f(x)$$

is the standardized variance of the forecast $\hat{y}_x$ for the response at x, based on the observation from design ξ_r. This relation implies that the addition of a trial at the site where $d(x, \xi_r)$ is maximum results in the largest possible increase of the information matrix. This relation can be used to approach the optimal design, e.g., by iteratively adding and deleting sites. In the potentially failing trial case, however, the effect of adding a trial is a weighted mean of the effects of adding the trial to all the subdesigns that can occur due to failing trials; the weights are functions of the probabilities $p_{\rho,j}^{(r)}$ and the information matrices $M(\xi_{\rho,j}^{(r)})$. As a consequence, the numerical effort in assessing possible candidates to be added or deleted is considerable.

3 Examples

On the basis of three examples, we investigate the effect of potentially failing trials on optimal design. For the probabilities of failing trials we choose three structures:

(a) constant probabilities

(b) constant probabilities and independence of failing trials

(c) increasing probabilities

We assume that the response variable Y is generated according to the quadratic polynomial

$$Y = \beta_0 + \beta_1 x + \beta_2 x^2 + u = f^T \beta + u \, ; \tag{7}$$

x is a non-random regressor variable with design space

$$\mathcal{X}_k = \left\{ 0, \frac{1}{k-1}, \dots, 1 \right\} \, ;$$

the error term u has expectation zero and variance σ^2 for all x in $\mathcal{X}_k$ and the error terms for any two observations are independent.

For the case where potentially failing trials are ignored, the D-optimal continuous design with $\mathcal{X} = [0,1]$ is $\xi_c^* = \{0, 0.5, 1; \frac{1}{3}, \frac{1}{3}, \frac{1}{3}\}$. Exact designs for the discrete design space $\mathcal{X}_k$ are similar in the sense that the observations are concentrated in sites that are close to the sites of ξ_c^*. *Table 1* gives corresponding designs for some values of k and sample sizes r together with the value of the optimality criterion in column "$p_0 = 1$"; the designs are indicated by "1" at sites where trials are planned and "." otherwise. The designs were found by comparing the optimality criteria of suitable candidates.

3.1 A Simple Probability Structure

Let us assume that at all sites at most one observation is taken; thus, the design of our experiment has the form $\xi_u = \{x_1, \dots, x_r; 1, \dots, 1\}$, where the $x_i \in \mathcal{X}_k$ are the selected sites. In modelling the probability structure we choose probabilities p_i, $i = 1, \dots, k$, for the events that a response is observed at all sites except at x_i: $p_1 = \dots = p_k = p$. Then, $p_0 = 1 - \sum_{i=1}^k p_i$ is the probability for the event that the responses of all sites are available or the responses of at least two sites are missing. Given these assumptions, we find for the expected loss or the covariance matrix of the LS-estimates $\hat{\beta}$

$$
\begin{aligned}
D_p(\hat{\beta}) &= \sum_{j=i}^r p_j D(\hat{\beta}_{-j}) + \left(1 - \sum_{j=1}^r p_j \right) D(\hat{\beta}) \\
&= \sigma^2 \left\{ \frac{1}{r-1} \sum_{j=1}^r p_j [M(\xi_{-j})]^{-1} + \frac{1}{r} \left(1 - \sum_{j=1}^r p_j \right) [M(\xi)]^{-1} \right\} \, ; \tag{8}
\end{aligned}
$$

here, ξ_{-j} and $\hat{\beta}_{-j}$ are the design and the LS-estimate, respectively, where the observation at site x_j is missing.

Table 1 shows the D-optimal designs for $p_0 = 1$, 0.8, and 0.2 for three values of k and some sample sizes r, together with the value of the optimality criterion. As it can be expected, for a given design the value of the optimality criterion increases with decreasing

p_0; this corresponds to the fact that with decreasing p_0, the weight of non-optimal designs increases which become effective due to failing trials. Even more interesting is the fact that the optimal design changes for small values of r with decreasing p_0. E.g., for $k = 8$ and $r = 4$, the optimal designs are (A) '1 . . 1 1 . . 1' and (B) '1 . 1 . . 1 . 1' for the case of certain response and potentially failing trials, respectively. The reason is that the latter design is more robust against failing trials: A missing observation in $x = 0$ (or $x = 1$) increases the optimality criterion by a factor 100 (from 8.17 to 817.0) in case of design A but only by 7.2 (from 18.2 to 130.72) in case B. The effect of omitting one of the inner observations is about the same: increase from 8.17 to 16.67 in case A, from 18.2 to 24.01 in case B.

Table 1: Exact D-optimal designs for experiments with constant probability for the trials to fail and $p_0 = 1$, 0.8, and 0.2; three sizes k of the design space $\mathcal{X}_k$ and various sample sizes r. In cases of unsymmetrical designs, the corresponding mirrored design has the same value of the optimality criterion. The respective values of the optimality criterion is given for three values of p_0; the minimal values of the optimality criterion are indicated by '∗'.

k	r	$\lvert D(\xi)\rvert$ $p_0 = 1$	$p_0 = 0.8$	$p_0 = 0.2$	ξ
6	4	8.35∗	65.12	649.36	1 . 1 1 . 1
		14.36	23.71∗	72.15∗	1 1 . . 1 1
	5	5.75∗	15.09	65.82	1 1 1 1 . 1
		6.34	9.46∗	23.95∗	1 1 1 . 1 1
	6	3.99	5.66	13.04	1 1 1 1 1 1
8	4	8.17∗	103.72	1169.13	1 . . 1 1 . . 1
		10.14	18.20∗	61.03∗	1 . 1 . . 1 . 1
	5	5.45∗	15.30	59.56	1 . . 1 1 . 1 1
		6.30	8.02∗	14.99∗	1 1 . 1 . . 1 1
	6	3.63	4.43	7.49	1 1 . 1 1 . 1 1
	7	2.71	3.26	5.32	1 1 . 1 1 1 1 1
	8	2.08	2.47	3.89	1 1 1 1 1 1 1 1
10	4	8.10∗	152.99	1855.79	1 . . . 1 1 . . . 1
		10.32	17.63∗	40.70∗	1 . 1 . . . 1 . . 1
	5	5.20∗	16.28	61.62	1 . . . 1 1 . . 1 1
		6.08	7.42∗	12.54∗	1 1 . . 1 . . . 1 1
	6	3.34	3.84	5.65	1 1 . . 1 1 . . 1 1
	7	2.41	2.75	3.94	1 1 . 1 1 1 . . 1 1
	8	1.83	2.08	2.93	1 1 . 1 1 1 1 . 1 1
	9	1.49	1.68	2.32	1 1 1 1 1 1 1 . 1 1
	10	1.22	1.36	1.84	1 1 1 1 1 1 1 1 1 1

3.2 Independently Failing Trials

We assume again that at all sites at most one observation is taken. For all sites the probability is p that a trial is failing, and the events that trials are failing are assumed to be independent. Thus, the probabilities $p_{\rho,j}^{(r)}$ in (6) are

$$p_{\rho,j}^{(r)} = \binom{r}{\rho} p^{\rho} (1-p)^{r-\rho}.$$

For the numerical illustration, two values for p were chosen out of the interval (B) [.86, .92] and (C) [.57, .73] so that the probability $P\{S < k-1\}$ is about 0.8 and 0.2, respectively, that the number S of non-failing trials is less that $k-1$. The corresponding probabilities are shown in the following table.

Table 2 shows the D-optimal designs for the cases (A) $p = 1$ and (B) and (C) as stated above, again for the three values of k and the sample sizes r, that were used in Section 3.1. In all reported cases, the penalty was set to zero ($\lambda = 0$). However, a nonzero penalty increases the value of the optimality criterion but only for very large values of λ the ranking of the designs. Contrary to the cases of Section 3.1, the value of the optimality criterion can be increased or decreased in the cases (B) and (C) as compared to the case $p = 1$. Particularly for small values of r, the optimality criterion of the optimal designs in cases (B) and (C) are considerably larger than that of the optimal design for $p = 1$. In agreement with the results in Section 3.1, the optimal designs tend to give weight to points that are away from the center with increasing p.

	k	p	$P\{S < k-1\}$	$P\{S = k-1\}$	$P\{S = k\}$	$P\{S \leq 2\}$
(B)	6	0.860	0.200	0.395	0.405	0.005
	8	0.895	0.202	0.386	0.412	0.000
	10	0.917	0.199	0.381	0.420	0.000
(C)	6	0.575	0.804	0.160	0.036	0.216
	8	0.670	0.799	0.160	0.041	0.019
	10	0.729	0.800	0.158	0.042	0.001

3.3 Increasing Probability for Failing Trials

Here, the probability structure is the same as that in Section 3.1 except that we let the probabilities grow: The probability, that a response is observed at all sites except at x_i, is now $p_i = p(i - 1)$, $i = 1, \ldots, k$, where p is chosen so that $p_0 = 1 - \sum_{i=1}^{k} p_i$ gets a prespecified value.

Table 3 shows the D-optimal designs for $p_0 = 1$, 0.8, and 0.2 for three values of k and some sample sizes r, together with the value of the optimality criterion. Here, in cases of unsymmetrical designs the optimality criterion of the mirrored designs is not the same. E.g., for $k = 8$, $r = 4$ and $p_0 = 0.2$, the optimality criteria for the optimal design '1 . 1 . . . 1 1' and the design '1 1 . . . 1 . 1' are 41.21 and 108.26, respectively. This is due to the fact that the loss of an observation is much more serious in the second case: A failing trial both at $x = 0$ in the former and at $x = 1$ in the latter case results in a design criterion of 294.12; the probabilities that this happens is 0 and 0.167 in the two cases; similarly, the design criterion is 66.69 for a failing trial in $x = 2/7$ and $x = 5/7$, and the corresponding probabilities are 0.038 and 0.111, respectively. As a consequence, the effect of potentially failing trials is reduced as compared to the case of constant probability of failing trials as discussed in Section 3.1. So, the ratio of the optimality criterion of the optimal designs for potentially failing trials ($p_0 = 0.2$) and certain response is decreased from 5.0 to 3.9 when $k = 10$ and $r = 4$.

Table 2: Exact D-optimal designs for experiments with uncertain response and probability for non-failing trial (A) $p = 1$, (B) $p \in [.86, .92]$, and (C) $p \in [.57, .73]$ and independent failing trials; three sizes k of the design space $\mathcal{X}_k$ and various sample sizes r. In cases of unsymmetrical designs, the corresponding mirrored design has the same value of the optimality criterion. The minimal values of the optimality criterion are indicated by '*'.

k	r	$\lvert D(\xi)\rvert$ (A)	(B)	(C)	ξ
6	4	8.35*	8.30	3.63	1 . 1 1 . 1
		14.36	1.02*	0.31*	1 1 . . 1 1
	5	5.75*	4.66	22.54	1 1 1 1 . 1
		6.34	3.47*	17.42*	1 1 1 . 1 1
	6	3.99	0.92	44.49	1 1 1 1 1 1
8	4	8.17*	19.21	24.45	1 . . 1 1 . . 1
		10.14	1.03*	0.90	1 . 1 . . 1 . 1
		22.08	1.17	0.82*	1 1 . . . . 1 1
	5	5.45*	3.39	41.79	1 1 . 1 1 . . 1
		6.30	1.38*	16.37*	1 1 . 1 . 1 . 1
	6	3.63	0.39	33.91	1 1 . 1 1 . 1 1
	7	2.71	0.10	31.56	1 1 . 1 1 1 1 1
	8	2.08	0.02	17.44	1 1 1 1 1 1 1 1
10	4	8.10*	34.59	79.79	1 . . . 1 1 . . . 1
		10.32	0.74*	0.90*	1 . 1 . . . . 1 . 1
	5	5.20*	3.45	76.81	1 . . . 1 1 . . 1 1
		6.08	0.69*	10.50*	1 . 1 . 1 . . 1 . 1
	6	3.34*	0.28	36.91	1 1 . . 1 1 . . 1 1
		7.17	0.19*	19.58	1 . 1 . 1 . 1 . 1 1
		7.39	0.23	19.05*	1 . 1 . 1 1 . 1 . 1
	7	2.41*	0.06	15.61	1 1 . 1 1 1 . . 1 1
		4.71	0.04*	14.37	1 1 . 1 1 . 1 . 1 1
		5.01	0.06	12.60*	1 1 . 1 1 1 . 1 . 1
	8	1.83*	0.01	4.22*	1 1 . 1 1 1 1 . 1 1
		3.66	0.01*	12.49	1 1 1 . 1 1 . 1 1 1
	9	1.49*	0.01	1.46*	1 1 1 1 1 1 1 . 1 1
		2.85	0.01*	2.21	1 1 1 1 1 1 . 1 1 1
	10	1.22	0.01	0.42	1 1 1 1 1 1 1 1 1 1

4 Concluding Remarks

When real-life experiments are performed it cannot be guaranteed that each trial will result in an observation. In the case where the data are to be used for estimating the model parameters and the experimental design is optimized, the loss of this uncertainty is twofold: A reduced number of observations (1) increases the standard error of the estimates, and (2) potentially decreases the efficiency of the design. The size of the effects depends on the probabilities that trials fail to give response at the sites of the design space. These effects must be expected to be serious for small and moderate sample sizes. This is demonstrated for simple cases in the examples of Section 3.

Table 3: Exact D-optimal designs for experiments with increasing probability for trials to fail and $p_0 = 1$, 0.8, and 0.2; three sizes k of the design space $\mathcal{X}_k$ and various sample sizes r. The respective values of the optimality criterion is given for three values of p_0; the minimal values of the optimality criterion are indicated by '*'.

		$\|D(\xi)\|$			
k	r	$p_0 = 1$	$p_0 = 0.8$	$p_0 = 0.2$	ξ
6	4	8.35*	51.97	397.93	1 . 1 1 . 1
		9.81	20.12*	83.19	1 . 1 . 1 1
		14.36	23.33	64.52*	1 1 . . 1 1
	5	5.75*	18.46	73.01	1 1 1 1 . 1
		6.34	9.11*	20.67*	1 1 1 . 1 1
	6	3.99	5.61	12.09	1 1 1 1 1 1
8	4	8.17*	71.90	585.93	1 . . 1 1 . . 1
		10.14	17.63*	50.35	1 . 1 . . 1 . 1
		12.56	17.75	41.21*	1 . 1 . . . 1 1
	5	5.45*	21.05	80.00	1 1 . 1 1 . . 1
		5.83	7.48*	14.01	1 . 1 . 1 . 1 1
		6.30	7.89	13.87*	1 1 . 1 . . 1 1
	6	3.63	4.42	7.24	1 1 . 1 1 . 1 1
	7	2.71	3.25	5.14	1 1 1 1 1 . 1 1
	8	2.08	2.46	3.80	1 1 1 1 1 1 1 1
10	4	8.10*	93.02	784.31	1 . . . 1 1 . . . 1
		9.11	14.48*	31.89*	1 . . 1 . . 1 . . 1
	5	5.20*	7.94	19.68	1 . . . 1 1 . . 1 1
		5.44	6.69*	11.49	1 . . 1 . 1 . . 1 1
		5.91	6.99	10.89*	1 . 1 . . 1 . . 1 1
	6	3.34	3.84	5.54	1 1 . . 1 1 . . 1 1
	7	2.41	2.73	3.84	1 1 . 1 1 1 . . 1 1
	8	1.83	2.07	2.85	1 1 . 1 1 1 1 . 1 1
	9	1.49	1.67	2.26	1 1 . 1 1 1 1 1 1 1
	10	1.22	1.36	1.82	1 1 1 1 1 1 1 1 1 1

Acknowledgement

The author is indebted to Michael Maderbacher for his assistance in the computations.

References

Hedayat, A.S. and Majumdar, D. (1984). Redesigning Experiments, pp. 113-140 in A.M. Abouammoh, E. El-Neweihi, E. Aly, M.A. Alosh, (Eds.), *Developments in Statistics and Its Applications. Proceedings of the First Saudi Symposium on Statistics and Its Applications.* Riyadh: King Saud University Library.

Das, A. and Sinha, B.K. (1994). Optimal Allocation of Observations Under Experimental Constraints: Resources Scarcity and External Force's Intervention. *Research Report*, Indian Statistical Institute, Calcutta.

Regression Design for One-Dimensional Subspaces

ABDELOUAFI IBRAHIMY and R. DENNIS COOK

Department of Applied Economics, National School of Agriculture, B.P. 40, Meknes, Morocco
and
Department of Applied Statistics, University of Minnesota, St. Paul, MN 55108, USA.

Abstract:

Consider a regression problem with univariate response y and $p \times 1$ vector of design variables x, and assume that the cumulative distribution function $F(y|x)$ depends on x only through the linear combination $\theta^T x$ so that $F(y|x) = F(y|\theta^T x)$ for all x in the design space. When the form of F is unknown, θ is not estimable. However, under certain conditions the subspace $S(\theta)$ of R^p spanned by θ is estimable. The goal of this paper is to begin investigating how to design an experiment so that standard methods of estimation may yield useful estimates of $S(\theta)$ when the family $F(y|\theta^T x)$ is unknown. This may provide a baseline for assessing the robustness of designs based on an assumed family F, in addition to allowing insight into model robust design.

1 Introduction

The choice of an experimental design can depend on many aspects of the phenomena under study, particularly the expected relationship between the response and the design variables. If the functional form of that relationship is unknown, then many standard design methods no longer apply. There are several papers dealing with designs for unknown models, among which is the pioneering paper by Box and Draper (1959). For a literature review, see Atkinson (1988). In this paper we approach the problem using a novel formulation based on subspaces.

Let y denote the scalar–valued response and let x denote the $p \times 1$ vector of design variables. The distribution function of y given x is represented by $F(y|x)$. The experiment consists of observing a value y_i of the response at each of n setting of the design variables, $x = x_i$, $i = 1,\ldots,n$. We assume that the y_i's are independently observed. Letting θ

---

*Kitsos, C.P., and Müller, W.G., Eds., *Proceedings of MODA4*, Physica Verlag, Heidelberg, 1995

denote an unknown $p \times 1$ parameter vector, we adopt the model form

$$F(y|x) = F(y|\theta^T x) \tag{1}$$

for all values of x in the design region. According to (1), if $\theta^T x$ were known and held fixed then no further information about the response could be obtained from x.

The family (1) includes many commonly used models. Models of the form $y|x = g(\theta^T x)+\varepsilon$ are included, for example. Response transformation models $y^{(\lambda)}|x = \theta_0+\theta^T x+\varepsilon$, transform-both-sides models (Carroll and Ruppert 1988), and generalized linear models are also covered by (1). The family of models (1) thus seems usefully large.

Without further restrictions, the length of θ is not generally estimable under (1). However, with suitable conditions (Li and Duan 1989), the subspace $S(\theta)$ spanned by θ is estimable and for this reason we assume without loss of generality that $\|\theta\| = 1$. This restriction does not change the relative magnitudes of the coordinates of θ so inferences on ratios are still possible. More importantly, a plot of y_i versus estimated linear combinations $\hat{\theta}^T x_i$ will contain relevant information on $F(y|\theta^T x)$ up to an estimation error. For further background on graphics, see Cook (1994). Li and Duan (1989) and Duan and Li (1991) describe estimation methods that can be used within the family (1). In this paper we restrict attention to ordinary least squares (OLS) estimation.

In Section 2, we introduce the design criteria and in Section 3, we present a class of small sample designs called equally spaced designs. To show the relative performance of these designs, a small case study is described in Section 4. Equally spaced designs have specific geometrical shapes that are discussed in Section 5. Concluding remarks are given in Section 6.

2 Design Criteria

The OLS estimate of θ is well-known, $\hat{\theta} = (X^T X)^{-1} X^T Y$, where X is the $n \times p$ centered design matrix with $x_i^T - \bar{x}^T$ at row i, and Y is the $n \times 1$ vector $(y_1, \ldots, y_n)^T$. In design problems based on known models, OLS estimates are often unbiased, at least asymptotically, and designs are chosen to minimize variability. However, both bias and variability must be addressed when designs are based on (1) because then $\hat{\theta}$ is generally a biased estimate of θ. The bias depends on the sample size n and will not necessarily decrease with large n. Under the key condition that $\mathrm{E}(x|\theta^T x)$ is a linear function of $\theta^T x$, Li and Duan (1989) showed that $\hat{\theta}$ is a consistent estimate of $S(\theta)$. In their development, Li and Duan assumed that x is random, but we are interested in situations where x can be set by design and thus all expectations involving the design variables will be with respect to the appropriate design measure.

Let P_θ be the projection operator onto $S(\theta)$ and let $Q_\theta = I - P_\theta$, where the usual inner product is used here and for all projection operators in this article. We define the bias in $\hat{\theta}$ to be $Q_\theta \mathrm{E}(\hat{\theta})$ which is the part of $\mathrm{E}(\hat{\theta})$ in the orthogonal complement of $S(\theta)$. For an arbitrary design $\xi_n = \{x_1, \ldots, x_n\}$ with design measure $\xi_n(x_i) = 1/n$, the bias can be written as

$$Q_\theta \mathrm{E}(\hat{\theta}) = Q_\theta \left(\frac{X^T X}{n} \right)^{-1} \frac{1}{n} \sum_{i=1}^{n} (x_i - \bar{x}) g(\theta^T x_i) \tag{2}$$

where the x_i's are chosen from the design region χ and $\mathrm{E}(y|x) = g(\theta^T x)$. One natural goal for the design problem is to choose a design that minimizes a functional of (2).

For large n, let the exact design ξ_n be generalized to a probability measure ξ on the design region χ. Then

$$
\begin{aligned}
Q_\theta\, \mathrm{E}(\hat{\theta}) &= Q_\theta M^{-1} \int_\chi (x - \mu) g(\theta^T x) \xi(dx) \\
&= Q_\theta M^{-1} E[g(\theta^T x)\, \mathrm{E}(x - \mu | \theta^T x)]
\end{aligned}
$$

where $\mu = \mathrm{E}(x)$, $M = E[(x - \mu)(x - \mu)^T]$, and all expectations involving x are taken with respect to the design measure ξ. Using Theorem 2.1 of Li and Duan (1989), the bias vanishes at a particular θ if $\mathrm{E}(x | \theta^T x)$ is a linear function of $\theta^T x$. Eaton (1986) has shown that $\mathrm{E}(x | \theta^T x)$ is a linear function of $\theta^T x$ for all values of θ if and only if the design points x are elliptically contoured. It follows that $\hat{\theta}$ is unbiased if x has uniform mass on each of a series of concentric ellipses inscribed in the design region. To minimize large sample bias then, we restrict possible large sample designs to be on elliptical contours. Without loss of generality, χ is now taken to be a hypersphere of radius unity. In sum, if (1) holds and ξ is the uniform measure on the hypersphere with radius 1, then the bias $Q_\theta\, \mathrm{E}(\hat{\theta}) = 0$ for all θ.

Assuming that $\mathrm{var}(y|x)$ is constant, the variability of $\hat{\theta}$ is

$$
\mathrm{var}(\hat{\theta}) \propto (X^T X)^{-1} \tag{3}
$$

In addition to minimizing bias, good designs should minimizes the determinant of (3) or, equivalently, maximize $|X^T X|$. Designs ξ^* which are restricted to the unit hypersphere and have information matrices of the form $M(\xi^*) = (X^T X)/n = I_p/p$ maximize $|X^T X|$. A distribution that ensures this property is again the uniform distribution on the unit hypersphere. These designs are known to be D-optimal for θ in linear models. The condition that $\mathrm{var}(y|x)$ is constant rules out heteroscedastic models, but (1) still contains substantial flexibility and the constant variance condition is unnecessary for considering bias reduction via (2), the main concern of this article.

3 Equally Spaced Designs

The discussion of the previous section led to designs with uniform mass on the unit hypersphere. While these designs might be approximated for large n and small p, adequate approximation may not otherwise be possible and thus investigating small sample designs that control bias seems necessary. We will need the following definition: Let ξ_n be an n-point design with support $S_n = \{x_1, \ldots, x_n\}$ and let

$$
c_{ij} = \{x | P_j x = P_j x_i \text{ with } x, x_i, x_j \in S_n\}
$$

where P_j is the projection matrix for $S(x_j)$. ξ_n is said to be an *equally spaced design* (ESD) if, for all i and $j = 1, \ldots, n$,

$$
\sum_{c_{ij}} x = a_{ij} x_j \tag{4}
$$

where the a's are constants which are not all zero.

For example, the support $S_8 = \{(\pm 1, 0), (0, \pm 1), (\pm 1/\sqrt{2}, \pm 1/\sqrt{2})\}$ with uniform mass defines an 8 points equally spaced two-dimensional design which we denote as ESD1 for later use. Figure 1 shows this design with numbers assigned to the design points for ease of reference. The design points are equally spaced points around the unit circle. Property (4) works as follows: Fix x_j to be any support point, say $x_j = x_1$. The sum of all support points with identical projections onto $S(x_1)$ must be in $S(x_1)$ and this holds for ESD1. For example, $c_{21} = \{x_2, x_8\}$ and $S(x_2 + x_8) = S(x_1)$. Additionally, the mean of the design points is zero and a_{i1} is either $\sqrt{2}$, 1 or -1 for all i. Generally, equally spaced designs force a form of equal spacing of the support points around the unit sphere in R^p. Examples of higher dimensional ESD's will be given in later sections.

Before discussing why ESD's may be useful for reducing bias, some general discussion of the sets c_{ij} is in order. For each fixed i and j, c_{ij} contains all points whose projection onto $S(x_j)$ is the same as that of x_i. Also, for each fixed j, the subsets define a partition of the support points. For example, the partition of ESD1 with respect to x_1 is $p_1 = (\{2,8\}, \{3,7\}, \{4,6\}, \{1\}, \{5\})$ and the sum of the support points in any member of this partition spans $S(x_1)$.

On the unit hypersphere, subsets c_{ij} forming the partition p_j reduce to

$$
\begin{aligned}
c_{ij} &= \{x | x_j^T x_i = x_j^T x, x \in S_n\} \\
&= \{x | \cos(\alpha_{ij}) = \cos(\alpha_{xj}), x \in S_n\}
\end{aligned}
$$

where α_{ij}, is the angle between x_i and x_j, and α_{xj} is the angle between x_j and x. Clearly, $x_k^T x_j = \cos(\alpha_{ij})$ for all $x_k \in c_{ij}$. In matrix form, this equality can be written as $X_{ij}^T x_j = \cos(\alpha_{ij})\mathbf{1}$, where X_{ij} is the $p \times n_{ij}$ matrix of the n_{ij} support points in c_{ij}, and $\mathbf{1}$ is the $n_{ij} \times 1$ vector of ones. It now follows from (4) that $a_{ij} = n_{ij} \cos(ij)$.

Propositions 3–6 in the Appendix (Section 7) provide some characterizations of ESD's. Briefly, for fixed j the subsets c_{ij} forming the partition p_j have mean vectors proportional to x_j. Given a c_{ij}, the points are in a subspace of dimension not greater than $p - 1$ and are orthogonal to x_j when centered by the conditional mean given in the Appendix. The mean support point $E(x) = 0$ and ESD's are invariant under the group of orthogonal transformations.

Once an ESD is found, the following Proposition applies.

Proposition 1 *If the support S_n forms an ESD with $(X^T X)^{-1} = kI$ for some constant k, and (1) holds for some θ such that $S(\theta) = S(x_j)$, where x_j is a support point, then the bias $Q_\theta E(\hat{\theta}) = 0$*

Justification:

$$
\begin{aligned}
Q_\theta E(\hat{\theta}) &= Q_\theta (X^T X)^{-1} \sum_{i=1}^{n} x_i g(\theta^T x_i) \\
&= \frac{1}{k} \sum_{c_{ij} \in p_j} \sum_{x \in c_{ij}} Q_{x_j} x g(x_j^T x)
\end{aligned}
$$

For a fixed j, $x_j^T x$ is constant for all x in c_{ij}. Thus, the bias is proportional to

$$
\sum_{c_{ij} \in p_j} Q_{x_j} g(x_j^T x) \sum_{x \in c_{ij}} x
$$

The conclusion now follows because

$$
\sum_{x \in c_{ij}} x \propto x_j \quad \Box
$$

According to Proposition 1, the bias vanishes at the support points of an ESD. Our experience indicates that ESD's may do well in practice when θ is not proportional to a support point, as will be illustrated a bit later.

Li (1993) has shown that the minimum information in estimating θ is obtained under the class of least favorable models with

$$g(\theta^T x) = \alpha + \theta^T x + \delta^T \eta(\theta^T x) \tag{5}$$

where $\eta(\theta^T x) = \mathrm{E}(x|\theta^T x)$ and δ is a p–dimensional nuisance parameter. In other words, this model is the most restrictive for estimating $S(\theta)$. In addition, Li determines I_{min}, the Fisher information matrix in estimating $S(\theta)$ under (5), and uses $I_{lf} - I_{min}$ as a measure of the information loss when using a linear fit instead of (5), where I_{lf} is the Fisher information matrix under the linear fit. For the design problem, the information loss is $I_{lf} - I_{min} = \mathrm{var}(\kappa_\theta(\theta^T x))$, where

$$\kappa_\theta(\theta^T x) = \mathrm{E}(x|\theta^T x) - L_\theta(\theta^T x)$$

and

$$L_\theta(\theta^T x) = \mu + (\mathrm{var}(\theta^T x))^{-1} M \theta \theta^T (x - \mu) \tag{6}$$

Here, μ and M are the mean and the covariance matrix of x with respect to the design measure ξ.

Proposition 2 *If $\theta = x_j$, where x_j is any point of the ESD support S_n, then $I_{lf} = I_{min}$.*

Justification: Assuming that S_n is an ESD with $M = kI$, it is known by Proposition 6 that $\mathrm{E}(x) = 0$. In this case (6) reduces to $L_\theta(\theta^T x) = \theta\theta^T x$ and $\kappa_\theta(\theta^T x) = \mathrm{E}(x|\theta^T x) - \theta\theta^T x$. To show that $I_{lf} = I_{min}$ over S_n, we need show only that $\mathrm{E}(x|\theta^T x) = \theta\theta^T x$ at $\theta = x_j$ for $j = 1, \ldots, n$. Setting $\theta = x_j$, the quantity $\mathrm{E}(x|x_j^T x)$ is the conditional design expectation of x given that $x_j^T x = \cos(\alpha_{xj})$, since x_j and x are on the unit hypersphere. But the values of x for which $x_j^T x = \cos(\alpha_{xj})$ define the subset c_{ij}. So $\mathrm{E}(x|x_j^T x = \cos(\alpha_{xj}))$ is the conditional mean of x over the subset which is given in Proposition 4: $\mathrm{E}(x|x_j^T x = \cos(\alpha_{xj})) = \cos(\alpha_{ij})x_j$. However, over c_{ij},

$$
\begin{aligned}
\mathrm{E}(x|x_j^T x = \cos(\alpha_{xj})) &= \mathrm{E}(x|x_j^T x = x_j^T x_i) \\
&= x_j x_j^T x_i
\end{aligned}
$$

which yields the desired result.□

This proposition indicates that even under the least favorable model, there is no loss of information in estimating $S(\theta)$ when $\theta = x_j$. This result reinforces the idea that g does not matter in estimating $S(\theta)$ at the points of an ESD.

4 A Two-Dimensional Example

To illustrate Proposition 1, we use the OLS estimate $\hat{\theta}$ and compute the norm of the bias (2) under the true model $\mathrm{E}(y|x) = \exp(\theta^T x)$, where θ is a vector on the unit circle. The vector θ is identified by its angle with the x–axis, varying from 0 to 2π in a counterclockwise direction. The design considered is ESD1 given in Figure 1. Figure 2 gives the norm of the bias (y–axis) versus θ expressed in radians (x–axis), and confirms that the bias vanishes at the design points labeled as "O". In addition, this norm remains fairly small at any other direction on the unit circle.

To illustrate the performance of an ESD with respect to the bias, we compare ESD1 to two other designs, labeled ER and W, of the same size: ER is an equiradial design known to be rotatable. This design is an 8–point ESD with 6 support points

$$\{(\cos(k\pi/3), \sin(k\pi/3)), k = 0, \ldots, 5\}$$

on the unit circle plus 2 center points. A more general discussion on rotatable designs is given in Section 5. The design W is a 3^2 factorial all–bias design (Welsh 1983) with the 8 support points $\{(\pm 1, \pm 1), (\pm 1, 0), (0, \pm 1)\}$. The comparisons are reported in Figure 3 which gives the norm of the bias (y–axis) versus the unit direction θ expressed in radians (x–axis). The relative magnitude of the bias induced by ESD1 is so small that it appears as a horizontal line. Moreover design W, having the support on the square, does not perform well in estimating $S(\theta)$ compared to the other designs. This underlines the importance of the shape of the design space. As expected, ER induces a fairly negligible bias outside the design supports, although it is greater than that of ESD1 because of the 2 center points. Under our rationale, center points are not essential because they do not contribute to bias reduction, although they might still be included to satisfy other criteria.

Another comparison we make is between designs ESD2 and BH: ESD2 an equally spaced design of size $n = 12$ with support points

$$\{(\cos(k\pi/6), \sin(k\pi/6), k = 0, \ldots, 11\}$$

while BH is an ESD of the same size with 5 points equally spaced around the unit circle and 7 points equally spaced around a concentric circle with radius about 0.44. Design BH is rotatable and has uniform precision (see Box and Hunter 1957, and Myers 1971, p. 160). As shown in Figure 4, the difference in the norm of the bias for ESD2 and BH is substantial.

5 Constructing Equally Distributed Designs

5.1 In 2 and 3 Dimensions

In 2D, for $n \geq 3$, designs with uniform mass on the support

$$S_n = \{\cos(\frac{2k\pi}{n}), \sin(\frac{2k\pi}{n}), k = 0, \ldots, n-1\}$$

are ESD's. These designs have information matrices equal to $I_2/2$ and constitute regular polygons on the unit circle. In addition, all designs are first-order rotatable designs (Box and Draper 1987). The polygon for $n = 8$ (ESD1) is a minimax design for estimating the stationary point of a quadratic model (Mandal and Heiligers 1992).

In 3D, designs tracing particular regular geometrical shapes can be ESD's. The support $\{(\pm 1, 0, 0), (0, \pm 1, 0), (0, 0, \pm 1)\}$ comprises an octahedron with vertices on the unit sphere. Adding points that are vertices of a regular cube results in an ESD with the 14 points of support,

$$\{(\pm 1, 0, 0), (0, \pm 1, 0), (0, 0, \pm 1), \frac{1}{\sqrt{3}}(\pm 1, \pm 1, \pm 1)\}$$

This design is a central composite design (CCD) with no center point.

Generally, the points of regular polyhedra are ESD's in 3D. The only known regular polyhedra are the tetrahedron ($n = 4$), the cube ($n = 8$), the octahedron ($n = 6$), the icosahedron ($n = 12$) and the dodecahedron ($n = 20$).

Certain 3D ESD designs can be constructed from 2D ESD's by choosing appropriate orthogonal transformations and using Proposition 7 in the Appendix. For example, begin with a 2D ESD having the 4 support points $(\pm 1/\sqrt{2}, \pm 1/\sqrt{2})$. Adding a third coordinate of 0 results in three 3D ESD's – $(0, \pm 1/\sqrt{2}, \pm 1/\sqrt{2})$, $(\pm 1/\sqrt{2}, 0, \pm 1/\sqrt{2})$ and $(\pm 1/\sqrt{2}, \pm 1/\sqrt{2}, 0)$ – each with 4 points. Combining all 12 points yields another 3D ESD that is a Box-Behnken design (Box and Draper 1987) with no center points and rescaled to the unit sphere. This 12 point design is rotatable. This example shows that ESD's can be combined to produce a new ESD, but this is not a general result: The 8 point design $\{(0, \pm 1/\sqrt{2}, \pm 1/\sqrt{2}), (\pm 1/\sqrt{2}, 0, \pm 1/\sqrt{2})\}$ is not an ESD.

5.2   In p Dimensions

The points of regular polyhedra are potential ESD's in any dimension. Simplex designs constitute regular polyhedra with $p(p + 1)/2$ edges of length $(2(p + 1)/p)^{-1/2}$ and define a class of small ESD's with $n = p + 1$. For $p = 2$ and $p = 3$, the corresponding shapes are an equilateral triangle (n = 3) and a tetrahedron ($n = 4$) respectively. To see the ESD property of this class, let

$$a_i = \left(\frac{p+1}{i(i+1)p} \right)^{1/2}$$

for $i = 1, \ldots, p$ and let $S_n = \{x_j\}$ contain the support points. The first support point x_1 is defined to be the $p \times 1$ vector with elements $-a_i$. The remaining support points are the $p \times 1$ vectors $(0, \ldots, 0, (j - 1)a_{j-1}, -a_{j+1}, \ldots, -a_p)^T$ for $j = 2, \ldots, p + 1$. It is easy to verify that (a) $\|x_i\| = 1$ for $i = 1, 2, \ldots, n$; (b) $x_i^T x_j = -1/p$ for all i and j with $i \neq j$ and (c)

$$X^T X = \sum_{i=1}^{n} x_i x_i^T = \frac{p+1}{p} I$$

The corresponding partition is

$$p_j = \{(x_1, \ldots, x_{j-1}, x_{j+1}, \ldots, x_n), (x_j)\}$$

for $j = 1, \ldots, n$.

Other ESD's can be constructed from the simplex designs as well. Clearly, $\{S_n, -S_n\}$ is an ESD with $2n$ support points. Other design points that can be added to a simplex design to produce another ESD are the midpoints of the edges joining 2 vertices. Similarly, the design with uniform mass on the support $S_n = \{\pm e_1, \ldots, \pm e_p\}$, where e_i is the $p \times 1$ unit vector with 1 at the i-th position and 0 elsewhere, is a regular polyhedron in p dimensions and can be shown to be an ESD with $n = 2p$. Also, the regular polyhedron with uniform mass on the support $S_n = \{1/\sqrt{p}(\pm 1, \ldots, \pm 1)\}$ is an ESD with $n = 2^p$. These two regular polyhedra considered altogether are respectively the star and the cube points of a rescaled CCD (Box and Draper, 1987) with no center point. It turns out that CCD's are ESD's as will be discussed briefly in the following section.

5.3   ESD's and Rotatability

A CCD is formed by 2^p points, $2p$ star points and a center point. Under our framework, the center point can be discarded. When considering the second-order rotatability criterion (see Box and Draper 1987, p. 480) and restricting the design space to a unit hypersphere, the points are on 2 concentric hyperspheres of radius unity and less than unity. The support is

$$S_n = \{(\pm v, \ldots, \pm v), (\pm 1, 0, \ldots, 0), \ldots, (0, \ldots, \pm 1)\}$$

where $v = 2^{-p/4}$. CCD's have information matrices proportional to the identity. These designs are ESD's and consequently the bias still vanishes at the support points. However, rotatability does not imply that a design is an ESD. Indeed several designs known to be rotatable are not ESD's. As an illustration, the non-regular icosahedron

$$\{(0, \pm 0.85, \pm 0.526), (\pm 0.85, 0, \pm 0.526), (\pm 0.85, \pm 0.526, 0)\}$$

is rotatable but is not an ESD.

6 Discussion

This study is intended as an initial investigation into the desirability of treating $S(\theta)$ as a design parameter for model robustness. Our results indicate that this approach may have merit. The developments so far are applicable (a) when model robustness is a primary issue and (b) when a target linear model is used for primary design, but some protection against model misspecification is desirable. In the latter case, standard linear model designs might be "moved" in the direction of an ESD to reduce the bias or preference might be given to standard rotatable designs which are also ESD's.

The bias as defined here depends on the true model, on the design space and on the sample size n. For large n, the bias vanishes if x is uniform on the unit sphere. Otherwise, ESD's with support on the largest hypersphere in the design region seem to perform well for estimating $S(\theta)$ under OLS estimation. The designs should also perform well when using other convex estimation methods (eg., M-estimates) based on a target linear model. The bias in $E(\hat{\theta})$ vanishes at the support points and our experience indicates that it is otherwise small for a large class of models. The bias can be reduced by increasing n but the amount of reduction depends on $F(y|\theta^T x)$. The nature of this dependence is under investigation.

7 Appendix: Characterization of Equally Distributed Designs

A series of properties concerning the subsets c_{ij} were claimed in Section 3. Those properties follow from Propositions 3 to 7.

Proposition 3 *For fixed j, the c_{ij}'s are mutually exclusive, $c_{ij} \cap c_{i'j} = \emptyset$ for $x_{i'} \notin c_{ij}$.*

Justification: Assume that $c_{ij} \cap c_{i'j} \neq \emptyset$. Then, there exists x in c_{ij} and in $c_{i'j}$ such that $P_{x_j} x = P_{x_j} x_i = P_{x_j} x_{i'}$. This implies that x_i and $x_{i'}$ are in the same subset c_{ij} which contradicts the initial assumption that $x_{i'} \notin c_{ij}$. □

Proposition 4 *For an ESD, the elements of c_{ij} have mean vector $\bar{x}_{ij} = \cos(\alpha_{ij}) x_j$.*

Justification: The justification is straightforward and omitted.□

Proposition 5 *The subspace $S(Z_{ij})$ is orthogonal to x_j where $Z_{ij} = X_{ij} - \bar{x}_{ij} \mathbf{1}^T$.*

Justification: Let $S_n = \{x_j\}$ be the support of an ESD, and center the points in c_{ij} by subtracting their mean $\bar{x}_{ij}$: $z = x - \bar{x}_{ij}$, where x belongs to c_{ij}. The centered elements of c_{ij} can be written in matrix form as $Z_{ij} = X_{ij} - \bar{x}_{ij}\mathbf{1}^T$. Using Proposition 4, $x_j^T Z_{ij} = 0$. $\square$

Proposition 5 implies that all subspaces $S(Z_{ij})$ are orthogonal to $S(x_j)$. If any two subspaces $S(Z_{ij})$ and $S(Z_{i'j})$ have the same dimension, then they must be equal. The corresponding affine subspaces must be parallel.

Proposition 6 *For an ESD with support $S_n = \{x_j\}$, $\bar{x} = 0$.*

Justification: For fixed j and j' with $j \neq j'$, let p_j and $p_{j'}$ be the partitions of the ESD with respect to x_j and $x_{j'}$, respectively. It follows that for some constants u and v,

$$\sum_{i=1}^{n} x_i = u x_j = v x_{j'}$$

But since x_j and $x_{j'}$ are not in the same manifold, we must have $u = v = 0$, which gives the desired result.$\square$

Proposition 7 *If $S_n = \{x_j\}$ is an ESD, then $PS_n = \{Px_j\}$ is also an ESD, where P is any orthogonal transformation.*

Justification: The result follows because orthogonal transformations preserve lengths and angles.$\square$

References

Atkinson, A.C. (1988). Recent development in the methods of optimum and related experimental designs. *International Statistical Review* 56, 99-115.

Box, G.E.P. and Hunter, J.S. (1957). Multifactor experimental designs for exploring response surfaces. *Annals of Mathematical Statistics* 28, 195-241.

Box, G.E.P. and Draper, N.R. (1959). A basis for the selection of a response surface design. *Journal of the American Statistical Association* 54, 622-654.

Box, G.E.P. and Draper, N.R. (1987). *Empirical Model-Building and Response Surfaces*. New York: Wiley.

Carroll, R.J. and Ruppert, D. (1988). *Transformations and Weighting in Regression*. New York: Chapman and Hall.

Cook, R.D. (1994). On the interpretation of regression plots. *Journal of the American Statistical Association* 89, 177-189.

Eaton, M.L. (1986). A characterization of spherical distributions. *Journal of Multivariate Analysis* 20, 272-276.

Duan, N. and Li, K.C. (1991). Slicing regression: A link-free regression method. *Annals of Statistics* 19, 505-530.

Li, K.C. (1993). Helices in high dimensional regression. Submitted.

Li, K.C. and Duan, N. (1989). Regression analysis under link violation. *Annals of Statistics* 17, 1009-1052.

Mandal, N.K and Heiligers, B. (1992). Minimax designs for estimating the optimum point in a quadratic response surface. *Journal of Statistical Planning and Inference* 31, 235-251.

134

Myers, R.H. (1971). *Response surface methodology.* Boston: Allyn and Bacon. (reprinted by Edwards Bros., Ann Arbor, MI).

Welch, W.J. (1983). A mean squared error criterion for the design of experiments. *Biometrika* 70, 205-213.

D-Optimal First Order Saturated Designs with $n \equiv 2\mathrm{mod}4$ Observations

CHRISTOS KOUKOUVINOS

Department of Mathematics, National Technical University of Athens, Zografou 15773, Athens, Greece.

Abstract:

In experimental situations where n two-level factors are involved and n observations are taken , then the D-optimal first order saturated design is an $n \times n$ ± 1 matrix with the maximum determinant. In this paper we discuss this problem for $n \equiv 2\mathrm{mod}4$, we summarize all the known results, and we give some new D-optimal designs.

1 Introduction

Consider an experimental situation in which a response y depends on k factors $x_1, \ldots, x_k$ with the first order relationship of the form $E(\underline{y}) = X\underline{\beta}$, where $\underline{y}$ is an $n \times 1$ vector of observations, the design matrix X is $n \times (k+1)$ whose jth row is of the form $(1, x_{j1}, x_{j2}, \ldots, x_{jk})$, $\quad j = 1, 2, \ldots, n$ and $\underline{\beta}$ is the $(k+1) \times 1$ vector of coefficients to be estimated. In a two-level factorial design, each x_i can be coded as ± 1. The design is then determined by the $n \times k$ matrix of elements ± 1. The ith column gives the sequence of factor levels for factor x_i, each row constitutes a run. When $k = n - 1$, the design is called a saturated design and the design matrix is an n×n square matrix. Note that n=k+1 is the minimal number of points (rows) required to estimate all coefficients of interest (the β_i's).

Several criteria have been advanced for the purpose of comparing designs and for constructing optimal designs. One of the most popular is the D-optimality criterion, which seeks to maximize $\det(X^T X)$, the determinant of the $X^T X$ matrix. Recall that for any $n \times n$ matrix, H, consisting entirely of elements ± 1, the maximum determinant possible is $\det(H^T H) = n^n$.

The construction of square $n \times n$ ± 1 matrices with maximum determinant is a difficult problem and its solution is not yet known even for small values of n. In the case $n \equiv 0$ mod4 it is known that Hadamard matrices of order n have maximum determinant. An

---

*Kitsos, C.P., and Müller, W.G., Eds., *Proceedings of MODA4*, Physica Verlag, Heidelberg, 1995

Hadamard matrix of order n is an $n \times n$ ± 1 matrix H satisfying $HH^T = H^TH = nI_n$ and its construction is known for many values of n. For more details see Seberry and Yamada (1992).

2 D-Optimal Designs for $n \equiv 2 \bmod 4$

If $n \equiv 2 \bmod 4$ and A, B are $\frac{n}{2} \times \frac{n}{2}$ commuting matrices, with elements ± 1, such that

$$AA^T + BB^T = (n-1)I_{\frac{n}{2}} + 2J_{\frac{n}{2}} \tag{1}$$

where $J_{\frac{n}{2}}$ is an $\frac{n}{2} \times \frac{n}{2}$ matrix of 1's, then the $n \times n$ matrix

$$R = \begin{pmatrix} A & B \\ -B^T & A^T \end{pmatrix}$$

has the maximum determinant (Ehlich (1964), Cohn (1989)) among all $n \times n$ ± 1 matrices. Such matrices are called D-optimal designs of order n. A particular case where this construction is used is when A and B are circulant matrices. If A, B are circulant matrices, then pre- and post-multiplying both sides of (1) by e^T and e respectively we obtain:

$$(\frac{n}{2} - 2r)^2 + (\frac{n}{2} - 2s)^2 = 2n - 2 \tag{2}$$

where e is the $\frac{n}{2} \times 1$ vector of 1's and r, s is the number of -1's in every row of A, B respectively.

If A, B satisfy (1) so do $\pm$ A, $\pm$B, i.e., we can always take $1 \le r \le s \le (n-2)/4$. Now form the two sets

$$P = \{p_1, p_2, \ldots, p_r\}, \quad Q = \{q_1, q_2, \ldots, q_s\}$$

where p_i, q_j denote the positions of -1's in the first row of A, B respectively. If the congruences

$$p_i - p_j \equiv a \bmod(\frac{n}{2}), \quad q_i - q_j \equiv a \bmod(\frac{n}{2})$$

have exactly $\lambda = r + s - (n-2)/4$ solutions for any $a \not\equiv 0 \bmod(\frac{n}{2})$, then P, Q are called supplementary difference sets, denoted by $2 - \{\frac{n}{2}; r, s; \lambda\}$.

We know (Chadjipantelis and Kounias (1985)) that:

Theorem 3 *(i) If P, Q are supplementary difference sets* $2 - \{\frac{n}{2}; r, s; \lambda\}$ *and A, B the corresponding incidence matrices, then*

$$AA^T + BB^T = 4(r + s - \lambda)I_{\frac{n}{2}} + 2(\frac{n}{2} - 2(r + s - \lambda))J_{\frac{n}{2}} \tag{3}$$

(ii) Given two $\frac{n}{2} \times \frac{n}{2}$ *circulant matrices A, B satisfying (3), then the corresponding sets P, Q are supplementary difference sets* $2 - \{\frac{n}{2}; r, s; \lambda\}$, *where r, s is the number of -1's in each row of A, B.*

Hence the construction of the two circulant matrices A, B satisfying (1) is equivalent to the construction of the corresponding supplementary difference sets.

This means that for $n < 200$, all such matrices are likely to be found except for $n = 22, 34, 58, 70, 78, 94, 106, 130, 134, 142, 162, 166, 178$ and 190, because $n - 1$ is not the sum of two squares.

D-optimal designs of circulant type and order n are constructed for the following values of n:

(i) $n = 2, 6, 10, 14, 18, 26, 30, 38$ by Ehlich (1964).

(ii) $n = 42, 46, 50, 54, 62, 66$ by Yang (1966a, 1966b, 1968, 1969, 1976).

(iii) $n = 86$ by Chadjipantelis and Kounias (1985).

(iv) $n = 74, 82, 90, 98, 102$ by Cohn (1989, 1992).

(v) $n = 114, 146, 182$ by Koukouvinos, Kounias and Seberry (1991).

(vi) $n = 122$ by Whiteman (1990).

(vii) $n = 126, 186$ by Dokovic (1991).

We should also mention that a D-optimal design of order 82 was exhibited by Kharaghani (1987) by using a symmetric block design with parameters $(41, 16, 6)$ constructed by Trung (1982). However Kharaghani's D-optimal design is not of circulant type. A symmetric block design with the mentioned parameters was also constructed by Bridges, Hall and Hayden (1981).

Recently two infinite series of D-optimal designs were discovered. The first series, due to Koukouvinos, Kounias and Seberry (1991), exists for $n = 2 (q^2 + q + 1)$ where q is a prime power. The D-optimal designs of the first series are all of circulant type. The second series, due to Whiteman (1990), exists for $n = 2(2q^2 + 2q + 1)$ were q is an odd prime power. The existence of this series was deduced by applying Kharaghani's observation to an infinite series of symmetric balanced incomplete block designs constructed by Brouwer (1983). The D-optimal designs of the second series are not of circulant type.

From the facts stated above it follows that the existence of D-optimal designs of order n with $n < 200$ is undecided only in the following cases:

$$n = 110, 118, 138, 150, 154, 158, 170, 174, 194, 198$$

3 Some New D-Optimal Designs

In this section we give some new D-optimal designs for $n = 90$ and $n = 226$. These are presented in the form of the corresponding supplementary difference sets.

1. $n = 90$;   2-{45; 16, 21; 15}

   (i) $A_1 = \{8, 9, 10, 15, 17, 20, 23, 25, 29, 31, 34, 35, 37, 38, 42, 44\}$

   $B_1 = \{9, 10, 11, 13, 14, 15, 18, 21, 22, 25, 26, 27, 29, 31, 32, 34, 35, 36, 41, 42, 44\}$

   (ii) $A_2 = \{9, 10, 12, 13, 14, 15, 17, 19, 21, 23, 28, 29, 33, 40, 41, 44\}$

   $B_2 = \{5, 6, 7, 9, 10, 12, 15, 16, 18, 19, 22, 27, 28, 30, 33, 35, 37, 38, 42, 43, 44\}$

2. $n = 226$;   2-{113; 49, 49; 42}

138

(i) $A_1 = \{1, 2, 3, 4, 5, 7, 16, 17, 19, 24, 27, 28, 30, 32, 34, 37, 39, 40, 42, 45, 46,$
$48, 49, 56, 58, 59, 60, 64, 66, 70, 75, 78, 80, 83, 84, 85, 90, 92, 93, 97, 98, 99,$
$101, 103, 105, 106, 107, 109, 112\}$

$B_1 = \{1, 2, 8, 12, 14, 15, 16, 17, 20, 21, 23, 24, 28, 29, 30, 32, 33, 35, 39, 40,$
$42, 45, 46, 49, 53, 56, 57, 58, 59, 60, 65, 66, 70, 75, 76, 79, 81, 86, 94, 98, 99,$
$103, 105, 106, 107, 108, 109, 110, 111\}$

(ii) $A_2 = \{1, 3, 4, 7, 8, 12, 14, 15, 16, 17, 18, 21, 23, 24, 28, 29, 30, 34, 41, 42, 45,$
$46, 48, 49, 52, 53, 57, 58, 62, 64, 65, 79, 81, 83, 84, 85, 88, 90, 91, 92, 97, 100,$
$101, 106, 107, 109, 110, 111, 112\}$

$B_2 = \{2, 4, 6, 7, 8, 9, 12, 14, 15, 17, 21, 23, 24, 26, 29, 31, 32, 42, 44, 45, 46,$
$50, 53, 55, 56, 57, 58, 60, 64, 65, 67, 68, 71, 77, 79, 81, 83, 85, 89, 96, 97, 98,$
$99, 102, 105, 107, 110, 111, 112\}$

We note that a D-optimal design for $n = 90$ was found by Cohn (1989), but the two designs given here are non-equivalent with Cohn's design. Also a D-optimal design for $n = 226$ was found by Whiteman (1990), but not of circulant type. The two designs given here are of circulant type and non-equivalent with Whiteman's design, and can be easily constructed by using the corresponding circulant matrices A and B in the design matrix R as described in section 2.

References

Bridges, G. W., Hall, M. Jr. and Hayden, L. J. (1981). Codes and designs. *J. Combin. Theory, Ser. A, 31, 155-174.*

Brouwer, A. E. (1983). An infinite series of symmetric designs. *Math. Centrum Amsterdam Report ZW 202/83.*

Chadjipantelis, T. and Kounias, S. (1985). Supplementary difference sets and D-optimal designs for $n \equiv 2 \bmod 4$. *Discrete Math., 57, 211-216.*

Cohn, J. H. E. (1989). On determinants with elements ± 1, II. *Bull. London Math. Soc., 21, 36-42.*

Cohn, J. H. E. (1992). A D-optimal design of order 102. *Discrete Math., 102, 61-65.*

Dokovic, D. Z. (1991). On maximal (1, -1)-matrices of order 2n, n odd. *Radovi Matematicki, 7, 371-378.*

Ehlich, H. (1964). Determinantenabschätzung für binäre matrizen. *Math. Z., 83, 123-132.*

Kharaghani, H. (1987). A construction of D-optimal designs for $n \equiv 2 \bmod 4$. *J. Combin. Theory, Ser A, 46, 156-158.*

Koukouvinos, C., Kounias, S. and Seberry, J. (1991). Supplementary difference sets and optimal designs. *Discrete Math., 88, 49-58.*

Seberry, J. and Yamada, M. (1992). Hadamard matrices, sequences and block designs. Contemporary Design Theory: A collection of surveys. Edited by J. Dinitz and D. R. Stinson. *J. Wiley and Sons, New York, 431-560.*

Trung, V. T. (1982). The existence of symmetric block designs with parameters (41, 16, 6) and (66, 26, 10). *J. Combin. Theory, Ser. A, 33, 201-204.*

Whiteman, A. L. (1990). A family of D-optimal designs. *Ars Combinatoria, 30, 23-26.*

Yang, C. H. (1966a). Some designs for maximal (+1,-1)-determinant of order $n \equiv 2 \mod 4$. *Math. Comp., 20, 147-148.*

Yang, C. H. (1966b). A construction of maximal (+1,-1)-matrix of order 54. *Bull. Amer. Math. Soc., 72, 293.*

Yang, C. H. (1968). On designs of maximal (+1,-1)-matrices of order $n \equiv 2 \mod 4$. *Math. Comp., 22, 174-180.*

Yang, C. H. (1969). On designs of maximal (+1,-1)-matrices of order $n \equiv 2 \mod 4$, II. *Math. Comp., 23, 201-205.*

Yang, C. H. (1976). Maximal binary matrices and sum of two squares. *Math. Comp., 30, 148-153.*

On Information Matrices for Fixed and Random Parameters in Generally Balanced Experimental Block Designs

BARBARA BOGACKA

Department of Mathematical and Statistical Methods, Agricultural University of Poznań, Wojska Polskiego 28, 60–637 Poznań, Poland.

Abstract:

> Information matrices are arguments of most of optimality criteria defined under fixed linear models and also for fixed effects in mixed linear models. However, in the context of mixed models interest often lies on variances of random effects as well as on fixed effects. In the paper the forms and some properties of the information matrix for fixed treatment parameters and for strata variances, in case of generally balanced block designs, are shown. A short discussion on optimality criteria is also presented.

1 The Model

The paper deals with experiments performed on one-way heterogeneous experimental material on which v treatments are to be compared. As an experiment and the related mathematical model of observations should give a basis to detach the variability of experimental units caused by different treatments from natural variability of these units, a kind of unification of the units should be done before applying the treatments. A one-way heterogeneous population of units can be divided into groups according to some common characteristics which may have an influence on final observations. Then, after double randomization, i.e. a randomization of the groups and a randomization of units inside the groups we will have an experimental material unified in a probabilistic sense. That is, the units inside the groups represent random variables having identical distributions with the same moments of the first and second order. This property is called "exchangeability" of the units (cf Kala, 1991). Then, the applied treatments can modify an expected value and the parameters of the dispersion matrix of the vector of potential observations, from the same group, in the same degree. So, the treatment assigning to the units is, to

---

*Kitsos, C.P., and Müller, W.G., Eds., *Proceedings of MODA4*, Physica Verlag, Heidelberg, 1995

some extend, arbitrary. Of course, it has to be done according to a design which should guarantee high efficiency, low costs etc. Let us note, that grouping and randomization of experimental units is not contradictory to an optimal choice of a design.

Many authors considered randomization models from estimation and testing point of view, and they underlined the adequacy and applicability of such models to practical experiments (cf Fisher, 1926; Neyman, Iwaszkiewicz, Kołodziejczyk, 1935; Nelder, 1965 a,b; Ogawa, Ikeda, 1973; Bailey, 1981,1991; Caliński, Kageyama, 1991; Kala, 1991).

We consider connected block designs for which an n dimensional observable random vector y can be written as follows (Caliński, Kageyama, 1991; Kala, 1991):

$$y = \Delta'\tau + D'\beta + \eta + e, \tag{1}$$

where τ denotes a v dimensional vector of treatment parameters, β stands for a b dimensional vector of random block effects, η is an n dimensional vector of random errors of experimental units and e is an n dimensional vector of random technical errors, Δ' and D' denote $(n \times v)$ and $(n \times b)$ dimensional design matrices for treatments and blocks, respectively, such that $\Delta D'$ is the incidence matrix, usually denoted by N. The expected value of the random vector y is $E(y) = \Delta'\tau$, and its dispersion matrix $Cov(y)$, denoted by V, is, in a general case, equal to:

$$V = (D'D - \frac{1}{N_B}J_n)\sigma_\beta^2 + (I_n - \frac{1}{K_H}D'D)\sigma_\eta^2 + I_n\sigma_e^2, \tag{2}$$

where σ_β^2, σ_η^2 and σ_e^2 denote variances of groups the population of units is divided into, of units within the groups and of technical error, respectively; scalars N_B and K_H denote the number of groups (potential blocks) and a weighted harmonic average of the group sizes K_i, respectively; J_n is the $(n \times n)$ dimensional matrix of ones and I_n denotes the identity matrix of order n.

Let us note that the number of blocks b, used in the experiment, does not have to be equal to the number of groups N_B, the population was initially divided into, and the block size k_i does not have to be equal to the group size K_i. The following relations have to be fulfilled:

$$b \leq N_B < \infty \quad \text{or} \quad b < N_B = \infty$$

$$\max_{i=1}^{b} k_i \leq \min_{i=1}^{N_B} K_i < \infty \quad \text{or} \quad \max_{i=1}^{b} k_i < \min_{i=1}^{N_B} K_i = \infty.$$

2 General Balance of the Designs

For further considerations we take into account proper block designs only, that is the designs with all block sizes equal, say k. As it is shown by Caliński (1993) proper block designs have very advantageous property, namely they are generally balanced. In case of the proper block designs under mixed model (1) the dispersion matrix (2) can be expressed as:

$$V = \xi_1 S_1 + \xi_2 S_2 + \xi_3 S_3, \tag{3}$$

where the matrices S_i, $i = 1, 2, 3$, are orthogonal projection operators onto three subspaces of R^n, so called strata. It means that $\oplus_i \mathcal{C}(S_i) = R^n$, where $\oplus$ denotes a direct sum of subspaces while $C(A)$ is the column space of A. The matrices S_i are of the following forms:

$$S_1 = I_n - \frac{1}{k}D'D, \quad S_2 = \frac{1}{k}D'D - \frac{1}{n}J_n, \quad S_3 = \frac{1}{n}J_n.$$

The coefficients ξ_i, $i = 1, 2, 3$, denote the strata variances and are equal to:

$$\xi_1 = \sigma_\eta^2 + \sigma_e^2, \quad \xi_2 = k\sigma_\beta^2 + (1 - \frac{k}{K_H})\sigma_\eta^2 + \sigma_e^2, \quad \xi_3 = (1 - \frac{b}{N_B})k\sigma_\beta^2 + (1 - \frac{k}{K_H})\sigma_\eta^2 + \sigma_e^2.$$

The strata variances depend on the way of grouping of experimental units and on the number of blocks and their sizes taken to the experiment. Some special cases are $K_H = \infty$ or $N_B = \infty$, $k = K_i$ for all i, i.e. $k = K_H$ or $b = N_B$. Then the coefficients ξ_i take much simpler forms.

The division of R^n is, in fact, performed in accordance with the stratification of the experimental units. Let us note that $S_i y$, $i = 1, 2, 3$, give submodels of the model (1). We have

$$y = y_1 + y_2 + y_3,$$

where

$$y_i = S_i y, \quad E(y_i) = S_i \Delta' \tau, \quad Cov(y_i) = \xi_i S_i, \quad i = 1, 2, 3. \tag{4}$$

Vectors $y_i \in C(S_i)$, $i = 1, 2, 3$, belong to the strata which are named intrablock, interblock and total-area stratum, respectively (Houtman and Speed, 1983; Bailey, 1991; Caliński, 1993).

Let us decompose the treatment space $T = C(\Delta')$ into v orthogonal subspaces $T_j = C(\Delta' s_j)$, where s_j are eigenvectors of the matrix $C_1 = \Delta S_1 \Delta'$ connected with the eigenvalues ϵ_j of C_1 with respect to $r^\delta = \Delta \Delta' = diag(r_1, \ldots, r_v)$, i.e. the vectors s_j, $j = 1, \ldots, v$, fulfill the conditions:

$$C_1 s_j = \epsilon_j r^\delta s_j, \quad s_i' r^\delta s_j = \delta_{ij}, \tag{5}$$

where δ_{ij} is the Kronecker delta. There is a strict connection between the structure of the dispersion matrix V in (3) and the treatment structure.

Lemma 1. (Caliński, 1993) *An experimental block design under model (1) with dispersion matrix (3) is generally balanced with respect to the decomposition of the treatment space $T = \oplus_j T_j$.*

Lemma 1 means that the matrices $C_i = \Delta S_i \Delta'$, $i = 1, 2, 3$, have common set of eigenvectors and the eigenvalues corresponding to them are summing up to 1 over $i = 1, 2, 3$. This fact is often used in theory of recovery of interblock information (cf Caliński, 1993; Bogacka, 1993).

The property of general balance (GB - for short) was introduced by Nelder (1965 a,b) and then considered by many authors, for instance: Bailey (1981, 1991), Houtman and Speed (1983), Payne and Tobias (1992), Mejza (1992), Caliński (1993). The GB property makes easier both estimation of treatment parameters functions and their variances, and also considerations on optimality of designs.

3 Information Matrices of the Design

In the mixed model case interest often lies on simultaneous estimation of τ and $\xi = (\xi_1, \xi_2, \xi_3)'$. Assuming a multivariate normal distribution of the vector y we can apply the Maximum Likelihood Method for estimating the parameters and use the likelihood function to derive the information matrix for the whole set of parameters, i.e. τ and ξ. The estimators $\hat{\tau}$ and $\hat{\xi}$ are asymptotically normally distributed, and if $n \to \infty$ we have (cf Searle, 1970; Hocking, 1985):

$$Cov(\hat{\tau}) \to (\Delta V^{-1}\Delta')^{-1},$$

$$Cov(\hat{\tau}, \hat{\xi}) \to 0,$$

$$Cov(\hat{\xi}) \to 2\left[(trace(V^{-1}S_iV^{-1}S_j))_{i,j=1,2,3}\right]^{-1}.$$

The inverses of the asymptotic dispersion matrices are the diagonal elements of the information matrix for τ and ξ, denoted here by Λ, namely

$$\Lambda = \begin{bmatrix} M & 0 \\ 0 & \Omega \end{bmatrix},$$

where

$$M = \Delta V^{-1}\Delta'$$

is the information matrix for τ, and

$$\Omega = \frac{1}{2}\left[(trace(V^{-1}S_iV^{-1}S_j))_{i,j=1,2,3}\right]$$

is the information matrix for ξ. In case of GB designs the matrix Λ takes simpler form and has some nice properties. First, let us look at the information matrix M for fixed parameters.

Theorem 1. (Bogacka, Mejza, 1994) *The information matrix M for parameters τ, under the model (1) with dispersion matrix given in (3), has the following properties:*
a) M is a $(v \times v)$ dimensional positive definite matrix,
b) M can be expressed as the linear combination

$$M = \sum_{i=1}^{3} \xi_i^{-1} C_i,$$

where $C_i = \Delta S_i\Delta'$, $i = 1,2,3$, are the strata information matrices for τ corresponding to submodels (4), respectively,
c) The eigenvalues ν_j, $j = 1,\ldots,v$, of M with respect to r^δ depend on the strata variances ξ_1, ξ_2 and ξ_3, and on the eigenvalues ϵ_j of the intrablock information matrix C_1 with respect to r^δ. That is, the spectral decomposition of M can be presented in the form:

$$M = r^\delta \sum_{j=1}^{v} \nu_j s_j s_j' r^\delta,$$

where

$$\nu_j = \frac{\epsilon_j\xi_2 + (1-\epsilon_j)\xi_1}{\xi_1\xi_2}, \quad j = 1,\ldots,v-1, \quad \nu_v = 1/\xi_3,$$

while $s_j, j = 1, \ldots, v$, are eigenvectors defined in (5).

The eigenvalues of the information matrix play an important role in statistical analysis. Let G denote the matrix of the form

$$G = [c_1, \ldots, c_v],\tag{6}$$

where $c'_j \tau = s'_j r^\delta \tau$, $j = 1, \ldots, v-1$, are the basic contrasts of the treatment parameters and $c'_v \tau = s'_v \tau = (1/\sqrt{n})1'_v r^\delta \tau$, where 1_v is the v dimensional vector of ones, is the weighted average of the treatment effects (Pearce, Caliński, Marshall, 1974). The information matrix $M_{G'\tau}$ for the set of functions $G'\tau$ is the inverse of the dispersion matrix of its estimator. Assuming that strata variances are known, we have

$$\hat{\tau}_{BLUE} = M^{-1}\Delta V^{-1}y$$
$$Cov(\widehat{G'\tau}) = Cov(G'M^{-1}\Delta V^{-1}y) = G'M^{-1}G.\tag{7}$$

There is the following relation between the information matrices for τ and for $G'\tau$, i.e. between M and $M_{G'\tau}$.

Theorem 2. *The eigenvalues of the information matrix $M_{G'\tau}$ for the set of treatment functions $G'\tau$, where G is given in (6), are equal to the eigenvalues ν_j of the information matrix M for τ with respect to r^δ.*

Proof. Let us note that the inverse of M is, on the basis of Theorem 1c) equal to $M = \sum_{j=1}^v \frac{1}{\nu_j} s_j s'_j$. Hence, denoting by $G' = L' r^\delta$ where $L = [s_1, \ldots, s_v]$, we have, by (7) and (5),

$$Cov(\widehat{G'\tau}) = L' r^\delta \left[\sum_{j=1}^v \frac{1}{\nu_j} s_j s'_j\right] r^\delta L = diag\left(\frac{1}{\nu_1}, \ldots, \frac{1}{\nu_v}\right).$$

So

$$M_{G'\tau} = \left[Cov(\hat{G'\tau})\right]^{-1} = diag(\nu_1, \ldots, \nu_v)$$

what proves the theorem. $\square$

Now, let us have a look at the information matrix for ξ.

Theorem 3. *The information matrix Ω for parameters ξ, under the model (1) with dispersion matrix of y given in (3), is of the following diagonal form*

$$\Omega = \frac{1}{2} diag\{\omega_1, \omega_2, \omega_3\},$$

where $\omega_1 = \frac{n-b}{\xi_1^2}$, $\omega_2 = \frac{b-1}{\xi_2^2}$, $\omega_3 = \frac{1}{\xi_3^2}$

Proof. Let us note that matrices S_i fulfill the conditions $S_i S_j = S_i \delta_{ij}$, $i,j = 1,2,3$, and the inverse of the matrix V given in (3) is $V^{-1} = \sum_{i=1}^3 \xi_i^{-1} S_i$. Then we can write

$$V^{-1}S_i V^{-1}S_j = \left(\sum_{l=1}^3 \xi_l^{-1} S_l\right) S_i \left(\sum_{p=1}^3 \xi_p^{-1} S_p\right) S_j = \frac{1}{\xi_i \xi_j} S_i S_j =$$

$$= \begin{cases} 0 & if \quad i \neq j \\ \frac{1}{\xi_i^2} S_i & if \quad i = j \end{cases}$$

Hence,

$$\Omega = \frac{1}{2} diag_{i=1}^3 \left\{ trace \left(\frac{1}{\xi_i^2} S_i \right) \right\} = \frac{1}{2} diag_{i=1}^3 \left\{ \frac{1}{\xi_i^2} rank(S_i) \right\} =$$

$$= \frac{1}{2} diag \left\{ \frac{n-b}{\xi_1^2}, \frac{b-1}{\xi_2^2}, \frac{1}{\xi_3^2} \right\}.$$

$\square$

Different statistical quantities describing "goodness" of a design, for instance efficiency factors defined for treatment parameters functions, are expressed by the eigenvalues of the information matrix $M_{G'T}$. Let us note that for $n \to \infty$, $j = 1, ..., v$, we have

$$var(\widehat{c_j'\tau}) \longrightarrow c_j' M^{-1} c_j = \frac{1}{\nu_j}.$$

Furthermore, the estimator of the contrast $c_j'\tau$, $j = 1, ..., v - 1$, is the following convex combination:

$$\widehat{c_j'\tau} = \frac{\epsilon_j}{\nu_j \xi_1} (\widehat{c_j'\tau})_1 + \frac{1 - \epsilon_j}{\nu_j \xi_2} (\widehat{c_j'\tau})_2,$$

where $(\widehat{c_j'\tau})_1$ and $(\widehat{c_j'\tau})_2$ are the intrablock stratum and the interblock stratum estimators of $c_j'\tau$, respectively. Hence, efficiency of the stratum estimation of the $j - th$ contrast can be of interst. Let us define it as a ratio of the variances of the estimators obtained in the whole space R^n and in the intrablock or interblock stratum $C(S_i)$, $i = 1, 2$, i.e. for $n \to \infty$, $j = 1, ..., v - 1$, we have

$$E_1 = \frac{var(\widehat{c_j'\tau})}{var(\widehat{c_j'\tau})_1} \longrightarrow \frac{1/\nu_j}{\xi_1/\epsilon_j},$$

$$E_2 = \frac{var(\widehat{c_j'\tau})}{var(\widehat{c_j'\tau})_2} \longrightarrow 1 - E_1.$$

The coefficient E_1 is very helpful in making decision concerning further statistical analysis, namely whether to perform the analysis in R^n or in a stratum $C(S_i)$, $i = 1, 2$.

Other way of looking at the estimator of $c_j'\tau$ is to compare the variance of this contrast in case of an orthogonal design (where $\epsilon_j = 1, j = 1, ..., v - 1$) to its variance in case of the considered design. Let us write

$$ef(\widehat{c_j'\tau}) = \frac{var_{ort}(\widehat{c_j'\tau})}{var(\widehat{c_j'\tau})} \longrightarrow \frac{\xi_1}{1/\nu_j}.$$

An efficiency factor of the design, denoted here by $Ef(d)$, defined as a harmonic mean of $ef(\widehat{c_j'\tau})$, $j = 1, ..., v - 1$, is also a good measure of the design.

$$Ef(d) = \frac{v - 1}{\sum_{j=1}^{v-1} \frac{1}{ef(\widehat{c_j'\tau})}} = \xi_1 \frac{v - 1}{\sum_{j=1}^{v-1} \frac{1}{\nu_j}}.$$

Let us note that all the efficiency factors mentioned above depend on the eigenvalues ν_j of $M_{G'\tau}$, so they depend both on the eigenvalues ϵ_j of C_1 with respect to r^δ and on the strata variances ξ_1 and ξ_2. Similarly, the general variance of the estimator of the set of contrasts or the volume of the elipsoid for estimators of these functions depend on ν_j, $j = 1, ..., v - 1$. Hence, considering an optimality criterion in the case of mixed model we should take into account the fact that the quantity we want to optimize can depend on unknown parameters.

In the case of considered here block designs we should examine at least two situations:
I. The optimization is being performed over the class $\mathcal{D}(v, b, k)$, where the number of treatments v, the number of blocks b, and the block size k are fixed. Then, the vector ξ can be treated as an unknown constant. It means that the most of the optimality criteria based on the information matrix for treatments, as it is in the fixed model case, can be applied in the mixed model.
II. The optimization is being performed over the class $\mathcal{D}(v, n)$, where $n = bk$, but both b and k can vary. Then, we can not assume that the variances ξ_i are constant in this class. It means that both the treatment effects and variance components should come within a definition of an optimality criterion. So, the criterion will be based on the information matrix Λ.

For instance, to optimize the efficiency factor of the design $Ef(d)$ we should define a criterion comprising the eigenvalues ϵ_j and the strata variances ξ_1 and ξ_2. As we know

$$var(\xi_i) \longrightarrow \frac{1}{\omega_i}, \quad i = 1, 2, 3.$$

So, to minimize the factor $Ef(d)$, and at the same time to minimize the varinces of the estimators of the strata variances, we can take as the optimality criterion the following function:

$$\Phi(\Lambda) = \sum_{j=1}^{v-1} \frac{1}{\nu_j} + \sum_{i=1}^{2} \frac{1}{\omega_i}.$$

In some cases we are interested only in strata variances. Let us take into account a linear function

$$f(\xi) = a'\xi,$$

where $a' = (a_1, a_2, a_3)$, $a_i \in R$, $i = 1, 2, 3$. Then, as $n \to \infty$, we have

$$var\left(f(\hat{\xi})\right) \longrightarrow a'Cov(\hat{\xi})a = a'\Omega^{-1}a.$$

In case of $a' = (1, 1, 1)$ we obtain the A-optimality criterion for strata variances defined as the sum of the asymptotic varinces of their estimators

$$\Phi_A(\Omega) = \sum_{i=1}^{3} \frac{1}{\omega_i},$$

where ω_i, $i = 1, 2, 3$, are the diagonal elements, and at the same time, the eigenvalues of Ω.

D-optimality criterion in the form $\lambda \log \det F_1 + (1 - \lambda) \log \det F_2$, where F_1 is the information matrix for fixed effects and F_2 is the information matrix for variance components, $\lambda \in [0, 1]$, was used by Giovagnoli and Sebastiani (1989). They have shown some

148

conditions, depending on the ratio of variance components, the balanced designs with one random factor and a technical error have to fulfill to be optimal with respect to this criterion. In their paper the balance means equal number of observations in subclasses.

Similarly, we can define other criteria. However, automatical transfer of definitions of optimality used formerly for treatment parameters only is not always reasonable. Other problem, much more difficult, is to state conditions a design has to fulfill to be optimal in such wide sense. This is still the problem to work on.

References

Bailey, R.A. (1981). A unified approach to design of experiments. *J. Roy. Statist. Soc. A* **144**, 214-223.

Bailey, R.A. (1991). Strata for randomized experiments. *J. Roy. Statist. Soc. B* **53**, 27-78.

Bogacka, B. (1993). Efektywność warstwowej estymacji kontrastów obiektowych w ogólnie zrównoważonych układach blokowych (Stratum estimation efficiency of treatment contrasts in generally balanced block designs), in Polish. *Materiały XXIII Colloquium Metodologicznego z Agrobiometrii, Lublin, 20-22.09.1993*, 160-174.

Bogacka, B. and Mejza, S. (1994). Optimality of generally balanced experimental block designs. *Proc. of International Conference on Linear Statistical Inference - LINSTAT'93, T.Caliński and R.Kala eds., Kluwer Academic Publishers, Netherlands*,185-194.

Caliński, T. (1993). The basic contrasts of a block experimental design with special reference to the notion of general balance. *Listy Biometryczne - Biometrical Letters* **30**, 13-38.

Caliński, T. and Kageyama, S. (1991). On the randomization theory of intra-block and inter-block analysis. *Listy Biometryczne - Biometrical Letters* **28**, 97-122.

Fisher, R.A. (1926). The arrangement of field experiment. *J. Min. Agr.* **23**, 503-513.

Giovagnoli, A. and Sebastiani, P. (1989). Experimental designs for mean and variance estimation in variance components models. *Computational Statistics and Data Analysis.* **8**, 21-28.

Hocking, R. R. (1985). *The Analysis of Linear Models.* Monterey. California: Brooks/Cole Publishing Company.

Houtman, A. M. and Speed, T. P. (1983). Balance in designed experiments with orthogonal block structure. *Ann. Statist.* **11**, 1069-1085.

Kala, R. (1991). Elementy teorii randomizacji. III. Randomizacja w doświadczeniach blokowych. *Listy Biometryczne - Biometrical Letters* **28**, 3–23.

Mejza, S. (1992). On some aspects of general balance in designed experiments. *Statistica* **2**, 263-278.

Nelder, J.A. (1965a). The analysis of randomized experiments with orthogonal block structure. I. Block structure and the null analysis of variance. *Proc. of the Royal Soc. A* **283**, 147-162.

Nelder, J.A. (1965b). The analysis of randomized experiments with orthogonal block structure. II. Treatment structure and the general analysis of variance. *Proc. of the Royal Soc. A* **283**, 163-178.

Neyman, J., Iwaszkiewicz, K. and Kołodziejczyk, S. (1935). Statistical problems in agricultural experimentation. *J. Roy. Statist. Soc. Suppl.* **2**, 107-154. Discussions *Ibid.* **2**, 154-180.

Ogawa, J. and Ikeda, S. (1973). On the randomization of block designs. In: J. N. Srivastava, Ed., *A Survey of Combinatorial Theory*. North Holland Publishing Company, Amsterdam, 335-347.

Payne, R.W. and Tobias, R.D. (1992). General balance, combination of information and the analysis of covariance. *Scandinavian J. Statist.*

Pearce, S. C., Caliński, T. and Marshall, T. F. (1974). The basic contrasts of an experimental design with special reference to the analysis of data. *Biometrika.* **63** 449-460.

Searle, S. R. (1970). Large sample variances of Maximum Likelihood Estimators of variance components using unbalanced data. *Biometrics.* **26**, 505-524.

Estimation of Parameters in Factorial Triallel Analysis for BIB Design — the Mixed Model

BRONISŁAW CERANKA and ZYGMUNT KACZMAREK

Department of Mathematical and Statistical Methods, Agricultural University of Poznań, Poland
and
Institute of Plant Genetics of the Polish Academy of Sciences, Poznań, Poland.

Abstract:

The paper presents the estimation some genetic parameters concerning hybrids obtained in factorial triallel crossing system. The hybrids are compared in a balanced incomplete block design in which the block effects are treated as random. The statistical analysis includes estimation (intra-block, inter-block and combining) of general line effects and two-line and three-line specific effects.

1 Introduction

Triallel cross hybrids are the products of three parental lines. Two of them are crossed in the first step and obtained hybrid is crossed with the third line. The basic paper on the theory of triallel analysis is given by Rawlings and Cockerham (1962). The main problem in the application of triallel crossing system is rapidly increasing number of three-way crosses with increasing number of parental lines. One possible solution in this situation is use of partial trialles based on the theory of partially incomplete balanced block designs (Hinkelman (1965)). Cockerham (1963, 1980) extended the factorial diallel crossing system to the factorial triallel crossing system. Then a greater number of lines can be tested. It means that factorial triallel system can be an alternative way for the plant breeders to get informations on the combining abilities of the parental lines.

Statistical analysis of factorial triallel crossing system for fixed model is given by Rasch at al (1990). The complete algorithms for the analysis of factorial triallel crossing system over environments for the fixed and random models are presented by Wolf (1988a, b).

In this paper the statistical analysis including combined estimation of general line effects and two-line and three-line specific effects from an experiment carried out in a balanced incomplete block design with random block effects is given.

---

*Kitsos, C.P., and Müller, W.G., Eds., *Proceedings of MODA4*, Physica Verlag, Heidelberg, 1995

2 Model of Observations

In a factorial triallel the basic set of parental lines is divided into three groups, designated P, Q, R with numbers p, q, r, respectively. Each line can be located only in one group. In the first step of triallel crossing p lines of first group are crossed with all q lines of second groups. After them pq single hybrids are crossed with all r lines of third group. As a result of this crossing $v = pqr$ three-line hybrids are obtained. The experiment with all hybrids is carried out in a balance incomplete block design. The described factorial triallel enables to estimate the general and specific effects of parental lines. In the linear model of observations obtained from this experiment is assumed that general and specific effects of parental line are fixed, but block effects are random. The general analysis of this model is given for example by Ceranka and Mejza (1988).

Let γ_{ijk} be the expected value of trait observed in the progeny of three-line cross. The following linear model for factorial triallel system can be formulated:

$$\gamma_{ijh} = \mu + g_i^P + g_j^Q + g_h^R + s_{ij}^{PQ} + s_{ih}^{PR} + s_{jh}^{QR} + s_{ijh}^{PQR}$$

$$i = 1, ..., p; j = 1, ..., q; h = 1, ..., r,$$

where μ is the general parameter, g_i^P is the general effect of the ith grandparental line from group P, g_j^Q is the general effect of the jth grandparental line from group Q, g_h^R is the general effect of the hth parental line from group R, s_{ij}^{PQ} is the specific effect for the pair of ith grandparental line from P and jth grandparental line from Q, s_{ih}^{PR} is the specific effect for the pair of ith grandparental line from P and hth parental line from R, s_{jh}^{QR} is the specific effect for the pair of jth grandparental line from Q and hth parental line from R and s_{ijh}^{PQR} is the specific effect for the combination of ith grandparental line from P, jth grandparental line from Q and hth parental line from R.

The general effect for ith grandparental line from group P can be defined as follows

$$g_i^P = \frac{1}{qr}\gamma_{i..} - \frac{1}{pqr}\gamma_{...}, \quad i = 1, ..., p,$$

where

$$\gamma_{i..} = \sum_{j=1}^{q} \sum_{h=1}^{r} \gamma_{ijh},$$

$$\gamma_{...} = \sum_{i=1}^{p} \sum_{j=1}^{q} \sum_{h=1}^{r} \gamma_{ijh}.$$

Similarly can be defined the general effects g_j^Q and g_h^R.

The specific effect for the pair of ith grandparental line from P and jth grandparental line from Q can be defined as follows

$$s_{ij}^{PQ} = \frac{1}{r}\gamma_{ij.} - g_i^P - g_j^Q - \frac{1}{pqr}\gamma_{...},$$

where

$$\gamma_{ij.} = \frac{1}{pq} \sum_{h=1}^{r} \gamma_{ijh}.$$

Similarly can be defined the specific effects s_{ih}^{PR} and s_{jh}^{QR}.

The specific effect for the combination of lines from particular groups P, Q and R can be written

$$s_{ijh}^{PQR} = \gamma_{ijh} - g_i^P - g_j^Q - g_h^R - s_{ij}^{PQ} - s_{ih}^{PR} - s_{jh}^{QR} - \frac{1}{r}\gamma_{...}.$$

The following restrictions are imposed on the parameters:

$\sum_{i=1}^{p} g_i^P = 0, \sum_{j=1}^{q} g_j^Q = 0, \sum_{h=1}^{r} g_h^R = 0,$

$\sum s_{ij}^{PQ} = 0$ for each i and each j,

$\sum s_{ih}^{PR} = 0$ for each i and each h,

$\sum s_{jh}^{QR} = 0$ for each j and each h

and $\sum s_{ijh}^{PQR} = 0$ for each pair (ij), each pair (ih) and each pair (jh).

The general and specific line effects can be treated as the contrasts between treatments which can be defined as follows. The vector γ is convinient to transform to a vector γ_1 with $v = pqr$ elements in such a way that (ijh)th element of γ receives in vector γ_1 the position $(i-1)qr + (j-1)r + h$, $i = 1, \ldots, p$, $j = 1, \ldots, q$, $h = 1, \ldots, r$. Then the general and specific effects of lines can be written in the vector form.

The general effect of the ith grandparental line from P is

$$g_i^P = \mathbf{c}_i'\gamma_1, \quad i = 1, \ldots, p,$$

where

$$\mathbf{c}_i' = \frac{1}{pqr}[c_{i1}, \ldots, c_{iv}],$$

and where

$$c_{im} = \begin{cases} p - 1, & \text{when} \quad m = (i-1)qr + (j-1)r + h, \\ & \qquad\qquad j = 1, \ldots, q, \ h = 1, \ldots, r, \\ -1, & \text{else.} \end{cases}$$

In similar way can be defined g_j^Q and g_h^R effects.

The specific effect for the pair of ith grandparental line from P and jth grandparental line from Q is

$$s_{ij}^{PQ} = \mathbf{c}_{ij}'\gamma_1, \quad i = 1, \ldots, q,$$

where

$$\mathbf{c}_{ij}' = \frac{1}{pqr}[c_{ij1}, \ldots, c_{ijv}],$$

and where

$$
c_{ijm} = \begin{cases}
(p-1)(q-1), & \text{when } m = (i-1)qr + (j-1)r + h, \\
& \qquad h = 1, \ldots, r \\
-(p-1), & \text{when } m = (i-1)qr + (t-1)r + h \\
& \qquad t = 1, \ldots, q,\, t \neq j, \\
-(q-1), & \text{when } m = (s-1)qr + (j-1)r + h, \\
& \qquad s = 1, \ldots, p,\, s \neq i, \\
1, & \text{else.}
\end{cases}
$$

The specific effects for the pair of ith grandparental line from P and hth parental line from R and for the pair of jth grandparental line from Q and hth parental line from R can be defined analogously.

The specific effect for the combination of ith grandparental line from P, jth grandparental line from Q and hth parental line from R is

$$
s_{ijh}^{PQR} = \mathbf{c}'_{ijh}\gamma_1, \quad i = 1, \ldots, p,\ j = \ldots, q,\ h = 1, \ldots, r,
$$

where

$$
\mathbf{c}'_{ijh} = \frac{1}{pqr}[c_{ijh1}, \ldots, c_{ijhv}],
$$

and where

$$
c_{ijhm} = \begin{cases}
(p-1)(q-1)(r-1), & \text{when } m = (i-1)qr + (j-1)r + h, \\
-(p-1)(q-1)', & \text{when } m = (i-1)qr + (j-1)r + u, \\
& \qquad u = 1, \ldots, r,\, u \neq h, \\
-(p-1)(r-1), & \text{when } m = (i-1)qr + (t-1)r + h \\
& \qquad t = 1, \ldots, q,\, t \neq j, \\
-(q-1)(r-1), & \text{when } m = (s-1)qr + (j-1)r + h, \\
& \qquad s = 1, \ldots, p,\, s \neq i, \\
p-1, & \text{when } m = (t-1)r + u; t = 1, \ldots, q, \\
& \qquad t \neq j, u = 1, \ldots, r,\, u \neq h, \\
q-1, & \text{when } m = (s-1)pq + u, s = 1, \ldots, p, \\
& \qquad s \neq i, u = 1, \ldots, r,\, u \neq h, \\
r-1, & \text{when } m = (s-1)pq + (t-1)r + h, \\
& \qquad s = 1, \ldots, p,\, s \neq i, t = 1, \ldots, q,\, t \neq j \\
-1, & \text{else.}
\end{cases}
$$

3 Estimation of Parameters

As mentioned before the estimation of general line effect and two-line and three-line specific effects will base on combining estimators. According to the method of combining estimators given by Mejza (1978), we can write the combining estimators of considered above parameters as follows.

The combined estimator of general effect of the ith grandparental line from P is

$$
\bar{g}_i^P = \hat{g}_i^P + (\tilde{g}_i^P - \hat{g}_i^P)\frac{k(p-3)(p-1)s_E^2}{\lambda p^2 q^2 r^2 \sum_{i=1}^p (\tilde{g}_i^P - \hat{g}_i^P)^2},
$$

where $\hat{g}_i^P$ is the intra-block estimator of g_i^P, of the from

$$
\hat{g}_i^P = \frac{1}{qr}\hat{\gamma}_{i..} - \frac{1}{pqr}\hat{\gamma}_{...},
$$

with

$$\hat{\gamma}_{i..} = \sum_{j=1}^{q}\sum_{h=1}^{r}\hat{\gamma}_{ijh}$$

and

$$\hat{\gamma}_{...} = \sum_{i=1}^{p}\sum_{j=1}^{q}\sum_{h=1}^{r}\hat{\gamma}_{ijh}$$

obtainable from $\hat{\gamma} = (k/\lambda pqr)\mathbf{Q} + (G/n)\mathbf{1}$, where $G = \mathbf{T}'\mathbf{1} = \mathbf{B}'\mathbf{1}, \mathbf{Q} = \mathbf{T} - \frac{1}{k}\mathbf{NB}$, $\mathbf{T}$ is the $pqr \times 1$ vector of treatment totals, $\mathbf{B}$ is the $b \times 1$ vector of block totals, $\mathbf{N}$ is the incidence matrix of BIB design, and where $\tilde{g}_i^P$ is the inter-block estimator of g_i^P, which can be obtained when replacing $\hat{\gamma}$ by $\tilde{\gamma} = ((bk^2 - \lambda v^2)/v)(\mathbf{NB} - \lambda v/n)G\mathbf{1}$ in the above formulae.

The combined estimators of general effects of the jth grandparental line from group Q and hth parental line from group R can be obtained analogously as the estimator of $\tilde{g}_i^P$.

The combined estimator of specific effect for the pair of ith grandparental line from P and jth grandparental line from Q is

$$\bar{s}_{ij}^{PQ} = \hat{s}_{ij}^{PQ} + (\tilde{s}_{ij}^{PQ} - \hat{s}_{ij}^{PQ})\frac{k(pq - p - q - 1)(p-1)(q-1)s_E^2}{\lambda p^2 q^2 r^2 \sum_{i=1}^{p}\sum_{j=1}^{q}(\tilde{s}_{ij}^{PQ} - \hat{s}_{ij}^{PQ})^2},$$

where for any pair i, j $(i = 1,\ldots,p; j = 1,\ldots,q)$ $\hat{s}_{ij}^{PQ}$ is the intra-block estimator of s_{ij}^{PQ} of the form

$$\hat{s}_{ij}^{PQ} = \frac{1}{r}\hat{\gamma}_{ij.} - \hat{g}_i^P - \hat{g}_j^Q - \frac{1}{pqr}\hat{\gamma}_{...}$$

and $\tilde{s}_{ij}^{PQ}$ is the inter-block estimator of s_{ij}^{PQ}, which can be obtained analogously as the estimator of $\tilde{g}_i^P$.

The combined estimators of specific effects for the pair of ith grandparental line from P and hth parental line from R and for the pair of jth grandparental line from Q and hth parental line from R can be obtained analogously as the estimator of $\bar{s}_{ij}^{PQ}$.

The combined estimator of specific effect for the combination of ith grandparental line from P, jth grandparental line from Q and hth parental line from R is

$$\bar{s}_{ijh}^{PQR} = \hat{s}_{ijh}^{PQR} + (\tilde{s}_{ijh}^{PQR} - \hat{s}_{ijh}^{PQR})\frac{k[(p-1)(q-1)(r-1) - 2](p-1)(q-1)(r-1)s_E^2}{\lambda p^2 q^2 r^2 \sum_{i=1}^{p}\sum_{j=1}^{q}\sum_{h=1}^{r}(\tilde{s}_{ijh}^{PQR} - \hat{s}_{ijh}^{PQR})^2},$$

where $\hat{s}_{ijh}^{PQR}$ is the intra-block estimator of s_{ijh}^{PQR} of the form

$$\hat{s}_{ijh}^{PQR} = \hat{\gamma}_{ijh} - \hat{g}_i^P - \hat{g}_j^Q - \hat{g}_h^R - \hat{s}_{ij}^{PQ} - \hat{s}_{ih}^{PR} - s_{jh}^{QR} + \frac{1}{pqr}\hat{\gamma}_{...},$$

and $\tilde{s}_{ijh}^{PQR}$ is the inter-block estimator of s_{ijh}^{PQR}, which can be obtained analogously as the estimator of $\tilde{g}_i^P$.

The factorial triallel analysis provides additional information regarding the effects of order in which the parents are involved in crosses. Comparing with diallel system crossing and line by tester system crossing, where only two genetic parameter effects can be estimated the factorial triallel design gives the possibility of estimation more number of genetic parameters.

156

References

Ceranka B., Mejza S. (1988): Analysis of diallel table experiments carried out in BIB designs - mixed model. Biometrical Journal 30, 3-16.

Cockerham C.C. (1963): Estimation of genetic variances. In: Statistical Genetics and Plant Breeding, ed. by W.D. Hanson and H.F. Robinson, Publ. 982 Natl. Acad. Sci. - Natl. Res. Counc., Washington. D.C., 53-93.

Cockerham C.C. (1980): Random and fixed effects in plant genetics. Theor. Appl. Genet. 56, 119-131.

Hinkelmann K. (1965): Partial triallel crosses. Sankhyā, Ind. J. Statistics, Ser. A., 27, 173-196.

Mejza S. (1978): Use of inter - block information to obtain uniformly better estimators of treatment contrasts. Mathematische Operationsforschung und Statistik, Series Statistics, 9, 335-341.

Rasch D. (ed.) (1990): Handbuch der Populationsgenetik und Zütchtungsmethodik. Deutscher Landwirtschaftsverlag Berlin.

Rawlings J. O., Cockerham C.C. (1962): Triallel analysis. Crop. Sci. 2, 228-231.

Wolf J. (1988a): Statistical analysis of factorial triallel design over environments. I. The fixed model. Arch. Züchtungsforsch. Berlin, 18, 261-268.

Wolf J. (1988b): Statistical analysis of factorial triallel design over environments. II. The random model. Arch. Züchtungsforsch. Berlin, 18, 320-335.

On the Optimality of Certain Nested Block Designs under a Mixed Effects Model

Stanisław Mejza and Sanpei Kageyama

Department of Mathematical and Statistical Methods, Agricultural University of Poznań, Poland
and
Department of Mathematics, Hiroshima University, Japan.

Abstract:

Some optimal statistical properties of C-designs in certain nested block designs under a mixed model are characterized.

1 Introduction

Preece (1967) introduced a class of designs called nested balanced incomplete block designs. In these designs, there are two systems of blocks, the second nested within the first (each block from the first system, called superblocks,

containing b blocks from the second) such that ignoring either system leaves a balanced incomplete block (BIB) design whose blocks are those of the other system. This idea has been generalized to nested partially balanced incomplete block designs by Homel and Robinson (1975), and Banerjee and Kageyama (1990, 1993). The various constructions of these designs have also been considered in Dey et al. (1986), and Jimbo and Kuriki (1983).

In the existing literature the concept of nested balanced block designs is mainly connected with the property of balance or partial balance of two associated block designs. In this paper, we consider the nested block design in a wider sense and use the name of nested block design (NBD for short) for the class of designs with the above described two systems of blocks, only. We do not require the design to be in some sense balanced.

Let v denote the number of treatments. The additional parameters of an NBD are: R, the number of superblocks, and k, the block size. This means that we consider a proper NBD with respect to two block systems.

The purpose of the paper is to give some characterizations of an NBD to be optimal in a class of designs where observations follow a mixed linear model with superblock and

---
*Kitsos, C.P., and Müller, W.G., Eds., *Proceedings of MODA4*, Physica Verlag, Heidelberg, 1995

158

sub-block effects random, while the treatment effects are treated as fixed.

There are some papers dealing with the optimal properties of block designs under a mixed effects model; see, for example, Shah and Sinha (1989), Khatri and Shah (1984), Bagchi (1987a, 1987b), Mukhopadhyay (1981), Jacroux (1989). The present linear mixed models are particular cases of the model considered by Bogacka and Mejza (1994). The same type of linear model, with dispersion structure typical for randomized experiments, will be considered here. Also, some results given by Bogacka and Mejza (1993, 1994) will be adapted to the NBDs case.

It seems to us that this paper is the first dealing with optimal properties of an NBD under a mixed effects model. In particular, we restrict our considerations to a subclass of NBDs in which the designs are orthogonal with respect to superblocks. Nevertheless, this calss includes many useful NBDs. To this subclass belong for example all $(\mu_1, \mu_2, ..., \mu_R)$-resolvable block designs.

Finally, we give characterizations of certain (superblock orthogonal) NBDs to be optimal with respect to the optimality criteria given by Cheng (1978) and Bondar (1983).

2 The Linear Model

Let us consider the mixed linear model of the form

$$\mathbf{y} = \Delta'\mu + G'\rho + D'\beta + \eta + e, \quad E(\mathbf{y}) = \Delta'\mu \tag{1}$$

where $\mathbf{y}$ is an $n \times 1$ vector of observations, $n = Rbk, \Delta', G'$ and D' are $n \times v, n \times R$ and $n \times Rb$ design matrices for treatments, superblocks and sub-blocks, μ, ρ and β are $v \times 1, R \times 1$ and $Rb \times 1$ vectors of treatment, superblock and sub-block effects, respectively, while η and e are $n \times 1$ vectors of unit and technical errors, respectively.

Assume that the dispersion matrix of $\mathbf{y}$ is of the form

$$V = \xi_0 P_0 + \xi_1 P_1 + \xi_2 P_2 + \xi_3 P_3, \tag{2}$$

where $\xi_0 = \sigma_e^2, \xi_1 = bk\sigma_\rho^2 + \sigma_e^2, \xi_2 = k\sigma_\beta^2 + \sigma_e^2, \xi_3 = \sigma_\eta^2 + \sigma_e^2$, and $\sigma_\rho^2, \sigma_\beta^2$ are superblock and sub-block variances, σ_η^2 and σ_e^2 are unit and technical error variances.

The matrices (projectors) $P_f, f = 0, 1, 2, 3$, are $P_0 = \frac{1}{n}J_n, \text{rank}(P_0) = 1, P_1 = \frac{1}{bk}G'G - \frac{1}{n}J_n, \text{rank}(P_1) = R-1, P_2 = \frac{1}{k}D'D - \frac{1}{bk}G'G, \text{rank}(P_2) = R(b-1), P_3 = I_n - \frac{1}{k}D'D, \text{rank}(P_3) = Rb(k-1), \sum_f P_f = I_n$, where $J_n = 1_n 1_n', 1_n$ stands for an n-dimensional column vector of ones, and I_n denotes the identity matrix of order n.

Note that model (1) with dispersion structure (2) is typical for randomization models for experiments with orthogonal block structure (cf. Nelder, 1965; Mejza, 1992).

The NBDs under consideration are called multistrata experiments (cf. Houtman and Speed, 1983). The information matrix for estimation of the treatment effects in model (1) is

$$\Lambda = \Delta V^{-1}\Delta' = \sum_{f=0}^{3} \xi_f^{-1} A_f \tag{3}$$

where $A_f = \Delta P_f \Delta', f = 0, 1, 2, 3$, are the respective strata information matrices. These matrices can be expressed as: $A_0 = \frac{1}{n}rr'$ (so called general area stratum), $A_1 = \frac{1}{bk}MM' - \frac{1}{n}rr'$ (inter-superblock stratum), $A_2 = \frac{1}{k}NN' - \frac{1}{bk}MM'$ (inter-subblock stratum), and $A_3 = \mathbf{r}^{-\delta} - \frac{1}{k}NN'$ (intra-subblock stratum), where $M = \Delta G'$ and

$N = \Delta D'$ are the incidence matrices for treatments vs superblocks and treatments vs sub-blocks, respectively, $\mathbf{r} = M\mathbf{1} = N\mathbf{1}$ denotes the vector of treatment replications, and $\mathbf{r}^\delta = \mathrm{diag}(r_1, r_2, ..., r_v)$.

Let $\mathcal{D}(v, Rb, k)$ denote a class of NBDs orthogonal with respect to superblocks, i.e. let $M = \frac{1}{R}\mathbf{r}\mathbf{1}'$ (*superblock orthogonal* NBD, for short). Then we have $A_1 = O, A_2 = \frac{1}{k}NN' - \frac{1}{n}\mathbf{r}\mathbf{r}', A_3 = \mathbf{r}^\delta - \frac{1}{k}NN'$.

Let $\epsilon_1, \epsilon_2, ..., \epsilon_v$ be eigenvalues corresponding to eigenvectors $\mathbf{w}_1, \mathbf{w}_2, ..., \mathbf{w}_v$ of matrix A_3 with respect to (w.r.t.) $\mathbf{r}^\delta$, i.e. $A_3\mathbf{w}_i = \epsilon_i\mathbf{r}^\delta\mathbf{w}_i, i = 1, 2, ..., v$. Since $A_3\mathbf{1} = \mathbf{0}$, let $\mathbf{w}_v = n^{-\frac{1}{2}}\mathbf{1}$ and $\mathbf{w}_i$ be $\mathbf{r}^\delta$-orthonormal in pairs, i.e. $\mathbf{w}_i\mathbf{r}^\delta\mathbf{w}_{i'} = \delta_{ii'}$. The matrices A_2 and A_3 mutually commute w.r.t. $\mathbf{r}^{-\delta}$ (i.e. $\mathcal{D}$ is a generally balanced NBD, cf. Mejza, 1992) and hence $1 - \epsilon_i, i = 1, 2, ..., v$ are the eigenvalues of the matrix A_2 corresponding to the eigenvector $\mathbf{w}_i$ w.r.t. $\mathbf{r}^\delta$. Then the information matrix Λ_d as in (3) can be expressed in a spectral decomposition as

$$\Lambda_d = \mathbf{r}^\delta \sum_{i=1}^{v} \nu_i \mathbf{w}_i \mathbf{w}_i' \mathbf{r}^\delta \tag{4}$$

where $\nu_i = (1 - \epsilon_i)/\xi_2 + \epsilon_i/\xi_3 = [(1 - \epsilon_i)\xi_3 + \epsilon_i\xi_2]/\xi_2\xi_3, \; i = 1, 2, ..., v-1, \; \nu_v = 1/\xi_0$.

3 Optimality Criteria

There are many criteria for optimality of a design. These criteria are mainly given for fixed effects linear models. However, they are general and can be adapted to the mixed-effects model (cf. Shah and Sinha, 1989; Bogacka and Mejza, 1994).

Some notation and theorems concerning optimality problems in a mixed effects model for block designs are taken from Bogacka and Mejza (1993, 1994).

Let $\lambda_j(\Lambda_d)$ denote the eigenvalue, ordinary or w.r.t. $\mathbf{r}^\delta$ (or w.r.t. another positive definite matrix) of the information matrix Λ_d associated with $\mathcal{D}$, $j = 1, 2, ..., v$. Moreover, let $\mathcal{B} = \{\Lambda \geq 0\}$ denote a set of nonnegative definite or positive definite matrices, let $\mathcal{M} = \{\Lambda_d : d \in \mathcal{D}\}$ stand for a set of information matrices for designs d belonging to a class $\mathcal{D}$, such that $\mathcal{M} \subset \mathcal{B}$. Furthermore, let $\lambda(\Lambda) = (\lambda_1(\Lambda), ..., \lambda_m(\Lambda))', 0 < m \leq v$, be the vector of positive eigenvalues of Λ, let $\lambda(\mathcal{B}) = \{\lambda(\Lambda) : \Lambda \in \mathcal{B}\}$, and let

$$\lambda^\delta(\Lambda) = \mathrm{diag}(\lambda_1(\Lambda), ..., \lambda_m(\Lambda)), \; \lambda_\square^\delta(\Lambda) = \mathrm{diag}(\lambda_{[1]}(\Lambda), ..., \lambda_{[m]}(\Lambda)),$$

where $\lambda_{[j]}(\Lambda)$ is the jth largest eigenvalue of Λ.

Definition 1. Given the function $\Phi : \mathcal{B} \mapsto \mathcal{R}$ (optimality criterion), a design $d^* \in \mathcal{D}$ is Φ-optimal in the class $\mathcal{D}$ if $\Phi(\Lambda_{d^*}) \leq \Phi(\Lambda_d)$ for all $d \in \mathcal{D}$.

Definition 2 (Cheng, 1978). A design $d^* \in \mathcal{D}$ is termed generally optimal in the class $\mathcal{D}$ if Λ_{d^*} minimizes in $\mathcal{M}$ every $\Phi : \mathcal{B} \mapsto \mathcal{R}$ of the form

$$\Phi(\Lambda) = \phi_j(\lambda(\Lambda)) = \sum_{j=1}^{m} f(\lambda_j(\Lambda)),$$

where the function $f : [0, x_0] \mapsto \mathcal{R}$, $x_0 = \max_{\Lambda \in \mathcal{M}} \mathrm{tr}(\Lambda)$, satisfies the following conditions:

(1) f is continuously differentiable on $(0, x_0)$,

(2) the first, second and third derivatives of f satisfy, respectively: $f' < 0, f'' > 0, f''' < 0$ on $(0, x_0)$,

(3) $f(0) = \lim_{x \to 0+} f(x) = \infty$.

The class of functions Φ used by Cheng includes A- and D-optimality criteria.

The more general definition of optimality as in Definition 2 was given by Bondar (1983). The design, optimal w.r.t. Bondar's definition, is said to be *BU-optimal*. That is,

Definition 3 (Bondar, 1983). A design $d^* \in \mathcal{D}$ is termed BU-optimal in the class $\mathcal{D}$ if Λ_{d^*} minimizes in $\mathcal{M}$ every $\Phi : \mathcal{B} \mapsto \mathcal{R}$ satisfying the following conditions:

(a) $\Phi(\Lambda) = \phi(\lambda(\Lambda))$, i.e. Φ is a function defined on the eigenvalues of $\Lambda \in \mathcal{B}$,

(b) $\phi(\lambda(\Lambda))$ is a Schur-convex function on the set $\lambda(\mathcal{B})$, i.e. $\lambda(\Lambda_1) \prec \lambda(\Lambda_2) \Rightarrow \phi(\Lambda_1) \leq \phi(\Lambda_2)$, where the symbol $\prec$ denotes the majorization of $\lambda(\Lambda_1)$ by $\lambda(\Lambda_2)$,

(c) $\lambda_\square^\delta(\Lambda_1) \geq_L \lambda_\square^\delta(\Lambda_2) \Rightarrow \Phi(\lambda_\square^\delta(\Lambda_1) \leq \Phi(\lambda_\square^\delta(\Lambda_2))$, where $\geq_L$ denotes the Loewner ordering of matrices.

Denoting by $\mathcal{F}_C$ and $\mathcal{F}_B$ the classes of functionals fulfilling the conditions of Definitions 1 and 2, respectively, it holds that $\mathcal{F}_C \subset \mathcal{F}_B$.

Following Cheng(1978), Shah and Sinha (1989), and Bogacka and Mejza (1994), we have the following.

Theorem 1. Let $d^* \in \mathcal{D}(v, Rb, k)$ be an experimental design under the mixed model (1) for which the matrix A_{3d^*} has two positive eigenvalues w.r.t. $r_{d^*}^\delta$, the greater one with multiplicity 1 and the smaller one with multiplicity $v - 2$. Furthermore, let:

$$\mathrm{tr}(\mathbf{r}_{d^*}^{-\delta} A_{3d^*}) = \max_{d \in \mathcal{D}} \mathrm{tr}(\mathbf{r}_d^{-\delta} A_{3d}),$$

$$\mathrm{tr}(\mathbf{r}_{d^*}^{-\delta} A_{3d^*}) - \gamma \, p(A_{3d^*}) \geq \mathrm{tr}(\mathbf{r}_d^{-\delta} A_{3d}) - \gamma \, p(A_{3d}) \text{ for all } d \in \mathcal{D},$$

where $p(A_{3d}) = \{\sum_{j=1}^{v-1}[\epsilon_j(A_{3d}) - \bar{\epsilon}(A_{3d})]^2\}^{\frac{1}{2}}$, $\gamma^2 = (v-1)/(v-2)$, and $\bar{\epsilon}(A_{3d})$ denotes the average of the eigenvalues $\epsilon_j(A_{3d})$. Then the design d^* minimizes every function

$$\phi_f(\nu(\Lambda_d)) = \sum_{j=1}^v f(\nu_j(\Lambda_d))$$

fulfilling the conditions of Cheng's definition.

Some of the so-called C-designs (Saha, 1976) satisfy the necessary conditions of Theorem 1. Hence, we have the following (see Bogacka and Mejza, 1994).

Corollary 1. Let $\mathcal{D}_0(v, Rb, k)$ denote a class of binary incomplete block designs under mixed model (1). Let $d^* \in \mathcal{D}_0(v, Rb, k)$ be an experimental design for which the matrix A_{3d^*} has two positive eigenvalues w.r.t. $r_{d^*}^\delta$, the greater one with multiplicity 1 and the smaller one with multiplicity $v - 2$. Then the design d^* is generally optimal in $\mathcal{D}_0$ if it is MS-optimal.

Similarly, following Bogacka and Mejza (1994), we can obtain the following theorem and corollaries.

Theorem 2. An experimental design $d^* \in \mathcal{D}(v, Rb, k)$ under the mixed model (1) is BU-optimal in the class $\mathcal{D}$ if and only if for all
$d \in \mathcal{D}$

$$\sum_{j=k}^{v-1} \epsilon_{[j]}(A_{3d}) \le \sum_{j=k}^{v-1} \epsilon_{[j]}(A_{3d^*}), \ k = 1, 2, ..., v-1,$$

where $\epsilon_{[j]}(A_{3d})$ is the jth largest positive eigenvalue of the matrix A_{3d} corresponding to the design d w.r.t. $\mathbf{r}_d^{\delta}$.

Corollary 2. If in a class $\mathcal{D}_1 = \{d \in \mathcal{D}(v, Rb, k) : \text{tr}(\mathbf{r}_d^{-\delta}\Lambda_d) = const.\}$ there exists a design d^* such that the eigenvalues of A_{3d^*} w.r.t. $\mathbf{r}_{d^*}^{\delta}$ are all equal, then, under mixed model (1), the design d^* is BU-optimal in the class $\mathcal{D}_1$.

Corollary 3. If in a class $\mathcal{D}_1$ there exists an $\mathbf{r}_d^{-\delta}$-balanced design (Caliński, 1977) then, under the mixed model (1), it is BU-optimal in the class $\mathcal{D}_1$.

A balanced block design in the sense of Kiefer (1975) is an example.

In many practical cases we are interested in applying only binary designs. Let $\mathcal{D}_2(v, Rb, k)$ denote the binary subclass of $\mathcal{D}_1(v, Rb, k)$. Because of the fact that BIB designs are the only $\mathbf{r}^{-\delta}$-balanced designs in the class $\mathcal{D}_1$ (when it exists: see Kageyama, 1980) we have the following:

Corollary 4. If in the class $D_2(v, Rb, k)$ a BIB design exists, then, under mixed model (1), it is BU-optimal in $\mathcal{D}_2$.

4 Characterization of $\mathcal{D}(v, Rb, k)$

Let us now consider NBDs in $\mathcal{D}(v, Rb, k)$ as candidates for optimality in one of the senses. Denote by $\mathcal{D}_i(v, b, k)$ a block preceding design of the ith superblock of the NBD and let N_i be its

incidence matrix for $i = 1, 2, ..., R$. Then $N = [N_1 : N_2 : \cdots : N_R]$ gives the incidence matrix of the NBD w.r.t. sub-blocks, while $M = [N_1 1 : N_2 1 : \cdots : N_R 1]$ is the incidence matrix of the NBD w.r.t. superblocks.

Under the present optimality criteria we will look for an NBD in the class $\mathcal{D}(v, Rb, k)$ which is a C-design or $\mathbf{r}^{-\delta}$-balanced w.r.t. sub-blocks (which are in fact particular cases of C-designs). This means that the previously considered statistical properties now always refer to a block design with sub-blocks as blocks.

We at first search for optimal designs in the class of α-resolvable block designs.

Definition 4.1. An NBD is said to be α-resolvable if each design D_i is α/R-resolvable for $i = 1, 2, ..., R$.

This restriction results from the fact that our designs are proper. From the α-resolvability of an NBD, the following can easily be obtained.

Proposition 4.1. An α-resolvable NBD is superblock orthogonal, i.e. belongs to the class $\mathcal{D}(v, Rb, k)$.

The class of α-resolvable designs is quite wide. There are many methods of constructing such designs, see, for example, Bose (1942), Ceranka, Kageyama and Mejza (1986), Kageyama (1973, 1984), Mukerjee and Kageyama (1985), Patterson and Williams (1976), Shrikhande and Raghavarao (1964).

Most of the α-resolvable designs in the literature are binary. Then, for the present class of proper designs only BIB designs are $\mathbf{r}^{-\delta}$-balanced (called also efficiency-balanced

162

or $\mathbf{r}^{-\delta}$-PEB(1)).

Proposition 4.2. If in the class of binary α-resolvable NBDs in $\mathcal{D}_1(v, Rb, k)$ a BIB design exists, then it is BU-optimal in the case of the mixed model (1).

The desirable properties of an α-resolvable NBD in $\mathcal{D}_1(v, Rb, k)$ and a BIB design are very useful in non-binary cases. Namely, using the treatment merging method, we can obtain new designs which preserve the unit structure and are also $\mathbf{r}^{-\delta}$-balanced with the same efficiency factor as the starting design (cf. Nigam et al., 1988). This observation leads to the following:

Proposition 4.3. If in the class of α-resolvable NBDs in $\mathcal{D}_1(v, Rb, k)$ an $\mathbf{r}^{-\delta}$-balanced design exists, then by the treatment merging method it is possible to obtain BU-optimal designs (under model (1)) in the class of NBDs in $\mathcal{D}_1(v^* < v, Rb, k)$.

Let us consider a subclass of C-designs (w.r.t. sub-blocks) in $\mathcal{D}_0(v, Rb, k)$. In the light of Theorem 3 it seems that α-resolvable C-designs are worth investigating. At the moment we can only refer to the methods of constructing α-resolvable C-designs (see Ceranka et al., 1986). Hence we do not check which of them are MS-optimal and finally generally optimal, following Corollary 1. Some of the α-resolvable C-designs, given by Ceranka et al. (1986), are so-called Regular Graph Designs (John and Mitchell, 1977) which are generally MS-optimal (cf. Shah and Sinha, 1989). These designs are generally optimal under model (1).

Acknowledgements

The paper was prepared while the first author visited the Hiroshima University under an exchange programme between PAN and JSPS in 1994. Thanks are due to these organizations for providing such opportunities for research.

References

[1] Bagchi, S. (1987a). On the E-optimality of certain asymetrical designs under mixed effects model. *Metrika* 37, 95-105.

[2] Bagchi, S. (1987b). On the optimalities of the $MBGDD$'s under mixed effects model. Report, Computer Science Unit, Indian Statistical Institute, Calcutta, India.

[3] Banerjee, S. and Kageyama, S. (1990). Existence of α-resolvable nested incomplete block designs. *Utilitas Math.* 38, 237-243.

[4] Banerjee, S. and Kageyama, S. (1993). Methods of constructing nested partially balanced incomplete block designs. *Utilitas Math.* 43, 3-6.

[5] Bogacka, B. and Mejza, S. (1993). On BU-optimality of block designs under mixed model. *Bulletin of the International Statistical Institute, Contributed Papers*, Book 1, 139-140.

[6] Bogacka, B. and Mejza, S. (1994). Optimality of generally balanced experimental block designs. *Proc. of the International Conference on Linear Statistical Inference, LINSTAT'93*, T.Caliński and R.Kala, eds, Kluwer Academic Publishers, Dordrecht, 185-194.

[7] Bondar, J. V. (1983). Universal optimality of experimental designs: definitions and a criterion. *Canad. J. Statist.* 11, 325-331.

[8] Bose, R. C. (1942). A note on the resolvability of balanced incomplete block designs. *Sankhyā* 6, 105-110.

[9] Caliński, T. (1977). On the notion of balance in block designs. In: J. R. Barra, F. Brodeau, G. Romier and B. van Cutsem, Eds., *Recent Developments in Statistics*. North-Holland, Amsterdam, 365-374.

[10] Ceranka, B., Kageyama, S. and Mejza, S. (1986). A new class of C-designs. *Sankhyā* B 48, 199-206.

[11] Cheng, C. S. (1978). Optimality of certain asymetrical experimental designs. *Ann. Statist.* 6, 1239-1261.

[12] Dey, A., Das, U. S. and Banerjee, A. K. (1986). Constructions of nested balanced incomplete block designs. *Calcutta Statist. Assoc. Bull.* 35, 161-167.

[13] Hormel, R. J. and Robinson, J. (1975). Nested partially balanced incomplete block designs. *Sankhyā* B 37, 201-210.

[14] Houtman, A. M. and Speed, T. P. (1983). Balance in designed experiments with orthogonal block structure. *Ann. Statist.* 11, 1069-1085.

[15] Jacroux, M. (1989). On the E-optimality of block designs under the assumption of random block effects. *Sankhyā B* 51, 1-12.

[16] Jimbo, M. and Kuriki, S. (1983). Constructions of nested designs. *Ars Combin.* 16, 275-285.

[17] John, J. A. and Mitchell, T. J. (1977). Optimal incomplete block designs. *J. Roy. Statist. Soc.* B 39, 39-43.

[18] Kageyama, S. (1973). On μ-resolvable and affine μ-resolvable balanced incomplete block designs. *Ann. Statist.* 1, 195-203.

[19] Kageyama, S. (1980). On properties of efficiency-balanced designs. *Commun. Statist.* A 9, 597-616.

[20] Kageyama, S. (1984). Some properties on resolvability of variance-balanced designs. *Geom. Dedicata* 15, 289-292.

[21] Khatri, C. G. and Shah, K. R. (1984). Optimality of block designs. In: *Proc. Indian Statistical Institute Golden Jubilee International Conference on Statistics: Applications and New Directions, Calcutta*, 326-332.

[22] Kiefer, J. (1975). Construction and optimality of generalized Youden designs. In: J. N. Srivastava, Ed., *A Survey of Statistical Design and Linear Models*. North-Holland, Amsterdam, 333-353.

[23] Mejza, S. (1992). On some aspects of general balance in designed experiments. *Statistica* 2, 263-278.

164

[24] Mukerjee, R. and Kageyama, S. (1985). On resolvable and affine resolvable variance-balanced designs. *Biometrika* 72, 165-172.

[25] Mukhopadhyay, S. (1981). On the optimality of block designs under mixed effects model. *Calcutta Statist. Assoc. Bull.* 30, 171-185.

[26] Nelder, J. A. (1965). The analysis of randomized experiments with orthogonal block structure. Block structure and the null analysis of variance. *Proc. Roy. Soc. London A* 283, 147-178.

[27] Nigam, A. K., Puri, P. D. and Gupta, V. K. (1988). *Characterizations and Analysis of Block Designs*. Wiley Eastern, New Delhi.

[28] Patterson, H. D. and Williams, E. R. (1976). A new class of resolvable incomplete block designs. *Biometrika* 63, 83-92.

[29] Preece, D. A. (1967). Nested balanced incomplete block designs. *Biometrika* 54, 479-486.

[30] Saha, G. M. (1976). On Calinski's patterns in block designs. *Sankhyā B* 38, 383-392.

[31] Shah, K. R. and Sinha, B. K. (1989). *Theory of Optimal Designs*. Springer-Verlag, Berlin.

[32] Shrikhande, S. S. and Raghavarao, D. (1964). Affine α-resolvable incomplete block designs. *Contributions to Statistics* (edited by C. R. Rao), pp.471-480. Pergamon Press, Statistical Publishing Society, Calcutta.

Construction of A-Optimum Cross-Over Designs

ALEXANDER N. DONEV and BYRON JONES

Department of Mathematics, The University, Southampton, SO17 5NH, UK
and
Department of Medical Statistics, School of Computing Sciences, De Monfort University,
The Gateway, Leicester, LE1 9BH, UK.

Abstract:

We describe an algorithmic approach to the construction of A-optimum repeated measurements designs. The algorithm is very flexible and can search for designs in non-standard situations. Some illustrative examples are given.

1 Introduction

In many medical experiments and clinical trials sequences of different treatments are given to a number of subjects. Each treatment is administered for a given period of time and a response is measured at the end of each period. Of most interest to the experimenter is the effect each treatment has during the period it is applied. However, the effect of a treatment may persist into the immediately following period and so the possibility of such a carry-over effect must be allowed for in the design. Sometimes an attempt is made to minimize the possibility of carry-over effects by including a wash-out period between each pair of active treatment periods. In the wash-out period no active treatment is given. The ideal length of the wash-out period is often difficult to determine and so it is advantageous to allow for the possible presence of carry-over effects.

The important advantage of using repeated measurement designs is that the treatments are compared within-subjects, rather than between subjects as would be the case if each subject received only a single treatment. When modelling the responses observed in such experiments we should therefore include terms in our model that account for differences between each of the following: subjects, periods, treatments and carry-over effects. The standard model used in the literature includes such terms in an additive way, and we will follow that model here. See Jones and Kenward (1989) for a thorough review. Jones and Donev (1996) consider models that are reasonable under other pharmacological assumptions. These authors also search for designs for non-standard combinations of numbers of

---

*Kitsos, C.P., and Müller, W.G., Eds., *Proceedings of MODA4*, Physica Verlag, Heidelberg, 1995

166

subjects and periods.

In this paper we will consider some non-standard situations and will use for the construction of designs the criterion of *A*-optimality. Under this criterion an optimum design minimizes the sum of the variances of all the estimated pairwise differences between the treatments, i.e. *A*-optimum designs provide highly efficient comparisons of the treatments. From a number of points of view this criterion is a natural one to consider and is easily understood by medical researchers. We will compare the properties of such designs with the corresponding designs obtained using the *D*-optimality criterion where the confidence ellipsoid of the model parameters is minimized. One of our aims is to show that it is no longer necessary for experimenters to try to make their experimental plans fit the designs available in the literature. In our approach the design is chosen to fit the experimental conditions, not vice versa.

2 Modelling Assumptions and Criterion of Optimality

We will first give the general form of the model and then consider a special case. Let $Y_u, u = 1, 2, \ldots, N$, denote the observations to be obtained in the experiment. If β_i, $i = 1, 2, \ldots, q$, denote the regression parameters to be included in a particular model, then the general model for Y_u is

$$Y_u = \sum_{i=1}^{q} \beta_i f_i + \epsilon_u , \qquad (1)$$

where ϵ_u is an independent random error with mean zero and variance σ^2. The functions f_i can be simple or complicated depending on the model required for a particular experiment. Here we will consider a simple but widely used model for repeated measurements experiments. In this model the response observed in period j on subject i is denoted by Y_{ij}, $i = 1, 2, \ldots, s$ and $j = 1, 2, \ldots, p$ (rather than by Y_u as in the general model). Then, if $d[i, j]$ is the treatment applied to subject i in period j, the model is

$$Y_{ij} = \mu + \tau_{d[i,j]} + \lambda_{d[i,j-1]} + \pi_j + s_i + \epsilon_{ij} , \qquad (2)$$

where μ is the general mean, π_j is the effect of period j, $\tau_{d[i,j]}$ is the effect of treatment $d[i, j]$, $\lambda_{d[i,j-1]}$ is the effect of the carry-over of treatment $d[i, j - 1]$, and $\lambda_{d[i,0]} = 0$, s_i is the effect of subject i and ϵ_{ij} is an independent random error with mean zero and variance σ^2. We assume that there are t treatments in the experiment. As model (2) is over-parameterised we have applied the following constraints:

$$\sum_{i=1}^{s} s_i = \sum_{i=1}^{t} \tau_i = \sum_{i=1}^{t} \lambda_i = \sum_{j=1}^{p} \pi_j = 0 .$$

We have set the last parameter in each set equal to minus the sum of the remaining parameters in the set and have exlcuded it from the model. The functions f_i in the general model (1) take the values -1, 0 and 1, and for observations in the first period we have set the carry-over parameter equal to zero.

The general model in matrix notation is

$$Y = F\beta + \epsilon .$$

The variance matrix of $\hat{\beta}$, the least squares estimator of the parameter vector β, is then

$$\text{var}(\hat{\beta}) = (F^T F)^{-1} \sigma^2 .$$

Let the information matrx $F^T F$ be partitioned as

$$\begin{pmatrix} F_{11} & F_{12} \\ F_{21} & F_{22} \end{pmatrix}$$

where F_{11} is the $(t-1) \times (t-1)$ matrix corresponding to the treatment effects. Taking into account the imposed parameterisation the ijth element of the matrix

$$F^{11} = (F_{11} - F_{12} F_{22}^{-1} F_{21})^{-1}$$

is

$$g_{ij} = \text{cov}(\hat{\tau}_i, \hat{\tau}_j)$$

for $i, j = 1, 2, ..., t\text{-}1$.

The criterion we will use to search for designs is A-optimality. That is, we require a design that minimises the sum of the variances of all estimated pairwise differences between the treatments. For example, if $t = 3$ we wish to minimise

$$A = \text{var}(\hat{\tau}_1 - \hat{\tau}_2) + \text{var}(\hat{\tau}_1 - \hat{\tau}_3) + \text{var}(\hat{\tau}_2 - \hat{\tau}_3),$$

where $\hat{\tau}_i$ is the least squares estimator of τ_i in model (2). Clearly

$$A = 2(\text{var}(\hat{\tau}_1) + \text{var}(\hat{\tau}_2) + \text{var}(\hat{\tau}_3) - \text{cov}(\hat{\tau}_1, \hat{\tau}_2) - \text{cov}(\hat{\tau}_1, \hat{\tau}_3) - \text{cov}(\hat{\tau}_2, \hat{\tau}_3)). \qquad (3)$$

The parameterisation implies that

$$\hat{\tau}_3 = -\hat{\tau}_1 - \hat{\tau}_2.$$

Therefore

$$\text{var}(\hat{\tau}_3) = \text{var}(\hat{\tau}_1) + \text{var}(\hat{\tau}_2) + 2\text{cov}(\hat{\tau}_1, \hat{\tau}_2)$$

$$\text{cov}(\hat{\tau}_1, \hat{\tau}_3) = -\text{var}(\hat{\tau}_1) - \text{cov}(\hat{\tau}_1, \hat{\tau}_2)$$

$$\text{cov}(\hat{\tau}_2, \hat{\tau}_3) = -\text{var}(\hat{\tau}_2) - \text{cov}(\hat{\tau}_1, \hat{\tau}_2)$$

When these results are substituted in (3) we obtain

$$A = 6(\text{var}(\hat{\tau}_1) + \text{var}(\hat{\tau}_2) + \text{cov}(\hat{\tau}_1, \hat{\tau}_2)).$$

In general

$$A = 2t(\sum_{i=1}^{t-1} \text{var}(\hat{\tau}_i) + \sum_{i<j}^{t-1} \text{cov}(\hat{\tau}_i, \hat{\tau}_j)).$$

It easy to see that the criterion is proportional to the sum of the elements in the upper triangle of the $(t-1) \times (t-1)$ matrix F^{11}. Therefore for every design we calculate

$$A = 2t \sum_{i=1}^{t-1} \sum_{j=i}^{t-1} g_{ij}.$$

Another criterion we might have considered using is the D-optimality criterion. Under this criterion an optimum design maximizes the determinant of $F^T F$. We can interpret this determinant as being proportional to the volume of a confidence ellipsoid for the estimated parameters. The determinant of the information matrix of any of the designs considered in this paper is proportional to the determinant of F^{11}. Therefore we have used the value of

$$D = \det(F^{11})$$

as our criterion of D-optimality when comparing designs.

3 An Algorithm

In general the number of possible experimental designs for a given situation is likely to be very large. This makes it impractical to search exhaustively for an A-optimum design. However, it is practical to use an algorithm that searches for an optimum design by iteratively improving a given initial design. The algorithm we have developed is a modification of the BLKL-algorithm of Atkinson and Donev (1989). This earlier algorithm constructed D-optimum experimental designs using an iterative search over a list of all candidate observations or points. At each iteration a point from the design is replaced by the point from the list of candidate points that gives the maximum improvement in the criterion of optimality. This continues until no further improvement is possible. Iterative search algorithms have been suggested many times in the literature, beginning with Fedorov (1972). However, it is not possible directly to use these previously suggested algorithms for the construction of repeated measurement designs. This is because the exchange of a point in these designs affects not only the point being removed but the immediately following point as well. This is a consequence of including carry-over effects in the model. To overcome this extra complication, we do not exchange single points, but complete treatment sequences. An additional minor complication is that when a design permits subjects to receive treatment sequences of different lengths, then we must ensure that only candidates with sequences of appropriate length are considered for exchange.

In the next section we give examples of applications of our algorithm. We have considered particularly simple, but important cases. Donev and Jones (1995) introduce the algorithmic approach to construction of cross-over trials and describe the algorithm in detail while Jones and Donev (1996) discuss a number of alternative models for three treatments and the flexibility which this algorithm supplies when solving practical design problems. In particular, the potential to use the number of periods and subjects that are most appropriate to the experiment should eliminate the practice of using text-book designs in inappropriate circumstances.

4 Examples and Discussion

To test the algorithm we used it first to search for designs with the following standard sets of parameters $(t, p, s) : (3, 3, 6), (3, 4, 6), (4, 4, 4)$ and $(4, 5, 4)$. In each case the design found was the optimum design listed in Jones and Kenward (1989, Table 5.11). The following examples illustrate the performance of the algorithm in some non-standard situations.

In our first example we used $t = 3$, $p = 4$, $s = 3$. Using the A-optimality criterion we found design D1 given in Table 1 to be optimum. The pairwise variances for this design are

$$\text{var}(\hat{\tau}_1 - \hat{\tau}_2) = 0.500$$
$$\text{var}(\hat{\tau}_1 - \hat{\tau}_3) = 0.696$$
$$\text{var}(\hat{\tau}_2 - \hat{\tau}_3) = 0.696$$

and the value of the optimality criterion is $A = 1.8930$.

Design D1	Design D2
$A \quad C \quad C \quad B$ $B \quad A \quad B \quad C$ $B \quad B \quad A \quad A$	$A \quad C \quad B \quad B$ $B \quad A \quad C \quad C$ $C \quad B \quad A \quad A$
$A = 1.8930$ $D = 0.7340 \times 10^6$	$A = 2.1816$ $D = 0.7939 \times 10^6$

Table 1. A-optimum design D1 and D-optimum design D2 for 3 treatments, 4 periods and 3 subjects.

It will be noted that design D1 is not variance-balanced as these three variances are not equal to each other. A balanced design does exist and was found by the algorithm using the D-optimality criterion. This is design D2 given in Table 1. It has pairwise variances equal to

$$\mathrm{var}(\hat{\tau}_1 - \hat{\tau}_2) = 0.727$$
$$\mathrm{var}(\hat{\tau}_1 - \hat{\tau}_3) = 0.727$$
$$\mathrm{var}(\hat{\tau}_2 - \hat{\tau}_3) = 0.727$$

In practice, the A-optimum design would be preferred because it provides much more precise estimates of the estimated pairwise comparisons of the treatments. These two designs provide an interesting example of how different designs for the same situation can be obtained by using different optimality criteria. A choice of design is therefore offered: one has higher precision and the other has a simpler variance structure.

If six periods are used, then the design that is A-optimum is also D-optimum. In fact, the algorithm found a number of designs that were equally good in terms of the optimality criterion. One of them is given in Table 2.

The optimality of this design is not surprising when one realises that it consists of two Latin squares that make up a Williams (1949) design put next to each other. In the Williams design each treatment follows every other treatment (excluding itself) an equal number of times. By joining the two squares together we also allow a treatment to follow itself.

An important feature of our approach is that it makes it easy to search for designs that have unequal numbers of periods. This may be required in practice if some subjects are unable to stay in the trial for as long a time as others and we need to use all the available subjects.

Design D3
$A \quad C \quad B \quad B \quad C \quad A$ $B \quad A \quad C \quad C \quad A \quad B$ $C \quad B \quad A \quad A \quad B \quad C$
$A = 1.0356$ $D = 0.1852 \times 10^9$

Table 2. A- and D-optimum design for 3 treatments, 6 periods and 3 subjects.

170

Suppose that five subjects are available to compare four treatments and three subjects can stay for six periods and two subjects can stay for three periods. The A-optimum design found by the algorithm is D4 given in Table 3. The D-optimum design is D5 (also given in Table 3) and has a slightly larger value of A.

Design D4						Design D5					
A	D	D	B	B	C	A	C	B	B	D	D
B	D	C	C	A	A	B	A	D	C	C	A
C	B	A	D	D	B	C	D	A	A	B	B
D	A	B				D	B	C			
A	C	D				A	C	D			
$A = 2.3128$						$A = 2.3144$					
$D = 0.1161 \times 10^{11}$						$D = 0.1265 \times 10^{11}$					

Table 3. A-optimum design D3 and D-optimum design D4 for 4 treatments, 6 periods and 5 subjects.

In a repeated measurement design it may be necessary to increase the number of periods after the experiment has started. This may be because data from some of the earlier periods may have been analyzed and the achieved precision of the experiment is found to be lower than expected. In another situation an experiment may have been completed and further observations on the same subjects need to be taken. Whatever the reason, we assume here that a given design is to be augmented by adding additional periods. Suppose four treatments have been compared using four subjects and four periods in a Williams (1949) design and this design is to be augmented by adding three more periods. The initial design and the augmentation found by the algorithm are given in Table 4. In this design, the variances of the estimated pairwise differences are all nearly equal to each other. Incidentally, this is also the design found using the D-optimality criterion.

Design D6						
Initial Design				Augmentation		
A	C	D	B	B	C	A
B	D	C	A	A	D	C
C	B	A	D	D	A	B
D	A	B	C	C	B	D
$A = 1.8112$						
$D = 0.7310 \times 10^{12}$						

Table 4. A- and D-optimum design for 4 treatments, 7 periods and 4 subjects obtained after augmentation of an initial design for 4 periods.

References

Atkinson A.C. and Donev (1989). The construction of exact D-optimum experimental designs with application to blocking response surface designs. *Biometrika*, **76**, 515-26.

Donev, A.N. and Jones, B. (1995). An algorithmic approach to constructing optimum cross-over designs. Submitted for publication.

Fedorov, V.V. (1972). *Theory of Optimal Experiments*, translated and edited by W.J. Studden and E.M. Klimko. New York: Academic Press.

Jones, B. and Donev (1996). Modelling and design of cross-over trials. Submitted for publication.

Jones, B. and Kenward, M.G. (1989). *The Design and Analysis of Cross-Over Trials*. London: Chapman and Hall.

Williams, E.J. (1949). Experimental designs balanced for the estimation of residual effects of treatments. *Australian Journal of Scientific Research*, **2**, 149-168.

An Algorithm for Sampling Optimization for Semivariogram Estimation

WERNER G. MÜLLER and DALE L. ZIMMERMAN

Department of Statistics, University of Economics, Augasse 2–6, A-1090 Vienna, Austria
and
Department of Statistics and Actuarial Science, University of Iowa, Iowa City, IA 52242, USA.

Abstract:

This paper describes an algorithm for the optimal selection of sampling locations for semivariogram estimation. We assume that the semivariogram is estimated by fitting a parametric function of separation distance between observation sites to a selected subset of the squared differences of original observations (thereby restricting ourselves to isotropic fields). We apply standard regression design theory to construct an optimal configuration of distances in the lag space, which is then mapped into the site space in such a way that dependence among the observations is minimized.

1 Introduction

In any spatial sampling problem, the choice of sampling locations can be an important consideration. The optimal choice depends, of course, on the goals of the investigation. If the goal is predicting (kriging) the value of the variable of study over a region then, assuming that the spatial dependence among observations is known, it is reasonable to choose the sampling locations so as to minimize some functional (e.g. the average) of the variance of prediction error (kriging variance) over the region. Several authors have considered design problems of this or similar type; see (Bras and Rodriguez-Iturbe (1976)), (McBratney et al. (1981)), (Barnes (1989)), and (Cressie et al. (1990)). On the other hand, if the primary goal is to estimate the second-order spatial dependence structure of the data then other criteria are called for. (Russo (1984)), (Warrick and Myers (1987)), and (Zimmerman and Homer (1991)) have proposed various design criteria for estimating spatial dependence as it is characterized by the semivariogram. However, these criteria can be criticized either for emphasizing aspects of the semivariogram estimation paradigm

*Kitsos, C.P., and Müller, W.G., Eds., *Proceedings of MODA4*, Physica Verlag, Heidelberg, 1995

that are not as crucial as others or for failing to account for the correlation between semivariogram estimates at different lags.

In this paper we present a design criterion for semivariogram estimation that does not suffer from either of these deficiencies. We also describe an algorithm for obtaining the optimal design with respect to this criterion. Our criterion corresponds to a semivariogram estimation paradigm that is slightly different than the classical paradigm in that it does not utilize all of the information contained in the data. This modified paradigm is required if we are to apply standard regression optimal design theory to our problem.

The paper is organized as follows. In Section 2 we describe the classical semivariogram estimation paradigm and propose our modification to it. In Section 3 we give our design criterion and discuss some other design considerations. An algorithm for obtaining the optimal design is described in Section 4. An example is given in Section 5 and we summarize our conclusions in the final section.

2 The Semivariogram and Its Estimation

Consider a two-dimensional random field $\{Z(\mathbf{s}): \mathbf{s} \in \mathcal{S} \subset \mathcal{R}^2\}$; we shall refer to $\mathcal{S}$ as the site space. The random field $Z(\cdot)$ is said to be intrinsically stationary if $E[Z(\mathbf{s})] \equiv \mu$ (a constant) and

$$\frac{1}{2} E\{[Z(\mathbf{s}+\mathbf{h}) - Z(\mathbf{s})]^2\} = \gamma(\mathbf{h}), \text{ for all } \mathbf{s}, \mathbf{s}+\mathbf{h} \in \mathcal{S},$$

that is, if expected squared differences depend only on relative (rather than absolute) locations of sites. The function $\gamma(\mathbf{h})$, called the semivariogram, is a popular mode of characterization of the second-order spatial dependence of random fields and plays an important role in kriging; see (Cressie (1991)). Its estimation is a major component of a geostatistical analysis. An additional property of $Z(\cdot)$ that simplifies modelling and inference is that of isotropy, which specifies that the semivariogram is a function of only the distance between sites. That is, an intrinsically stationary random process with semivariogram $\gamma(\cdot)$ is said to be isotropic if $\gamma(\mathbf{h}) = \gamma(r)$, where $r = \|\mathbf{h}\| = (\mathbf{h}'\mathbf{h})^{1/2}$. The set $\{r : r = \|\mathbf{s} - \mathbf{s}'\|, \mathbf{s}, \mathbf{s}' \in \mathcal{S}\}$ is called the lag space. We initially restrict attention to isotropic processes and to semivariograms with finite "range" of dependence, i.e., those for which a number $a \geq 0$ exists such that

$$\gamma(r) = \sigma^2 \equiv \text{var}\,[Z(\mathbf{s})] \qquad \text{for } r > a. \tag{1}$$

Perhaps the most widely used parametric semivariogram that satisfies (1) is the so-called spherical semivariogram

$$\gamma_S(r, \theta) = \begin{cases} 0, & r = 0 \\ \theta_1 + \theta_2 \cdot \left\{\frac{3}{2}\left(\frac{r}{\theta_3}\right) - \frac{1}{2}\left(\frac{r}{\theta_3}\right)^3\right\}, & 0 < r \leq \theta_3 \\ \theta_1 + \theta_2, & r > \theta_3. \end{cases} \tag{2}$$

This semivariogram increases monotonically from a "nugget effect" of θ_1 near the origin to a "sill value" of $\theta_1 + \theta_2$, which is attained at the "range" $a = \theta_3$. These features correspond to spatial correlation that monotonically decreases as the distance between sites increases and vanishes beyond a distance of θ_3 units.

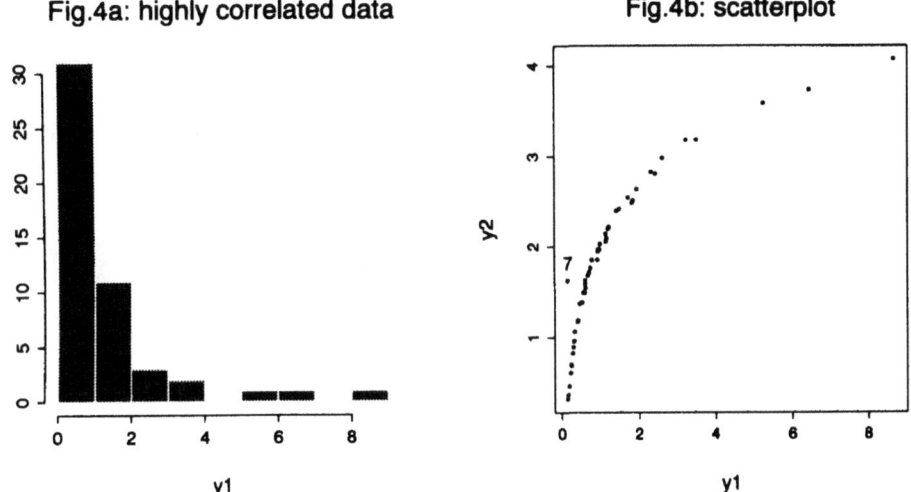

Fig4: Example 3: highly correlated data. (a) histogram of y_1; (b) scatterplot of y_1 and y_2. Observation 7 is jointly outlying but not marginally outlying.

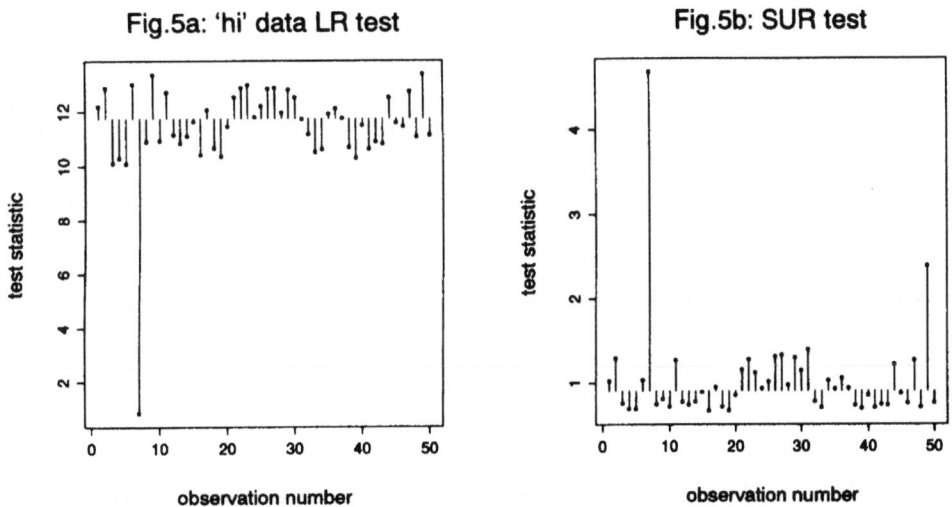

Fig5: Example 3: highly correlated data. Index plot of deletion tests for $\lambda_0 = (0,1)$, with observation 7 an outlier: (a) $T_{LR(i)}$; (b) $T_{SU(i)}$:

Example 2 Continued: Lightly Correlated Data

Fig.6a shows an index plot of the statistic $T_{SU(i)}$, for testing the parameter values $(0, 1)$, when there is deletion in the sums in (11), but no deletion in the estimate $\hat{\Sigma}_W$. The plot is virtually indistinguishable from the full deletion plot in Fig.3a, with the true parameter

176

between two points in the site space, our problem is to find the best set of distances between site pairs. After that set $\xi^*(r)$ (the optimal design in the lag space) has been found, we need to map it into the site space to yield the optimal sampling design.

As (3) is generally a nonlinear model it is natural to base our evaluation of precision on the asymptotic information matrix $\mathcal{M}(\xi,\theta) = \int_0^\infty \frac{\partial\gamma(r,\theta)}{\partial\theta}(\frac{\partial\gamma(r,\theta)}{\partial\theta})^T \gamma^{-2}(r,\theta)\xi(r)dr$, which for the spherical semivariogram (2) is given by

$$\int_0^{\theta_3} \begin{pmatrix} 1 & \frac{r}{2\theta_3}(3 - \frac{r^2}{\theta_3^2}) & \frac{3\theta_2 r}{2\theta_3^2}(\frac{r^2}{\theta_3^2} - 1) \\ \frac{r}{2\theta_3}(3 - \frac{r^2}{\theta_3^2}) & \frac{r^2}{4\theta_3^2}(3 - \frac{r^2}{\theta_3^2})^2 & \frac{3\theta_2 r^2}{4\theta_3^3}(3 - \frac{r^2}{\theta_3^2})(\frac{r^2}{\theta_3^2} - 1) \\ \frac{3\theta_2 r}{2\theta_3^2}(\frac{r^2}{\theta_3^2} - 1) & \frac{3\theta_2 r^2}{4\theta_3^3}(3 - \frac{r^2}{\theta_3^2})(\frac{r^2}{\theta_3^2} - 1) & \frac{9\theta_2^2 r^2}{4\theta_3^4}(\frac{r^2}{\theta_3^2} - 1)^2 \end{pmatrix} \gamma_S^{-2}(r,\theta)\xi(r)dr$$

$$+ \frac{1}{(\theta_1 + \theta_2)^2} \int_{\theta_3}^\infty \begin{pmatrix} 1 & 1 & 0 \\ 1 & 1 & 0 \\ 0 & 0 & 0 \end{pmatrix} \xi(r)dr. \tag{5}$$

It is unfortunate that in this model (as in all nonlinear models) $\mathcal{M}(\cdot)$ depends upon the true values of the parameters, with the consequence that all scalar design criteria based on $\mathcal{M}$ are affected. Therefore, if we are looking for the set ξ^* that maximizes (4) in some sense the best we can do is to find a solution that holds locally for a specific prior guess $\hat\theta$. However, the determination of solutions for various guesses will allow us to assess whether the design is robust with respect to the values of the parameters. As a scalar design criterion we will use the determinant of $\mathcal{M}(\cdot)$, the popular D-criterion.

Furthermore, it is noteworthy that optimal design theory requires that the $\hat\gamma_{ij}$'s be uncorrelated. Indeed, this is why we have supposed that the semivariogram will be estimated using only a portion of the lags. Hence, when we remap our optimal regressor values into the site space we should attempt to satisfy this condition to the greatest extent possible. The only way this can be accomplished is to separate pairs of sampling sites sufficiently far apart (ideally farther than the "range" θ_3).

4 An Algorithm

The proposed algorithm can be best described as a method of distributing a collection of pins of various lengths on a given surface, say a table, in such a way that the distances between them are as large as possible. In other words, for a given number N of possible original observations, we can define $\frac{N}{2}$ line segments r_i with lengths and frequency approximately corresponding to the optimal design $\xi^*(r)$ given in the previous section.

(0) Initialize the procedure by randomly distributing the segments r_i, $i = 1,\ldots,\frac{N}{2}$ within the region S, thereby assigning endpoints $s_i = (s_{i1}, s_{i2})$ and $s_i' = (s_{i1}', s_{i2}')$ to each segment.

(1) Construct the matrix $\mathcal{D} = \mathcal{D}_{kl}, k = 1,\ldots,N, l = 1,\ldots,N-2$ with the distances between all endpoints, i.e. with the entries $((s_{i1} - s_{j1})^2 + (s_{i2} - s_{j2})^2)^{\frac{1}{2}}$, $((s_{i1} - s_{j1}')^2 + (s_{i2} - s_{j2}')^2)^{\frac{1}{2}}$, $((s_{i1}' - s_{j1}')^2 + (s_{i2}' - s_{j2}')^2)^{\frac{1}{2}}$ and $((s_{i1}' - s_{j1})^2 + (s_{i2}' - s_{j2})^2)^{\frac{1}{2}}$ for $i \neq j$. Find the smallest entry $\Delta = \min \mathcal{D}_{kl}$ and identify the two points s_k^* and s_l^* corresponding to it.

(2) Select randomly one of the two points s^* and use its counterpart s'^* as an anchor. Randomly select a point $\dot s$ within S that lies approximately the same distance from s'^* as s^* does (lying on a circle centered at s'^*).

(3) Check if the distance to the closest of the other points is larger than Δ. If yes, exchange s^* by $\dot{s}$; if no, repeat (2) until all points on the circle have been checked.

(4) Repeat (1) until either $\Delta > \hat{\theta}_3$ or Δ cannot be further increased.

Note that the algorithm provides a measure of how well the independence criterion is met, namely $\hat{\kappa} = \frac{\gamma(\Delta)}{\gamma(\hat{\theta}_3)}$.

5 The Example

Before presenting our example, we note that there are several factors that might affect the applicability of the given approach. The region might be too small to accommodate the point pairs and their respective distances, i.e. the "independence" measure $\hat{\kappa}$ depends upon θ_3, the size (and shape) of the region and the number of observations N. Therefore, and to make our example comparable with the design presented in (Warrick and Myers (1987)), we take the site space to be the unit square, we assume that the locations of $N = 30$ sites are to be determined, and we take the parameter values to be $\hat{\theta}_3 = 0.09$; $\hat{\theta}_2 = 0.25, 1, 4$; and (since only the ratio of "sill" to "nugget" effect is of interest) $\hat{\theta}_1 = 1$.

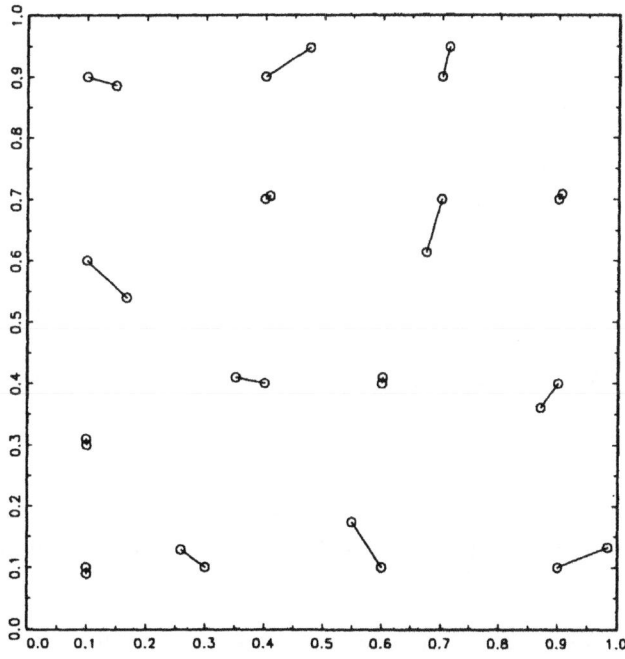

Figure 1: One possible optimal design on a unit square.

A standard algorithm for finding optimal designs for nonlinear regression models, as for instance described in (Silvey (1980)), was applied to obtain $\xi^*(r)$. It yielded a design with 5 observation pairs as close as possible to each other, 5 pairs at a distance of $r = 0.0521$ and 5 at any distance larger than the range $r = 0.09$ (to improve the performance of the algorithm one might want to choose a distance just beyond that). If we now start the algorithm from Section 4 with a slightly modified step (0): i.e. by choosing the initial points on a regular grid, then it yields a design as given in Figure 1 immediately.

178

6 Conclusions

Since classical optimal design theory assumes independent observations, the locations of site pairs should guarantee independence. This can only be achieved if the semivariogram has a finite range and the site space is large enough with respect to this range and the sample size N. The algorithm presented here separates point pairs as far as possible, thus giving the closest approximation to (or, ideally meeting the condition of) independence. We have noted that instead of using the information from $N(N-1)/2$ possible observations only $N/2$ are used here. It is not clear, *a priori*, if the deliberate sacrifice of a proportion of observations that large, though they are highly correlated, results in improved estimation as opposed to a straightforward technique such as that of (Warrick and Myers (1987)).

The usefulness of the algorithm can only be assessed in two ways: via simulation, or by comparing it to a technique that directly employs ideas for optimal designs for correlated observations. We are proceeding with this second approach; see (Müller and Zimmerman (1995)).

References

Barnes, R.J. (1989). Sample design for geological site characterization. In Armstrong, M., editor, *Geostatistics*, pages 809–822. Kluwer, Dordrecht.

Bras, R.L. and Rodriguez-Iturbe, I. (1976). Network design for the estimation of areal mean of rainfall events. *Water Resources Research*, 12:1185–1195.

Cressie, N., Gotway, C.A. and Grononda, M.O. (1990). Spatial prediction from networks. *Chemometrics and Intelligent Laboratory Systems*, 7:251–271.

Cressie, N. (1985). Fitting variogram models by weighted least squares. *Mathematical Geology*, 17:563–586.

Cressie, N. (1991). *Statistics for Spatial Data*. John Wiley and Sons, New York.

McBratney, A.B., Webster, R. and Burgess, T.M. (1981). The design of optimal schampling schemes for local estimation and mapping of regionalized variables - 1. *Computers and Geosciences*, 7:331–334.

Müller, W.G. and Zimmerman, D.L. (1995). Optimal design for semivariogram estimation. Technical report, University of Economics, Department of Statistics, Vienna, forthcoming.

Russo, D. (1984). Design of an optimal sampling network for estimating the variogram. *Soil Science Society of America Journal*, 48:708–716.

Silvey, S.D. (1980). *Optimal Design*. Chapman and Hall, London.

Warrick, A.W. and Myers, D.E. (1987). Optimization of sampling locations for variogram calculations. *Water Resources Research*, 23:496–500.

Zimmerman, D.L. and Homer, K. (1991). A network design criterion for estimating selected attributes of the semivariogram. *Environmetrics*, 2:425–441.

Part II
Estimation and Optimization

Multivariate Transformations, Regression Diagnostics and Seemingly Unrelated Regression

ANTHONY C. ATKINSON

Department of Statistical and Mathematical Sciences, The London School of Economics, London WC2A 2AE, UK.

Abstract:

The assumption of multivariate normality provides the customary powerful and convenient way of analysing multivariate data: if data are not normal, the analysis may often be simplified by an appropriate transformation. The paper derives deletion diagnostics for the effect of individual observations on the estimated transformation to normality, using the parametric family of power transformations of Box and Cox. The likelihood ratio test is compared with a seemingly unrelated regression test using constructed variables. The examples include both unstructured data and multivariate multiple regression. They indicate that the likelihood ratio test is more informative in the presence of appreciable correlation in the data than the seemingly unrelated regression test. Numerical results are given for the effect of deletion in the two main stages in the construction of deletion diagnostics for seemingly unrelated regression models.

1 Introduction

Univariate data have long been transformed to approximate normality, with the consequence that standard methods, such as regression, can then be used for their analysis. It has been argued that the development of more recent methods of analysis, such as those based on generalized linear models (McCullagh and Nelder 1989, §11.3.3), has reduced the importance of transformations to normality. However, for multivariate data, there is a paucity of alternatives to models based on multivariate normality so that the transformation of such data to normality is correspondingly more attractive. This paper develops diagnostic methods for the effect of individual observations on the evidence for a transformation of multivariate data and on the estimation of the vector transformation

*Kitsos, C.P., and Müller, W.G., Eds., *Proceedings of MODA4*, Physica Verlag, Heidelberg, 1995

parameter. Methods for the diagnostic regression analysis of univariate data are described in the books of Cook and Weisberg (1982) and of Atkinson (1985). The review paper of Davison and Tsai (1992) describes developments for generalized linear models and their extensions.

The parametric family of power transformations introduced by Box and Cox (1964) was extended to multivariate data by Andrews, Gnanadesikan, and Warner (1971) and by Gnanadesikan (1977). Velilla (1993) compares marginal and joint transformations and gives further references to related work. In these papers concern is with likelihood analysis of the data, that is with a procedure using statistics aggregated over all observations. Methods for detecting the influence of individual observations on transformations of univariate data are described by Atkinson (1985) who uses the deletion of individual observations to determine effects on parameter estimation and test statistics. An approach which is relatively computationally intense is to delete each observation in turn and to recalculate the likelihood. Computational savings, in the univariate case, are made through the use of constructed variables, when the problem becomes one of standard deletion diagnostics in regression. The present paper makes a start on investigating similar techniques for multivariate transformations.

The multivariate transformation is described in §2 and an example given, for which deletion diagnostics are calculated, in §3. For n observations on the m dimensional response this requires $n + 1$ numerical optimzations in m dimensions. Deletion diagnostics using constructed variables are derived in §4. These lead to a multivariate multiple regression formulation in which the parameters are different for each response, the estimates being related solely through the correlation structure of the responses, that is the seemingly unrelated regression model. Examples of transformation diagnostics derived from seemingly unrelated regression are given in §5. In §6 some properties of the seemingly unrelated regression solution are investigated. In the examples up to this point there is no structure in the multivariate means. However the method is more general and §7 presents an example in which the multivariate data arise from a designed experiment. The paper concludes in §8 with a few comments on unresolved points.

2 Multivariate Transformations to Normality

For multivariate data let y_i be the $m \times 1$ vector of responses at observation i with y_{ij} the observation on response j. In the extension of the Box and Cox (1964) family to multivariate responses the normalized transformation of y_{ij} is

$$
\begin{aligned}
z_{ij}(\lambda_j) &= (y_{ij}^{\lambda_j} - 1)/\lambda_j \dot{y}_j^{\lambda_j - 1} & (\lambda \neq 0) \\
&= \dot{y}_j \log y_{ij} & (\lambda = 0),
\end{aligned}
\tag{1}
$$

where $\dot{y}_j$ is the geometric mean of the jth response. The value $\lambda_j = 1$ $(j = 1, \ldots, m)$ corresponds to no transformation of any of the responses. If the transformed observations are normally distributed with mean μ_i for the ith observation and covariance matrix Σ, twice the profile loglikelihood of the observations is given by

$$
\begin{aligned}
2L_{max}(\lambda) &= \text{const} - n \log |\hat{\Sigma}(\lambda)| - \sum_{i=1}^{n} \{z_i(\lambda) - \hat{\mu}(\lambda)\}^T \hat{\Sigma}^{-1}(\lambda)\{z_i(\lambda) - \hat{\mu}_i(\lambda)\} \\
&= \text{const} - n \log |\hat{\Sigma}(\lambda)| - \sum_{i=1}^{n} e_i(\lambda)^T \hat{\Sigma}(\lambda)^{-1} e_i(\lambda).
\end{aligned}
\tag{2}
$$

In (2) $\hat{\mu}_i(\lambda)$ and $\hat{\Sigma}(\lambda)$ are derived from least squares estimates for fixed λ and $e_i(\lambda)$ is the $m \times 1$ vector of residuals.

The calculation of $\hat{\mu}_i(\lambda)$ and $\hat{\Sigma}(\lambda)$ is simplified when, as in the examples of maximum likelihood estimation in this paper, the matrix of explanatory variables X is the same for all responses. As a result, the least squares estimates are found by independent regression on each response, yielding the $m \times p$ matrix of parameter estimates $\hat{\beta}(\lambda) = (X^T X)^{-1} X^T z(\lambda)$. Then, in the usual way,

$$
\begin{aligned}
(n-p)\hat{\Sigma}(\lambda) &= \sum_{i=1}^{n} e_i e_i^T \\
&= \{z(\lambda) - X\hat{\beta}(\lambda)\}^T \{z(\lambda) - X\hat{\beta}(\lambda)\}.
\end{aligned}
\tag{3}
$$

When these estimates are substituted in (2), the profile loglikelihood reduces to

$$
2L_{max}(\lambda) = \text{const}' - n \log |\hat{\Sigma}(\lambda)|.
\tag{4}
$$

So, to test the hypothesis $\lambda = \lambda_0$, the statistic

$$
T_{LR} = n \log\{|\hat{\Sigma}(\lambda_0)|/|\hat{\Sigma}(\hat{\lambda})|\}
\tag{5}
$$

is compared with the χ^2 distribution on m degrees of freedom. In (5) $\hat{\lambda}$ is the vector of m parameter estimates maximising (2), which is found by numerical search.

Example 1: Soil Data

As a first example we take 57 readings on five properties of soil samples given by Mulira (1992). The first two variables are measurements of pH, which are highly correlated. The other three are measures of available phosphorus, potassium and magnesium: the marginal distributions of these three responses show appreciable skewness. There are no explanatory variables and so no regression structure.

The maximum likelihood estimates of the 5 transformation parameters are $\hat{\lambda}_1 = -0.24$, $\hat{\lambda}_2 = -0.03$, $\hat{\lambda}_3 = 0.01$, $\hat{\lambda}_4 = -0.86$ and $\hat{\lambda}_5 = -0.25$. The null hypothesis of no transformation, that is all $\lambda_j = 1$, yields a value of 140.1 for the likelihood ratio statistic (5), so the data should be transformed. For the hypothesis of the log transformation for all five variables the corresponding value is 12.1, compared with 11.07 for the 5% point of χ^2 on 5 degrees of freedom. If λ_4 is taken as -1, with all other y_j logged, the statistic has the value 2.1.

There are two puzzling features of this analysis. One is that the values of pH are already the logarithms of hydrogen ion concentration, so it is surprising that they should be logged again. The second feature is that the reciprocal transformation is indicated for just one of the three concentrations y_3, y_4 and y_5, which might be expected to have rather similar properties.

3   Deletion Statistics for the Likelihood Ratio Test

The analysis of the soil data in the previous section depends solely on aggregate statistics. The effect of deletion of observations on the value of the likelihood statistic T_{LR} (5) for fixed λ_0 can be found either by application of the standard deletion formulae (Cook and Weisberg 1982, §2.2; Atkinson 1985 also §2.2) to the calculation of the elements of

Σ or by repeating the m multiple regressions with observation i deleted for $i = 1, \ldots, n$. The computationally intensive step is calculation of the deletion estimate $\hat{\lambda}_{(i)}$. The seemingly unrelated regression method of the next section provides a computationally simpler alternative.

However the deletion calculations are performed, they lead to n values of the deletion statistic

$$T_{LR(i)} = n \log\{|\hat{\Sigma}_{(i)}(\lambda_0)|/|\hat{\Sigma}_{(i)}(\hat{\lambda}_{(i)})|\} \tag{6}$$

Example 1 Continued: Soil Data

An index plot of the deletion statistic $T_{LR(i)}$ is given in Fig.1a for the logarithmic transformation of all five variables. Fig.1b is the same plot but with $\lambda_4 = -1$. The most important observation is 20, deletion of which increases the value of both statistics, confirming the rejection of zero as a value for λ_4 while the statistic for $\lambda_4 = -1$ only increases to 4.37, not sufficient for rejection of this value. Fig.2 shows similar plots for the deletion maximum likelihood estimates of λ_1 and λ_4. Deletion causes appreciable fluctuations in $\hat{\lambda}_{1(i)}$, a relatively imprecisely estimated parameter, and much smaller fluctuations in the more problematic estimate of λ_4. The plots show that observation 20 is just the most important observation for $\hat{\lambda}_1$, but that it has a negligible effect on $\hat{\lambda}_4$. However, the plots fail to reveal that any one observation is having a strong effect on the conclusions drawn from the data about the correct transformation.

4 Constructed Variables and Seemingly Unrelated Regression

Although the nonlinear optimization required to find $T_{LR(i)}$ is not difficult, it is repeated n times for each diagnostic analysis, which can be for several models and a variety of values of the vector λ_0. An advantage of the procedure of this section is that it brings the diagnostic method into the framework of multivariate regression analysis, for which sufficiently flexible software is available to allow uncomplicated model development.

Constructed variables for the univariate Box-Cox transformation are described in detail in Atkinson (1985), particularly Chapter 6. The model is linearized by Taylor expansion which leads to inclusion in the regression of the constructed variable

$$
\begin{aligned}
w_{ij}(\lambda_0) &= \frac{\partial z_{ij}(\lambda)}{\partial \lambda_j}\Big|_{\lambda=\lambda_0} \\
&= y_{ij}^{\lambda_{0j}} \log y_{ij}/(\lambda_{0j}\dot{y}_j^{\lambda_{0j}-1}) - z_{ij}(\lambda_0)(1/\lambda_{0j} + \log \dot{y}_j).
\end{aligned} \tag{7}
$$

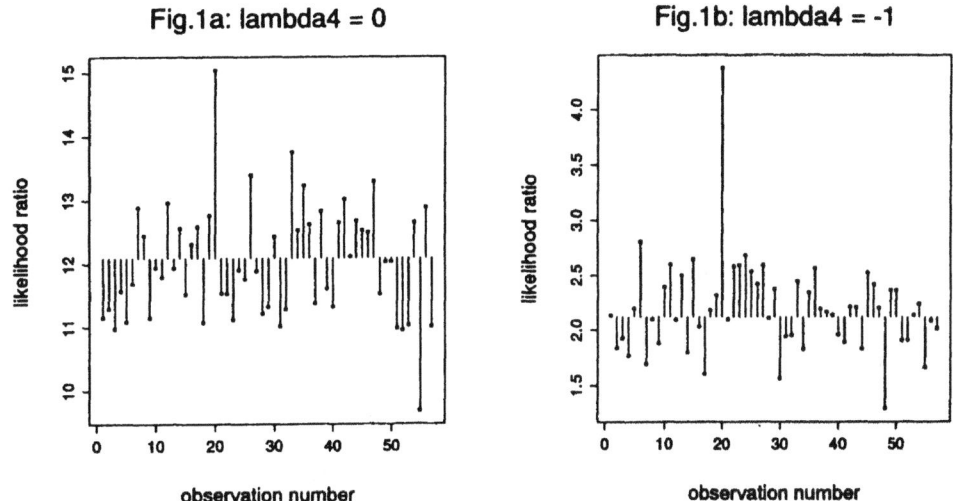

Fig.1: Example 1: soil data. Index plot of deletion likelihood ratio test $T_{LR(i)}$ for the simultaneous power transformation of all 5 responses: (a) null hypothesis the logarithmic transformation $(\lambda_j = 0, j = 1, \dots, 5)$; (b) as (a) but $\lambda_4 = -1$.

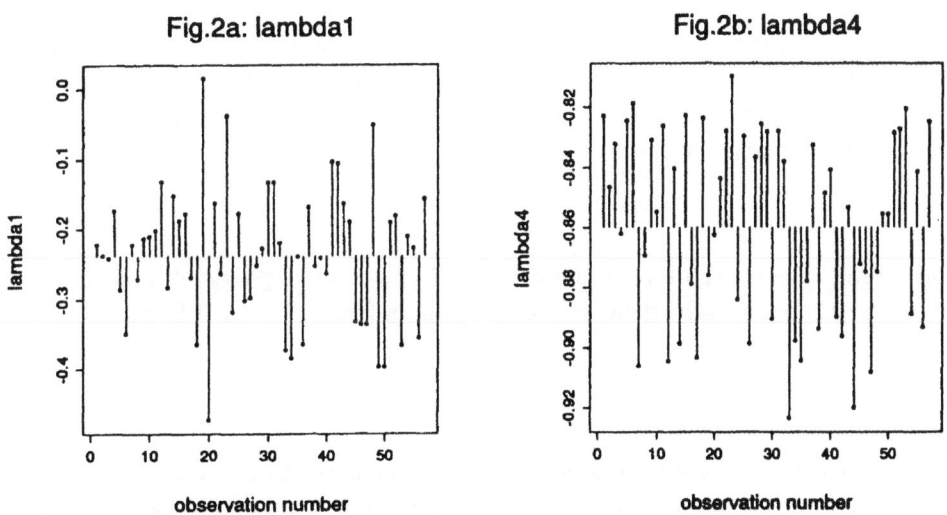

Fig.2: Example 1: soil data. Index plot of deletion maximum likelihood estimates of transformation parameters: (a) $\hat{\lambda}_{1(i)}$; (b) $\hat{\lambda}_{4(i)}$.

Provided the model for $z(\lambda)$ contains a constant, regression on (7) is equivalent, in the special cases of $\lambda = 1$ and 0, to regression on the variables

$$
\begin{array}{rll}
w(1) & = & y\{\log(y/\dot{y}) - 1\} \qquad\qquad (\lambda = 1) \\
w(0) & = & \dot{y}\log y(\log y/2 - \log \dot{y}) \quad (\lambda = 0).
\end{array}
\qquad (8)
$$

In (8) the subscripts i and j have been omitted for typographic clarity.

For multivariate transformation the regression model for the jth response when the constructed variable is included is

$$
z_j(\lambda_0) = X\beta_j(\lambda_0) + w_j(\lambda_0)\gamma_j.
\qquad (9)
$$

Testing that λ_0 is the correct transformation of the response is equivalent to testing that the γ_j in (9) are zero. Because the explanatory variables are no longer the same for all responses the simplification of the regression in §2 no longer holds: the covariance Σ between the m responses has to be allowed for in estimation and independent least squares is replaced by generalized least squares. In the particular form (9) the parameters for each response are different, the estimates being related only through covariances of the y_j. This special structure is known as seemingly unrelated regression (Zellner 1962).

In all there are nm observations on y. If the data are stacked the $nm \times nm$ covariance matrix of y is block diagonal with n blocks of the $m \times m$ matrix Σ. As a result of the block diagonal structure the calculation of the parameters β and γ can be achieved without inversion of an $nm \times nm$ matrix.

Since X is $n \times p$, there are in all $p^* = m(p+1)$ parameters in (9). Let X^* be the $n \times p^*$ matrix of explanatory variables formed by copying each column of X m times - first all the constants, then m columns of x_{1i} and so on up to the m columns of the various w_j. If β^* is the $p^* \times 1$ vector of parameters, calculation of the least squares estimates requires the $p^* \times p^*$ covariance matrix Ψ. Let J be a $(p+1) \times 1$ vector of ones. Then

$$
\Psi^{-1} = J^T J \otimes \Sigma^{-1},
\qquad (10)
$$

a matrix containing $(p+1) \times (p+1)$ copies of Σ^{-1}. The vector of parameter estimates can then be written in the seemingly standard least squares form $\hat{\beta}^* = A^{-1}B$ where

$$
\begin{array}{rll}
A_{jk} & = & \sum_{i=1}^{n} x_{ij}^* \Psi_{jk}^{-1} x_{ik}^* \\
B_j & = & \sum_{i=1}^{n}\sum_{k=1}^{m} x_{ij}^* \Psi_{jk}^{-1} z_{ik}.
\end{array}
\qquad (11)
$$

Although the pattern is clear, the matrices do not combine according to dimensions and the summations are over the n observations rather than the nm of the stacked data. Discussion of seemingly unrelated regression is to be found in many textbooks on econometrics, for example §2.9 of Harvey (1990).

Because (11) contains Ψ, estimation of Ψ, or equivalently Σ, is required for the procedure to be operational. The estimation proceeds in two steps:

[1] Obtain $\hat{\Sigma}_W$, an estimate of Σ, from the independent regressions as in §2, but with X augmented to include w_j for the jth response.

[2] Seemingly unrelated regression using (11) with $\hat{\Psi}^{-1}$ in (10) calculated using $\hat{\Sigma}_W^{-1}$.

Iteration in the estimation of Σ is possible, but it is not usual and is not explored here.

The likelihood ratio test for evidence of a transformation in (9), that is whether all elements of γ are zero, is found from (2). If $e_0(\lambda_0)$ is the vector of residuals from regression on X and $e_1(\lambda_0)$ the corresponding vector from regression on X and $w(\lambda_0)$ with estimated covariance matrix $\hat{\Sigma}_W(\lambda_0)$, the difference in loglikelihoods is

$$\sum_{i=1}^{n} e_{0i}(\lambda_0)^T \hat{\Sigma}_W^{-1}(\lambda_0) e_{0i}(\lambda_0) - \sum_{i=1}^{n} e_{1i}(\lambda_0)^T \hat{\Sigma}_W^{-1}(\lambda_0) e_{1i}(\lambda_0) = R_0(\lambda_0) - R_1(\lambda_0). \qquad (12)$$

The difference in form from (5) arises because in (12) the same estimate of Σ is used both under the null and the alternative hypotheses. Since $\hat{\Sigma}_W$ is not the estimated covariance matrix from the generalised least squares regression, the null distribution of (12) may not be close to χ_m^2. An improvement is obtained by standardizing by the sum of squares $R_1(\lambda_0)$ to yield

$$T_{SU} = \{R_0(\lambda_0) - R_1(\lambda_0)\}(n - p)/R_1(\lambda_0). \qquad (13)$$

The set of n deletion statistics $T_{SU(i)}$ is obtained in two steps analogous to those for the aggregate statistic. In the first step independent regressions on the $n - 1$ observations remaining after deletion of the ith yield the covariance estimate $\hat{\Sigma}_{W(i)}$. This is then used in (10) as the estimated covariance matrix for the estimation of $\beta_{(i)}^*$ in (11) by summation over all observations except i.

5 Examples of Seemingly Unrelated Regression Diagnostics

As a first stage in the investigation of the properties of the proposed procedure, two simulated data sets are used to compare the likelihood statistic and the approximation using seemingly unrelated regression.

Example 2: Lightly Correlated Data

For both examples $n = 50$. Let Z_1 and Z_2 be independent standard normal random variables. The observations were formed as $Y_1 = \exp(Z_1)$ and $Y_2 = 0.5Z_1 + 0.5Z_2 + 2$, so that the transformation to normality is $\lambda_1 = 0$ and $\lambda_2 = 1$. For the particular sample analysed here observation 7 was an outlier with $y_{1,7} = 0.0079$ and the next smallest value of $y_1 = 0.1527$, so that 7 is marginally extreme. The effect of this too small value is that a negative outlier is formed when the data are logged, so that $\hat{\lambda}_1$ is increased above zero.

Fig.3 shows index plots of $T_{LR(i)}$ and of $T_{SU(i)}$. The parameter values 0 and 1 which generated the data are rejected by both statistics with values of 8.75 and 8.55, to be compared with 5.99 for the 5% point of χ_2^2. The value of $\hat{\lambda}_1$ is 0.21 reflecting the effect of the small outlier. However, index plots show that when observation 7 is deleted the true value of the parameters is no longer rejected. Fig.3b is the index plot for the null hypothesis $\lambda_1 = \lambda_2 = 0$, which is rejected by both statistics. Deletion of observation 7 in this case reduces the size of the statistic, since the logarithmic transformation of y_1 is now acceptable, although both statistics remain significant. Perhaps more importantly, the plot calls attention to the anomalous observation. Fig.3b also shows that the values of $T_{SU(i)}$, the statistic using constructed variables, are greater than those of $T_{LR(i)}$. This is in line with the results of Atkinson and Lawrance (1989) comparing likelihood ratio and constructed variable statistics for univariate transformations.

188

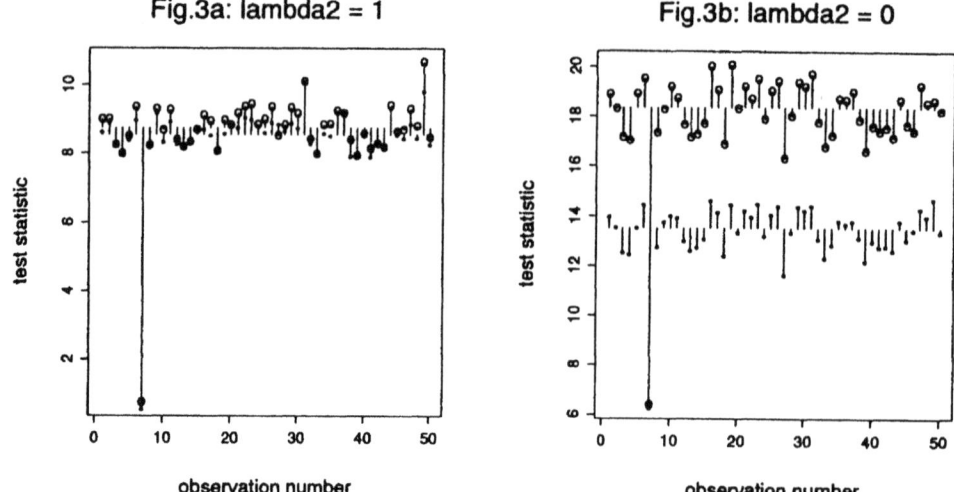

Fig3: Example 2: lightly correlated data. Index plot of deletion tests with observation 7 an outlier: o $T_{SU(i)}$; • $T_{LR(i)}$: (a) null hypothesis $\lambda_0 = (0,1)$, the data generation value; (b) $\lambda_0 = (0,0)$.

Example 3: Highly Correlated Data

This example is like Example 2 except that $Y_2 = 0.95Z_1 + 0.05Z_2 + 2$. Observation 7 is again an outlier with a small value of y_1. The scatterplot of Fig.4b shows the way in which observation 7 is outlying in the two dimensional space of the observations. The histogram of values of y_1 in Fig.4a shows that observation 7 is not marginally outlying - in fact it has the next to smallest value of y_1.

The likelihood ratio test of $\lambda_1 = 0, \lambda_2 = 1$ has the value 11.75. As Fig.5a shows, this drops to 0.83 when observation 7 is deleted. Thus the deletion statistic $T_{LR(i)}$ correctly identifies the effect of observation 7 on inference about the correct transformation. However the seemingly unrelated regression test fails, as shown in Fig.5b, for these highly correlated data. The value of 0.90 for the aggregate statistic indicates that $(0,1)$ is acceptable for λ, but when observation 7 is deleted, the value increases to 4.68. Although this value is not significant, the plot indicates that deletion of observation 7 seems to make the correct transformation less favourable rather than more so.

6   Structure of the Seemingly Unrelated Regression Test

Deletion of observations affects the statistics in two ways, through the summations in (11) and through replacement of the estimate $\hat{\Sigma}_W$ by the deletion estimate $\hat{\Sigma}_{W(i)}$. The effect of deletion on $\hat{\beta}^*$ for fixed $\hat{\Sigma}_W$ can be found by methods similar to those for standard regression diagnostics, but the effect of deletion in the covariance matrix is more complicated. A Taylor series approximation is given by de Gruttola, Ware, and Louis (1987). In this section we exhibit the effects of these two deletions numerically for the lightly correlated example of the previous section.

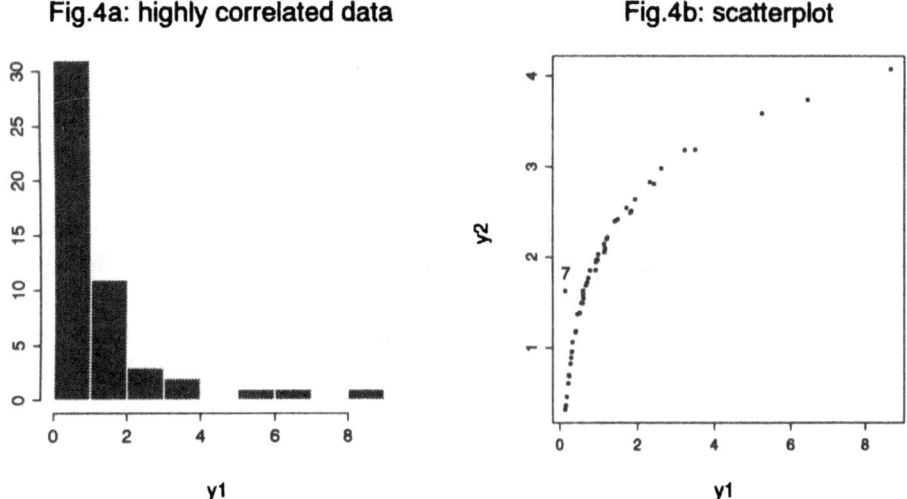

Fig4: Example 3: highly correlated data. (a) histogram of y_1; (b) scatterplot of y_1 and y_2. Observation 7 is jointly outlying but not marginally outlying.

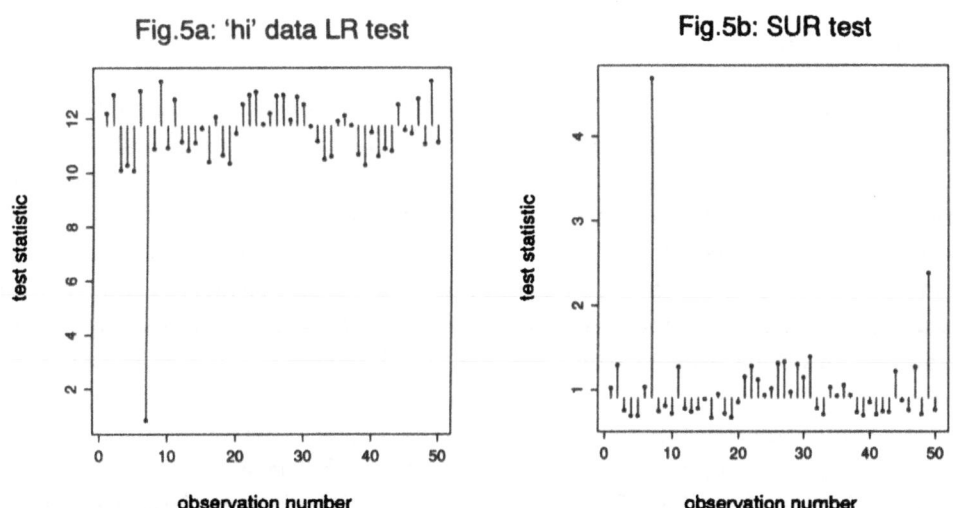

Fig5: Example 3: highly correlated data. Index plot of deletion tests for $\lambda_0 = (0, 1)$, with observation 7 an outlier: (a) $T_{LR(i)}$; (b) $T_{SU(i)}$:

Example 2 Continued: Lightly Correlated Data

Fig.6a shows an index plot of the statistic $T_{SU(i)}$, for testing the parameter values $(0, 1)$, when there is deletion in the sums in (11), but no deletion in the estimate $\hat{\Sigma}_W$. The plot is virtually indistinguishable from the full deletion plot in Fig.3a, with the true parameter value accepted by the test only when observation 7 is deleted. Fig.6b is very different. This was calculated with deletion only in the estimate $\hat{\Sigma}_{W(i)}$, the sums in (11) being taken over all observations. For the deletion of all observations except 7 the plot is much like that of Fig.6a, particularly in the values furthest from the aggregate value of 8.75. But when observation 7 is deleted, the statistic increases to 11.76, a stronger rejection of the true parameter value than that given by the aggregate statistic.

The general conclusion from this analysis is that the effect of deletion on estimation of the covariance matrix is of secondary importance compared with deletion in the sufficient statistics in (11). But the behaviour of the statistic in Fig.6b needs further investigation, reminiscent as it is of the behaviour of $T_{SU(i)}$ for the highly correlated data in Fig.5b, which also increased rather than decreased on deletion of the outlying observation.

7 Regression

All examples treated so far are unstructured multivariate samples. To stress that the method is equally applicable to data fitted to a regression model, some data from a designed experiment are analysed.

Example 4: Baby Food

Box and Draper (1987), p.265, give results from an experiment on the storage of baby food. The 27 trial five-factor design consists of a 2^{5-1} fractional factorial, with 10 star points and one centre point. The four responses are measurements of viscosity over time. Viscosity has a notoriously skewed distribution and, in the analysis given by Box and Draper, there is a discussion of data transformation leading to the choice of the logarithmic transformation. The fitted model contains terms in x_2, x_3 and x_5 as well as, surprisingly, the interaction $x_3 x_4$ in the absence of x_4. This model is used here but the data are modified by changing the four responses for observation 11 from (9.3, 5.8, 5.0, 12.5) to (0.3, 0.8, 0.5, 2.5). Fig.7a shows the resulting index plot of $T_{LR(i)}$ for testing the logarithmic null hypothesis, namely $\lambda_1 = \lambda_2 = 0$. For all observations the statistic equals 20.61 and individual deletions, apart from observation 11, give values fluctuating about this level. When the altered observation 11 is deleted the value of 4.34 admits the logarithmic transformation. Fig.7b shows a similar pronounced effect on the estimation of λ_1, the estimates of the other transformation parameters being and remaining closer to zero. In this example the altered observation 11 is again a marginal outlier but, because of the effect of regression, it is not clear that it will cause the parameter estimate to increase although here it does.

8 Discussion

Gnanadesikan (1977) describes three forms of transformation: marginal, joint, which has been the subject of this paper, and transformation to normality in a specified direction. A question, answered in some special cases by Velilla (1993), is under what conditions, and by how much, is the power of the tests increased by the use of joint transformations?

191

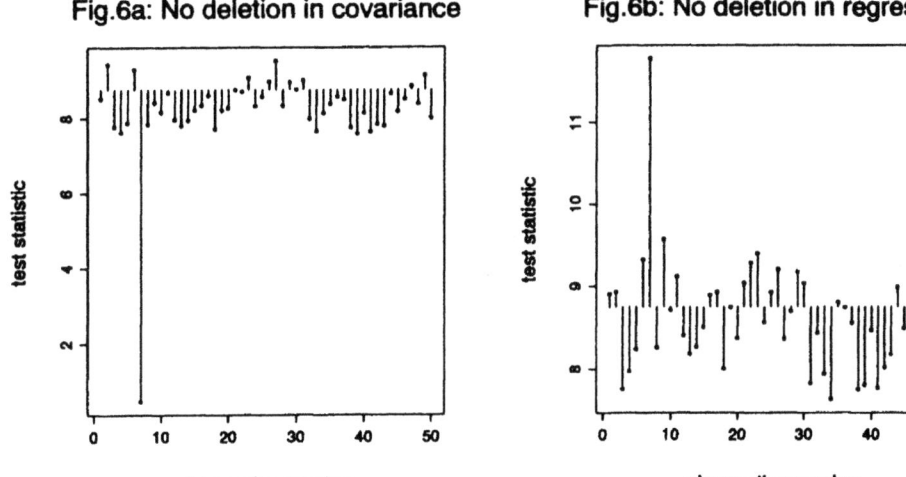

Fig6: Example 2: lightly correlated data. Structure of seemingly unrelated regression test: index plots of $T_{SU(i)}$: (a) no deletion in estimation of the covariances matrix Σ; (b) no deletion in regression calculations (11).

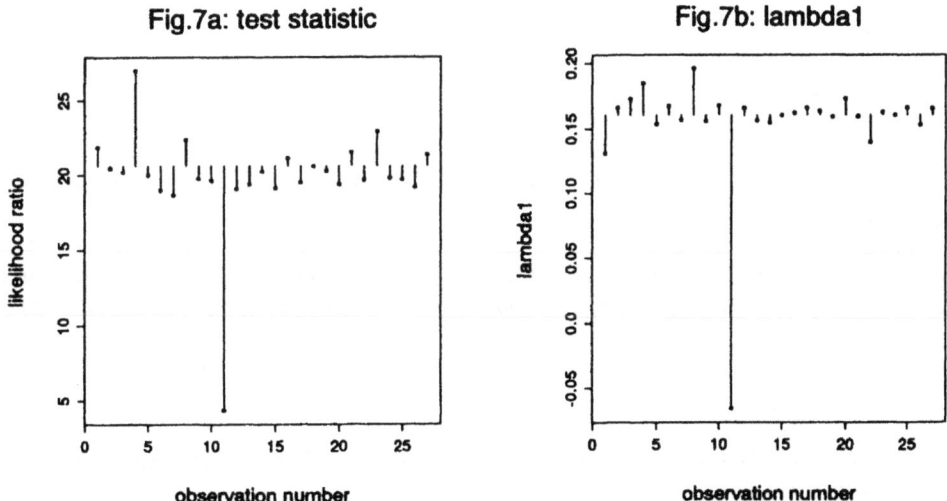

Fig 7: Example 4: modified baby food data. Index plots: (a) $T_{LR(i)}$ for $\lambda_0 = (0, 0, 0, 0)$; (b) $\hat{\lambda}_{1(i)}$.

192

The question relates to the comparison of the likelihood ratio and seemingly unrelated regression statistics of this paper, since the latter procedure begins with independent regressions. The examples suggest that, both for the aggregate and diagnostic statistics, the procedure using constructed variables can fail. It is natural to wonder whether it can be improved by iteration in the estimation of Σ. The answer will reflect more generally on seemingly unrelated regression procedures. Questions about diagnostics for seemingly unrelated regression are raised by the example in §6 of the effects of deletion in only a part of the diagnostic calculation.

References

Andrews, D. F., R. Gnanadesikan, and J. L. Warner (1971). Transformations of multivariate data. *Biometrics 27*, 825–840.

Atkinson, A. C. (1985). *Plots, Transformations, and Regression*. Oxford: Oxford University Press.

Atkinson, A. C. and A. J. Lawrance (1989). A comparison of asymptotically equivalent tests of regression transformation. *Biometrika 76*, 223–229.

Box, G. E. P. and D. R. Cox (1964). An analysis of transformations (with discussion). *Journal of the Royal Statistical Society, Series B 26*, 211–246.

Box, G. E. P. and N. R. Draper (1987). *Empirical Model-Building and Response Surfaces*. New York: Wiley.

Cook, R. D. and S. Weisberg (1982). *Residuals and Influence in Regression*. London: Chapman and Hall.

Davison, A. C. and C.-L. Tsai (1992). Regression model diagnostics. *International Statistical Review 60*, 337–353.

de Gruttola, V., J. H. Ware, and T. A. Louis (1987). Influence analysis of generalized least squares estimators. *Journal of the American Statistical Association 82*, 911–917.

Gnanadesikan, R. (1977). *Methods for Statistical Data Analysis of Multivariate Observations*. New York: Wiley.

Harvey, A. C. (1990). *The Econometric Analysis of Time Series (2nd edition)*. Cambridge, Mass.: MIT Press.

McCullagh, P. and J. A. Nelder (1989). *Generalized Linear Models (2nd edition)*. London: Chapman and Hall.

Mulira, H.-M. (1992). *Computational Methods for Transformations to Multivariate normality*. Ph. D. thesis, Department of Statistical and Mathematical Sciences, London School of Economics.

Velilla, S. (1993). A note on the multivariate Box-Cox transformation to normality. *Statistics and Probability Letters 17*, 259–263.

Zellner, A. (1962). An efficient method of estimating seemingly unrelated regressions and test of aggregation bids. *Journal of the American Statistical Association 57*, 348–368.

Regression Rank Scores: Asymptotic Linearity and RR-Estimators

JANA JUREČKOVÁ

Department of Probability and Statistics, Charles University, Sokolovská 83,
186 00 Prague 8, Czech Republic.

Abstract:

The uniform asymptotic linearity of regression rank scores process, proved by
the author in 1992, is extended to a broad class of distributions of the errors
including the Cauchy. This extends the applicability of RR-estimators in the
linear model and has various other applications.

1 Introduction

Consider the linear regression model

$$\mathbf{Y}_n = \mathbf{X}_n \beta + \mathbf{E}_n \qquad (1)$$

with vector of observations $\mathbf{Y} = \mathbf{Y} = (Y_{n1}, ..., Y_{nn})'$, known design matrix $\mathbf{X}_n = \mathbf{X}$ of
order $(n \times p)$, unknown parameter $\beta = (\beta_1, ..., \beta_p)'$ and with vector $\mathbf{E}_n = \mathbf{E} = (E_1, ..., E_n)'$
of *i.i.d.* errors with joint (unknown) distribution function F. We shall assume throughout
that β_1 is an *intercept*, i.e. $x_{i1} = 1, i = 1, ..., n$.

The *regression rank scores* (RR's) of model (1) were introduced in Gutenbrunner and
Jurečková (1992) as components of the optimal solution $\hat{\mathbf{a}}(\alpha)$, $0 < \alpha < 1$, of the following
parametric linear programming problem:

$$
\begin{aligned}
\mathbf{Y}'\hat{\mathbf{a}}(\alpha) &= \max \\
\mathbf{X}'(\hat{\mathbf{a}}(\alpha) - (1-\alpha))\mathbf{1}_n &= \mathbf{0} \\
\hat{\mathbf{a}}(\alpha) &\in [0,1]^n, \quad 0 < \alpha < 1.
\end{aligned}
\qquad (2)
$$

The regression rank scores are dual to the *regression quantiles* introduced by Koenker and
Bassett (1978); in fact, the α-regression quantile $\hat{\beta}(\alpha)$ of model (1) is the component $\hat{\beta}$

---

*Kitsos, C.P., and Müller, W.G., Eds., *Proceedings of MODA4*, Physica Verlag, Heidelberg, 1995

of the optimal solution $(\hat{\beta}, \mathbf{r}^+, \mathbf{r}^-)$ of the linear program

$$\alpha \mathbf{1}_n' \mathbf{r}^+ + (1 - \alpha) \mathbf{1}_n' \mathbf{r}^- = \min$$
$$\mathbf{X}\hat{\beta} + \mathbf{r}^+ - \mathbf{r}^- = \mathbf{Y} \qquad (3)$$
$$(\hat{\beta}, \mathbf{r}^+, \mathbf{r}^-) \quad \epsilon \quad \mathbf{R}^p \times \mathbf{R}_+^n \times \mathbf{R}_+^n, \quad 0 < \alpha < 1.$$

Regression rank scores represent an extension of ranks while the regression quantiles are counterparts of order statistics. The duality of (2) and (3) also extends the duality of ranks and of order statistics from the location to the linear regression model. Following this analog, one naturally expects that the regression quantiles are convenient mainly for estimation while the regression rank scores are convenient for testing various hypotheses in the model (1). Really, the L-estimators of β based on regression quantiles are already well-known [see Koenker and Bassett (1978), Ruppert and Carroll (1980), Jurečková (1983, 1984), Koenker and Portnoy (1987), Gutenbrunner and Jurečková (1992), among others]; among these estimators, the trimmed least squares estimator is a very natural and computionally appealing counterpart of the trimmed mean. On the other hand, a general class of linear tests based on regression rank scores (nonlinear RR-tests) was constructed in Gutenbrunner at al. (1993) and extended to a broad class of distributions of errors in Jurečková (1994). Nonlinear RR-tests of the Kolmogorov-Smirnov type were constructed by Jurečková (1991). The regression quantiles are simple and their idea is natural, acceptable even for practitioners; finding all regression quantiles is a straightforward application of *parametric* linear programming. The path of $\hat{\beta}(.)$ is a piecewise constant function from $[0, 1]$ to $\mathbf{R}^p$. Detailed descriptions of algorithms for regression quantiles may be found in Koenker and d'Orey (1987, 1994) and Osborne (1992). The regression rank scores process may be essentially computed as a byproduct of the regression quantile computation requiring only some additional storage. See Koenker and d'Orey (1994) for algorithmic details.

The main advantage of regression rank scores is apparently their *regression invariance* which means that

$$\hat{\mathbf{a}}(\alpha, \mathbf{Y} + \mathbf{Xb}) = \hat{\mathbf{a}}(\alpha, \mathbf{Y}) \quad \forall \mathbf{b} \epsilon \mathbf{R}^p. \qquad (4)$$

Hence, provided β is the nuisance parameter while our inference concerns something else and is based on $\hat{\mathbf{a}}(\alpha)$, β need not be estimated. An example is the extended linear model

$$\mathbf{Y} = \mathbf{X}\beta + \mathbf{Z}\delta + \mathbf{E} \qquad (5)$$

where $\mathbf{X}$ is of order $(n \times p)$ and $\mathbf{Z}$ of order $(n \times q)$ and the inference concerns δ while β is a nuisance parameter. The linear RR-tests of $H_0 : \delta = \mathbf{0}$ and of related hypotheses were constructed in Gutenbrunner at al. (1993) and Jurečková (1994). However, we could also construct estimators of δ based on RR's, analogous to R-estimators (estimators based on ranks) and this is the primary goal of the present paper.

In the case of $\beta = 0$, the rank-based estimators of δ were constructed by Adichie (1967), Jurečková (1971), Koul (1971) and Jaeckel (1972) and later on studied by many authors. If $\beta \neq \mathbf{0}$ but unknown and one wishes to estimate δ by an R-estimator, then, due to the nonlinearity of rank-based methods, we must estimate the whole $(\beta', \delta')'$ and then estimate δ by the δ-components of the R-estimator. The challenging property of invariance of RR's to the $\mathbf{X}$-regression means that inference based on RR's under the submodel (1) is analogously invariant and hence independent of the nuisance parameter β. Without the necessity of estimation β we do not risk an eventual breakdown of the estimator caused by leverage points in $\mathbf{X}$; moreover, computional aspects also play a role.

The RR-estimator of δ (of Jaeckel's type) with nuisance β was proposed by Jurečková (1992a), but its consistency and asymptotic normality were proved under more restrictive conditions. Jurečková and Sen (1993) showed that the RR-estimator of δ is asymptotically equivalent to the componentwise RR-estimator (i.e., the estimator calculated in such a way that while estimating every single component of δ, one considers all other components as nuisance). It is also asymptotically equivalent to the δ-part of the ordinary R-estimator (in either of the Jurečková, Koul or Jaeckel versions). The asymptotic equivalence is meant in the first order sense: Two estimators $\mathbf{T}_{n1}$ and $\mathbf{T}_{n2}$ are called the first order asymptotically equivalent provided

$$n^{1/2}\|\mathbf{T}_{n1} - \mathbf{T}_{n2}\| = o_p(1). \tag{6}$$

From the first order asymptotic equivalence point of view, we do not distinguish between the three above possible versions of estimate of δ. However, differences appear in finite sample case where the computational aspects also play a role. Moreover, differences appear in the higher order asymptotic relations when (6) is supplemented by an asymptotic (nonnormal) distribution of properly standardized difference of $\mathbf{T}_{n1}$ and $\mathbf{T}_{n2}$. A replacement of the nuisance parameter by an estimate apparently leads to weaker second order properties comparing with an estimator invariant to β; this interesting problem deserves a special study. An analogous phenomenon we observe in testing a hypothesis about δ with nuisance β in model (5): While the RR-tests need not an estimate of β, they are asymptotically equivalent to the aligned rank tests which replace β by an estimate. However, the numerical evidence clearly shows that the quality of an aligned test heavily depends on the estimator of β while, theoretically, a difference appears only in the higher order asymptotics.

The basic tool in the proof of consistency and asymptotic normality of RR-estimator is the asymptotic linearity of the RR-process and of linear RR-statistics, proved in Jurečková (1992b). The basic tool for the proof of the uniform asymptotic linearity is the approximation the regression quantiles and regression rank scores processes by weighted empirical processes, proved in Gutenbrunner at al. (1993) under more restrictive conditions excluding densities with havier tails than those of t-distribution with 5 $d.f.$; these conditions were relaxed in Jurečková (1994) to cover a broad class of densities including Cauchy. In the present paper, we shall prove the consistency and asymptotic normality of the RR-estimator of δ in model (5) for a broad class of distributions of errors.

2 RR-Estimators

Let $\hat{\mathbf{a}}_n(\alpha, t)$ denote the regression rank scores calculated for the pseudo-observations $\mathbf{Y}_n - \mathbf{Z}_n t$, $t \epsilon \mathbf{R}^q$, in the submodel (1) of (5), corresponding to $\delta = \mathbf{0}$. In other words, $\hat{\mathbf{a}}_n(\alpha, \mathbf{t})$ is an optimal solution $\hat{\mathbf{a}}_n$ of the linear program

$$(\mathbf{Y}_n - \mathbf{Z}_n t)'\hat{\mathbf{a}}_n : \ = \max$$
$$\mathbf{X}_n'(\hat{\mathbf{a}}_n - (1-\alpha))\mathbf{1}_n) = \mathbf{0}, \quad \hat{\mathbf{a}}_n \epsilon [0,1]^n, \quad 0 \le \alpha \le 1. \tag{7}$$

Following Jurečková (1994), we shall impose the following conditions on the distribution function F of the errors and on the matrices $\mathbf{X}$ and $\mathbf{Z}$:

(F.1) *We assume that F is absolutely continuous with absolutely continuous, positive and bounded density $f(x)$, $\underline{x} < x < \overline{x}$, and that the derivative f' of f is bounded a.e. in $(\underline{x}, \overline{x})$, where $\underline{x} = \sup\{x : F(x) = 0\}$ and $\overline{x} = \inf\{x : F(x) = 1\}$.*

(F.2) *We assume that f is monotonically decreasing to 0 as $x \to \underline{x}+$ and $x \to \bar{x}-$ and there exist A and a, $0 \leq A \leq a < \infty$ such that*

$$\lim_{\alpha \to 0,1} \frac{f(F^{-1}(\alpha))}{(\alpha(1-\alpha))^{1+A}} = 1 \tag{8}$$

and

$$\sup_{0 < \alpha < 1} \left\{ \alpha(1-\alpha) \left| \frac{f'(F^{-1}(\alpha))}{f^2(F^{-1}(\alpha))} \right| \right\} \leq 1+a. \tag{9}$$

Fix b satisfying $0 < \delta \leq b - a \leq a + \delta$ for some $\delta > 0$ and denote

$$\alpha_n^* = n^{-1/(1+2b)} \tag{10}$$

and

$$\sigma_\alpha = \frac{(\alpha(1-\alpha))^{1/2}}{f(F^{-1}(\alpha(1-\alpha)))}, \quad 0 < \alpha < 1. \tag{11}$$

Moreover, we impose the following conditions on $\mathbf{X}_n$ and $\mathbf{Z}_n$:

(X.1) $x_{i1} = 1, \quad i = 1, ..., n.$

(X.2) $\lim_{n \to \infty} \mathbf{D}_n = \mathbf{D}$, where $\mathbf{D}_n = n^{-1} \mathbf{X}_n' \mathbf{X}_n$ and $\mathbf{D}$ is a positive definite $(p \times p)$ matrix.

(X.3) $n^{-1} \sum_{i=1}^n \|\mathbf{x}_{ni}\|^4 = O(1)$ as $n \to \infty.$

(X.4) $\max_{1 \leq i \leq n} \|\mathbf{x}_{ni}\| = O(n^\Delta)$ as $n \to \infty$ where

$$\Delta = \frac{b - a - \delta}{1 + 2b} \wedge \frac{1}{4} \tag{12}$$

with a, b, δ fixed in (F.2) and (10).

(Z.1) $\max_{1 \leq i \leq n} \|\mathbf{z}_{ni}\| = O(n^\Delta)$ as $n \to \infty.$

(Z.2) $\lim_{n \to \infty} \mathbf{Q}_n = \mathbf{Q}$ where $\mathbf{Q}$ is a positive definite $(q \times q)$ matrix and

$$\mathbf{Q}_n = n^{-1}(\mathbf{Z}_n - \hat{\mathbf{Z}}_n), \quad \hat{\mathbf{Z}}_n = \mathbf{X}_n(\mathbf{X}_n' \mathbf{X}_n)^{-1} \mathbf{X}_n' \mathbf{Z}_n. \tag{13}$$

Take a score-generating function $\varphi(u) : (0,1) \to \mathbf{R}^1$, nondecreasing and square-integrable and such that $\varphi'(u)$ exists for $0 < u < \alpha_0$ and for $1 - \alpha_0 < u < 1$, $0 < \alpha < \frac{1}{2}$ and in this domain it satisfies the *Chernoff-Savage type condition*

$$|\varphi'(u)| \leq c(u(1-u))^{-1-(\delta^*/2)} \tag{14}$$

for some $c > 0$ and $0 < \delta < \delta^*$. Put

$$\varphi_n(u) = \begin{cases} \varphi(\alpha_n^*) & \dots \ 0 \leq u < \alpha_n^* \\ \varphi(u) & \dots \ \alpha_n^* \leq u \leq 1 - \alpha_n^* \\ \varphi(1 - \alpha_n^*) & \dots \ 1 - \alpha_n^* \leq u < 1 \end{cases} \tag{15}$$

and denote

$$\hat{b}_{ni}(t) = -\int_0^1 \varphi_n(u) d\hat{a}_{ni}(u, t), \quad t \epsilon \mathbf{R}^q, \quad i = 1, ..., n. \tag{16}$$

Following Jaeckel (1972), introduce the *measure of rank dispersion*

$$D_n(\mathbf{t}) = \sum_{i=1}^{n} (Y_i - \mathbf{z}_i'\mathbf{t})[\hat{b}_{ni}(\mathbf{t}) - \overline{\varphi}_n] \tag{17}$$

where

$$\begin{aligned}
\overline{\varphi}_n &= \int_0^1 \varphi_n(u)d(u) = -\int_0^1 \varphi_n(u)d(1-u) \\
&= -\int_0^1 \varphi_n(u)d(n^{-1}\sum_{i=1}^n \hat{a}_{ni}(u,\mathbf{t})) = n^{-1}\sum_{i=1}^n \hat{b}_{ni}(\mathbf{t}).
\end{aligned} \tag{18}$$

By (7) and (18), the measure $D_n(\mathbf{t})$ is invariant to the $\mathbf{X}$-regression. It is continuous, piecewise linear, convex function of $\mathbf{t}\epsilon\mathbf{R}^q$. We propose to estimate δ by minimizing $D_n(\mathbf{t})$ with respect to $\mathbf{t}\epsilon\mathbf{R}^q$; more precisely, denote

$$\mathcal{D}_n = \{\mathbf{t}\epsilon\mathbf{R}^q : D_n(\mathbf{t}) = \min\} \tag{19}$$

and consider any point of $\mathcal{D}_n$ as an estimator $\hat{\delta}$ of δ. The following theorem shows that, under the above conditions, $\hat{\delta}$ is consistent and asymptotically normal estimator of δ.

Theorem 2.1 *Under the conditions (F.1), (F.2), (X.1)-(X.3), (Z.1)-(Z.3) and (14), $\hat{\delta}_n$ is consistent estimator of δ in model (5) and $\sqrt{n}(\hat{\delta}_n-\delta)$ is asymptotically normally distributed*

$$\mathcal{N}_q(\mathbf{0}, \mathbf{Q}^{-1}A^2(\varphi)/\gamma^2(\varphi, F)) \tag{20}$$

where

$$\gamma(\varphi, F) = \int_0^1 \varphi(u)d(F^{-1}(u)) \neq 0 \tag{21}$$

and

$$A^2(\varphi) = \int_0^1 (\varphi(u) - \overline{\varphi})^2 du, \quad \overline{\varphi} = \int_0^1 \varphi(u)du. \tag{22}$$

Moreover, $\sqrt{n}(\hat{\delta}_n - \delta)$ admits the asymptotic representation

$$\sqrt{n}(\hat{\delta}_n - \delta) = n^{-1/2}\gamma^{-1}\mathbf{Q}^{-1}(\mathbf{Z}_n - \hat{\mathbf{Z}}_n)'\varphi(F(\mathbf{E}_n)) + o_p(1), \tag{23}$$

where

$$\varphi(F(\mathbf{E}_n)) = (\varphi(E_1), ..., \varphi(E_n))'. \tag{24}$$

Proof. The proof is postponed to Section 3. We shall first prove the uniform asymptotic linearity of the regression rank scores process which, in turn, yields a uniform approximation of dispersion measure $D_n(\mathbf{t})$ by a quadratic function of $\mathbf{t}$.

3 Uniform Asymptotic Linearity of Regression Rank Scores Process

We shall keep the notation of Section 2 and consider the process of regression rank scores

$$\begin{aligned}
A_n(\alpha, n^{-1/2}\mathbf{t}) &= n^{-1/2}\sum_{i=1}^n d_{ni}\hat{a}_{ni}(\alpha, n^{-1/2}\mathbf{t}) \\
&= n^{-1/2}\sum_{i=1}^n d_{ni}\hat{a}_{ni}(\alpha, \mathbf{Y}_n - n^{-1/2}\mathbf{Z}_n\mathbf{t}), \quad \mathbf{t}\epsilon\mathbf{R}^q
\end{aligned} \tag{25}$$

where $\mathbf{d}_n = (d_{n1}, ..., d_{nn})'$ are vectors satisfying the conditions

(D.1) $\mathbf{X}_n'\mathbf{d}_n = \mathbf{0}$, $\quad n^{-1}\sum_{i=1}^n d_{ni}^2 \to \Gamma^2$ *as* $n \to \infty$, $\quad 0 < \Gamma < \infty$

198

(D.2) $n^{-1} \sum_{i=1}^{n} |d_{ni}|^3 = O(1)$ as $n \to \infty$

(D.3) $\max_{1 \le i \le n} |d_{ni}| = O(n^{\Delta})$ with Δ given in (12).

The following theorem shows that the process (25) is asymptotically linear in **t**, uniformly in $\|\mathbf{t}\| \le C_n = C(\log \log n)^{1/2}$ for any fixed C, $0 < C < \infty$, and $\alpha_n^* \le \alpha \le 1 - \alpha_n^*$.

Theorem 3.1 *Under the conditions (F.1), (F.2), (X.1)-(X.3), (Z.1)-(Z.3), (D.1)-(D.3) and (14),*

$$\sup \left\{ \left| A_n(\alpha, n^{-1/2}\mathbf{t}) - A_n(\alpha, 0) + n^{-1}\mathbf{d}_n' \mathbf{Z}_n \mathbf{t} f(F^{-1}(\alpha)) \right| : \right.$$
$$\left. \|\mathbf{t}\| \le C_n, \alpha_n^* \le \alpha \le 1 - \alpha_n^* \right\} = o_p \left(n^{-\frac{1}{2} + \frac{b-(\delta/2)}{1+2b}} \log n \right) \tag{26}$$

as $n \to \infty$, where $C_n = C(\log \log n)^{1/2}$, $0 < C < \infty$.

Proof. Denote

$$\mathcal{C}_n = \left\{ (\mathbf{t}, \alpha) : \|\mathbf{t}\| \le C_n, \alpha_n^* \le \alpha \le 1 - \alpha_n^* \right\}. \tag{27}$$

Consider the function

$$r_n(\mathbf{t}, u, \mathbf{v}, \alpha) = \sigma_\alpha^{-1} \sum_{i=1}^{n} [\rho_\alpha(E_{i\alpha} - n^{-1/2}\sigma_\alpha(\mathbf{x}_i'\mathbf{v} + \mathbf{z}_i'\mathbf{t} + d_i u)) - \rho_\alpha(E_{i\alpha})]$$
$$+ n^{-1/2} \sum_{i=1}^{n} (\mathbf{x}_i'\mathbf{v} + \mathbf{z}_i'\mathbf{t} + d_i u)\psi_\alpha(E_{i\alpha}) \tag{28}$$
$$- \tfrac{1}{2}(\alpha(1-\alpha))^{1/2}[n^{-1}\mathbf{v}'\mathbf{D}_n\mathbf{v} + n^{-1}\mathbf{t}'\mathbf{Z}_n'\mathbf{Z}_n\mathbf{t} + \Gamma^2 u^2 + 2n^{-1}\mathbf{v}'\mathbf{X}_n'\mathbf{Z}_n\mathbf{t}],$$

$\mathbf{t} \epsilon \mathbf{R}^p$, $\mathbf{v} \epsilon \mathbf{R}^q$, $u \epsilon \mathbf{R}^1$. Then, the conditions of Lemma 3.1 in Jurečková (1994) are fulfilled for the model

$$Y_i = \sum_{j=1}^{p} x_{ij}\beta_j + \sum_{k=1}^{q} z_{ik}\delta_k + d_i\eta + E_i, \quad i = 1, ..., n \tag{29}$$

with parameters $\beta \epsilon \mathbf{R}^p$, $\delta \epsilon \mathbf{R}^q$, $\eta \epsilon \mathbf{R}^1$ and hence

$$\sup\{|r_n(\mathbf{t}, u, \mathbf{v}, \alpha)| : (\mathbf{t}, u, \mathbf{v}, \alpha) \epsilon \mathcal{B}_n\} = O_p \left(n^{-\frac{1}{2} + \frac{b-(\delta/2)}{1+2b}} \right) \tag{30}$$

as $n \to \infty$, where

$$\mathcal{B}_n = \{ (\mathbf{t}, u, \mathbf{v}, \alpha) : \|\mathbf{t}\| \le C_n, |u| \le C_n, \|\mathbf{v}\| \le C_n, \alpha_n^* \le \alpha \le 1 - \alpha_n^* \}. \tag{31}$$

Let $\hat{\beta}_n(\alpha, n^{-1/2}\mathbf{t})$ denote the α-regression quantile calculated for the pseudo-observations $\mathbf{Y}_n - n^{-1/2}\mathbf{Z}_n\mathbf{t}$, i.e., the solution of the minimization

$$\sum_{i=1}^{n} \rho_\alpha(Y_i - \mathbf{x}_i'\mathbf{v} - n^{-1/2}\mathbf{z}_i'\mathbf{t}) := \min \tag{32}$$

with respect to $\mathbf{v} \epsilon \mathbf{R}^p$. Then we have

Lemma 3.1 *Under the conditions of Theorem 3.1,*

$$\sup \left\{ n^{1/2}\sigma_\alpha^{-1} \|\hat{\beta}_n(\alpha, n^{-1/2}\sigma_\alpha\mathbf{t}) - \beta(\alpha)\| : (\mathbf{t}, \alpha) \epsilon \mathcal{C}_n \right\} = O_p((\log \log n)^{1/2}) \tag{33}$$

as $n \to \infty$.

Proof. Consider the function

$$G_{n\alpha}(\mathbf{t}, \mathbf{v}) = \sigma_\alpha^{-1} \sum_{i=1}^{n} [\rho_\alpha(E_{i\alpha} - n^{-1/2}\sigma_\alpha(\mathbf{x}_i'\mathbf{v} + \mathbf{z}_i'\mathbf{t})) - \rho_\alpha(E_{i\alpha} - n^{-1/2}\sigma_\alpha\mathbf{z}_i'\mathbf{t})] \qquad (34)$$

which is convex in $\mathbf{v}$ and is minimized by

$$\mathbf{T}_{n\alpha}(\mathbf{t}) = n^{1/2}\sigma_\alpha^{-1}[\hat{\beta}_n(\alpha, n^{-1/2}\sigma_\alpha\mathbf{t}) - \beta(\alpha] \quad \epsilon\mathbf{R}^p. \qquad (35)$$

Using the Pollard (1991) convexity argument as in the proof of Theorem 3.1 in Gutenbrunner et al. (1993), we get

$$\sup\{\|\mathbf{T}_{n\alpha}(\mathbf{t}) - \mathbf{U}_{n\alpha}(\mathbf{t})\| : (\mathbf{t}, \alpha)\epsilon\mathcal{C}_n\} \xrightarrow{p} 0 \qquad (36)$$

as $n \to \infty$, where

$$\mathbf{U}_{n\alpha}(\mathbf{t}) = n^{-1/2}(\alpha(1-\alpha))^{-1/2}\mathbf{D}_n^{-1} \sum_{i=1}^{n} \mathbf{x}_i[\psi_\alpha(E_{i\alpha}) - n^{-1/2}(\alpha(1-\alpha))^{1/2}\mathbf{z}_i'\mathbf{t}] \qquad (37)$$

which implies (33).

Completing the proof of Theorem 3.1. Differentiating (30) with respect to u at u=0 we obtain

$$\begin{aligned} &\sup\left\{ \left| n^{-1/2}\sum_{i=1}^{n} d_{ni}[\psi_\alpha(E_{i\alpha} - n^{-1/2}\sigma_\alpha(\mathbf{x}_i'\mathbf{v} + \mathbf{z}_i'\mathbf{t})) - \psi_\alpha(E_{i\alpha})] \right.\right. \\ &\left.\left. +(\alpha(1-\alpha))^{1/2}n^{-1}\mathbf{d}_n'\mathbf{Z}_n\mathbf{t} \right| : \|\mathbf{v}\| \leq C_n, \|\mathbf{t}\| \leq C_n, \alpha_n^* \leq \alpha \leq 1 - \alpha_n^* \right\} \\ &= o_p\left(n^{-\frac{1}{2}+\frac{b-(\delta/2)}{1+2b}} \log n \right). \end{aligned} \qquad (38)$$

Inserting $n^{1/2}\sigma_\alpha^{-1}[\hat{\beta}_n(\alpha, n^{-1/2}\sigma_\alpha\mathbf{t}) - \beta(\alpha)]$ for $\mathbf{v}$ in (38), we obtain that for $n \geq n_0$

$$\begin{aligned} &\sup\left\{ \left| n^{-1/2}\sum_{i=1}^{n} d_{ni}\left(I[Y_i - n^{-1/2}\sigma_\alpha\mathbf{z}_i'\mathbf{t} > \mathbf{x}_i'\hat{\beta}(\alpha, n^{-1/2}\sigma_\alpha\mathbf{t})] - I[E_i > F^{-1}(\alpha)] \right) \right.\right. \\ &\left.\left. +(\alpha(1-\alpha))^{1/2}n^{-1}\mathbf{d}_n'\mathbf{Z}_n\mathbf{t} \right| : (\mathbf{t}, \alpha)\epsilon\mathcal{C}_n \right\} \\ &= o_p\left(n^{-\frac{1}{2}+\frac{b-(\delta/2)}{1+2b}} \log n \right) \end{aligned} \qquad (39)$$

and hence under $\mathbf{t} = \mathbf{0}$

$$\begin{aligned} &\sup\left\{ \left| n^{-1/2}\sum_{i=1}^{n} \left(I[Y_i > \mathbf{x}_i'\hat{\beta}(\alpha)] - I[E_i > F^{-1}(\alpha)] \right) \right| : \alpha_n^* \leq \alpha \leq 1 - \alpha_n^* \right\} \\ &= o_p\left(n^{-\frac{1}{2}+\frac{b-(\delta/2)}{1+2b}} \log n \right). \end{aligned} \qquad (40)$$

On the other hand, by (D.3),

$$\begin{aligned} &\sup\left\{ \left| n^{-1/2}\sum_{i=1}^{n} d_{ni}I[Y_i - n^{-1/2}\sigma_\alpha\mathbf{z}_i'\mathbf{t} = \mathbf{x}_i'\hat{\beta}_n(\alpha, n^{-1/2}\sigma_\alpha\mathbf{t})] \right| : (\mathbf{t}, \alpha)\epsilon\mathcal{C}_n \right\} \\ &= O_p\left(n^{-\frac{1}{2}+\Delta} \log n \right). \end{aligned} \qquad (41)$$

Combining (39) - (41), we obtain

$$\begin{aligned} &\sup\left\{ \left| A_n(\alpha, \sigma_\alpha\mathbf{t}) - A_n(\alpha, \mathbf{0}) + (\alpha(1-\alpha))^{1/2}n^{-1}\mathbf{d}_n'\mathbf{Z}\mathbf{t} \right| : (\mathbf{t}, \alpha)\epsilon\mathcal{C}_n \right\} \\ &= o_p\left(n^{-\frac{1}{2}+\frac{b-(\delta/2)}{1+2b}} \log n \right). \end{aligned} \qquad (42)$$

In view of (F.2), (42) implies for $\mathbf{u} = \sigma_\alpha \mathbf{t}$

$$\sup\left\{\left|A_n(\alpha, \mathbf{u}) - A_n(\alpha, \mathbf{0}) + f(F^{-1}(\alpha))n^{-1}\mathbf{d}_n'\mathbf{Z}\mathbf{u}\right| : (\mathbf{u}, \alpha)\epsilon\mathcal{C}_n\right\}$$
$$= o_p\left(n^{-\frac{1}{2} + \frac{b-(\delta/2)}{1+2b}}\log n\right). \qquad \square \tag{43}$$

Theorem 3.1 enables us to prove the uniform asymptotic linearity of linear RR-statistics which, in turn, leads to the asymptotic distribution of the estimator $\hat{\delta}_n$; however, it has various other applications as does the asymptotic linearity of ordinary linear rank statistics.

Let $\varphi : (0,1) \to \mathbf{R}^1$ be a score-generating function such that $\varphi'(\alpha)$ exists for $0 < \alpha < \alpha_0$ and for $1 - \alpha_0 < \alpha < 1$ where it satisfies the Chernoff-Savage type condition (14). Consider the *linear RR-statistic*

$$S_n(n^{-1/2}\mathbf{t}) = n^{-1/2}\sum_{i=1}^n d_{ni}\hat{b}_{ni}(n^{-1/2}\mathbf{t}), \quad \mathbf{t}\epsilon\mathbf{R}^q \tag{44}$$

($\hat{b}_{ni}$ was defined in (15) and (16)).

Theorem 3.2 *Assume the conditions of Theorem 3.1 and that*

$$\gamma = \gamma(\varphi, F) = -\int_0^1 \varphi(\alpha)df(F^{-1}(\alpha) \neq 0. \tag{45}$$

Then, as $n \to \infty$,

$$\sup\{|S_n(n^{-1/2}\mathbf{t}) - S_n(0) + \gamma n^{-1}\mathbf{d}_n'\mathbf{Z}_n\mathbf{t}| : \|\mathbf{t}\| \leq C_n\} \xrightarrow{p} 0. \tag{46}$$

Proof. Denote $\gamma_n = -\int_0^1 \varphi_n(\alpha)df(F^{-1}(\alpha))$; then $\lim_{n\to\infty}\gamma_n = \gamma$. We may write

$$S_n(n^{-1/2}\mathbf{t}) - S_n(0) + \gamma_n^{-1}\mathbf{d}_n'\mathbf{Z}_n\mathbf{t}$$
$$= -\int_0^1 \varphi_n(\alpha)d[A_n(\alpha, n^{-1/2}\mathbf{t}) - A_n(\alpha, \mathbf{0}) + f(F^{-1}(\alpha))n^{-1}\mathbf{d}_n'\mathbf{Z}_n\mathbf{t}] + o_p(1) \tag{47}$$
$$= \int_{\alpha_n^*}^{1-\alpha_n^*}[A_n(\alpha, n^{-1/2}\mathbf{t}) - A_n(\alpha, \mathbf{0}) + f(F^{-1}(\alpha))n^{-1}\mathbf{d}_n'\mathbf{Z}_n\mathbf{t}]d\varphi_n(\alpha) + o_p(1).$$

We split the integration domain into the intervals

$$(\alpha_n^*, \alpha_0), \quad [\alpha_0, 1 - \alpha_0], \quad (1 - \alpha_0, 1 - \alpha_n^*)$$

and denote the respective integrals by $I_1(\mathbf{t})$, $I_2(\mathbf{t})$ and $I_3(\mathbf{t})$. Theorem 3.1 and the dominated convergence theorem immediately imply that

$$\sup_{\|\mathbf{t}\|\leq C_n} |I_2(\mathbf{t})| = o_p(1). \tag{48}$$

Further, by (14),

$$|I_1(\mathbf{t})| = \left|\int_{\alpha_n^*}^{\alpha_0} \varphi'(\alpha)[A_n(\alpha, n^{-1/2}\mathbf{t} - A_n(\alpha, \mathbf{0}) + f(F^{-1}(\alpha))n^{-1}\mathbf{d}_n'\mathbf{Z}\mathbf{t}]d\alpha\right|$$
$$\leq o_p\left(n^{-\frac{1}{2}+\frac{b-(\delta/2)}{1+2b}}\log n\right) \cdot \int_{\alpha_n^*}^{\alpha_0}(\alpha(1-\alpha))^{-1-\delta^*}d\alpha \tag{49}$$
$$= o_p\left(n^{-\frac{1}{2}-\frac{(\delta-\delta^*)}{2(1+2b)}}\log n\right)$$

uniformly in $\|\mathbf{t}\| \leq C_n$. We obtain an analogous bound for $I_3(\mathbf{t})$. $\qquad \square$

By means of Theorem 3.2, we shall prove the following approximation of the dispersion function (17):

Lemma 3.2 *Under the conditions of Theorem 3.2,*

$$sup_{\|t\|\le C_n} |D_n(\delta + n^{-1/2}t) - D_n(\delta) + t'S_n(\delta) - \frac{1}{2}\gamma t'Q_n t| \xrightarrow{p} 0 \qquad (50)$$

as $n \to \infty$ *where*

$$S_n(\delta) = n^{-1/2}(Z_n - \hat{Z}_n)\hat{b}_n(\delta). \qquad (51)$$

Proof of Lemma 3.2. We may put $\beta = 0$ without loss of generality. The function $D_n(t)$, being continuous, convex and piecewise linear, is differentiable in t *a.e.* and

$$\frac{\partial}{\partial t}D_n(\delta + n^{-1/2}t)\big]_{t_0} = -S_n(\delta + n^{-1/2}t_0) \qquad (52)$$

at any point t_0 of differentiability. By Theorem 3.2,

$$\sup_{\|t\|\le C_n} \|S_n(\delta + n^{-1/2}t) - S_n(0) + \gamma Qt\| \xrightarrow{p} 0 \qquad (53)$$

and this implies the Lemma. $\qquad \square$

Completing the proof of Theorem 2.1. Proceeding as in the proof of Theorem 1 in Jurečková (1992a) and using (30) and the convexity of D_n, we obtain

$$n^{1/2}(\hat{\delta}_n - \delta) = U_n + o_p(1) \qquad (54)$$

where

$$U_n = \arg\min_{t\epsilon R^q}\left\{-t'S_n(\delta) + \frac{1}{2}\gamma t'Q_n t\right\} = \gamma^{-1}Q_n^{-1}S_n(\delta). \qquad (55)$$

Hence, by Theorem 4.2 of Jurečková (1994),

$$\begin{aligned}U_n &= n^{-1/2}\gamma^{-1}Q_n^{-1}(Z_n - \hat{Z}_n')\varphi(F(E_n)) + o_p(1)\\ \varphi(F(E_n)) &= (\varphi(F(E_1)), ..., \varphi(F(E_n)))'\end{aligned} \qquad (56)$$

which gives the asymptotic distribution of $n^{1/2}(\hat{\delta}_n - \delta)$ as well as its asymptotic representation. The proof of Theorem 2.1 in complete. $\qquad \square$

Remark. In the case of known β, the RR-estimator coincides with Jaeckel's R-estimator (Jaeckel (1972)) and this is asymptotically equivalent to R-estimator constructed by Jurečková (1971) and Koul (1971). The R-estimator of the parameters of a regression line was proposed by Adichie (1967). The RR-estimator is convenient when β is unknown and a nuisance. The whole family of RR-estimators (including the R-estimators) provides us with great flexibility: Jurečko- vá and Sen (1993) proved that estimating δ componentwise by means of RR-estimators is asymptotically equivalent to estimating δ as a whole. This has a computational as well as various other aspects.

Acknowledgement

The author thanks the referee for valuable comments to the text and to the language.
The research was supported by Czech Republic Grant GAČR 2168.

202

References

- Adichie,J.N. (1967). Estimate of regression parameters based on rank tests. *Ann. Math. Statist.* 38 894-904.

- Antoch,J. and Jurečková,J. (1985). Trimmed LSE resistant to the leverage points. *Comp. Statist. Quaterly* 4 329-339.

- Bassett,G. and Koenker,R. (1978). Asymptotic theory of least absolute error regression. *J.Amer. Statist. Assoc.* 73 618-622.

- Gutenbrunner,C. (1986). Zur Asymptotik von Regressions Quantil Prozessen und abgeleiten Statistiken. *PhD Thesis, Universität Freiburg.*

- Gutenbrunner,C. and Jurečková,J. (1993). Regression rank scores and regression quantiles. *Ann. Statist.* 20 305-330.

- Gutenbrunner,C., Jurečková,J., Koenker,R. and Portnoy,S. (1993). Tests of linear hypotheses based on regression rank scores. *Nonpar. Statist.* 2 307-331.

- Jaeckel,L.A. (1972). Estimating regression coefficients by minimizing the dispersion of residuals. *Ann. Math. Statist.* 43 1449-1458.

- de Jongh,P.J., de Wet,T. and Welsh.A.H. (1988). Mallows type bounded influence regression trimmed means *J.Amer. Statist. Assoc.* 83 805-810.

- Jurečková,J. (1969). Asymptotic linearity of a rank statistic in regression parameter. *Ann. Math. Statist.* 40 1889-1900.

- Jurečková (1971). Nonparametric estimate of regression coefficients. *Ann. Math. Statist.* 42 1328-1338.

- Jurečková,J. (1983). Robust estimators of location and regression parameters and their second order asymptotic relations. *Trans. 9th Prague Cont. on Inform. Theory, Statist. Decis. Functions and Random Processes* (J.Á.Víšek, ed.), pp.19-21. *Academia, Prague & Reidel, Dordrecht.*

- Jurečková,J. (1984). Regression quantiles and trimmed least squares estimator under a general design. *Kybernetika* 20 345-357.

- Jurečková,J. (1991). Tests of Kolmogorov-Smirnov type based on regression rank scores. *Trans. 11th Prague Conf. on Inform.Theory, Statist. Decis. Functions and Random Processes.* (J.Á.Víšek, ed), pp. 41-49. *Academia, Prague.*

- Jurečková,J. (1992a). Estimation in a linear model based on regression rank scores. *Nonpar. Statist.* 1 197-203.

- Jurečková (1992b). Uniform asymptotic linearity of regression rank scores process. *Nonparametric Statistics and Related Topics* (A.K.Md.E.Saleh, ed.), pp.217-229. *Elsevier Science Publishers.*

- Jurečková,J. (1994). Regression rank scores tests applied to heavy-tailed distributions. Submitted.

- Jurečková,J. and Sen.P.K. (1993). Asymptotic equivalence of regression rank scores estimators and R-estimators in linear model. *Statistics and Probability: A R.R.Bahadur Festschrift* (J.K.Ghosh, S.K.Mitra, K.R.Partha- sarathy and B.L.S.Prakasa Rao, eds.), pp.279-292. *Wiley Eastern, New Delhi.*

- Koenker,R. and Bassett,G. (1978). Regression quantiles. *Econometrica* 46 33-50.

- Koenker,R. and Portnoy,S. (1987). L-estimation for linear models. *J.Amer. Statist. Assoc.* 82 851-857.

- Koenker,R. and d'Orey,V. (1987). Computing regression quantiles. *Applied Statistics* 36 383-393.

- Koenker,R. and d'Orey,V. (1994). A remark on algorithm AS 229: Computing dual regression quantiles and regression rank scores. *Applied Statistics* 43 410-414.

- Koul,H.L. (1971). Asymptotic behavior of a class of confidence regions based on ranks in regression. *Ann. Math. Statist.* 42 466-476.

- Osborne,M.R. (1992). An effective method for computing regression quantiles. *IMA Journal of Numerical Analysis* 12 151-166.

- Pollard,D. (1991). Asymptotics for least absolute deviation regression estimation. *Econometric Theory* 7 186-200.

- Ruppert,D. and Carroll,R.J. (1980). Trimmed least squares estimation in the linear model. *J.Amer. Statist. Assoc.* 75 828-838.

The Asymptotic Distribution of Regression Parameters

ANNE-MAI PARRING

Institute of Mathematical Statistics, University of Tartu, J.Liivi 2, EE2400 Estonia.

1 The Problem

The regression analysis is undoubtly one of the most popular methods of mathematical statistics. The classical model for the linear regression assumes the dependent variable X_0 to be a linear function with the same coefficients $\alpha_0, \alpha_1, ..., \alpha_p$ for each set of arguments

$$X_0 = \alpha_0 + \alpha_1 x_{i1} + ... + \alpha_p x_{ip} + \epsilon \qquad (1).$$

If we fix the matrix of arguments $\aleph$ and measure the values of dependent variable X_0, we will get the vector of measured values $\mathbf{x}_0$ and can calculate the least squares estimates (LSE) of coefficients

$$\hat{\alpha}_* = (\aleph'\aleph)^{-1}\aleph'\mathbf{x}_0 \qquad (2)$$

and the variance matrix of these estimates

$$D\hat{\alpha}_* = s^2(\aleph'\aleph)^{-1}. \qquad (3)$$

These results are connected with the fixed matrix $\aleph$. The variance matrix (3) gives us the variability of coefficients' estimates for that fixed matrix $\aleph$.

In the most of the real problems we observe the random vector

$$\mathbf{X}_* = (X_0, X_1, X_2, ..., X_p)' = [\, X_0 \vdots \mathbf{X}' \,]'$$

and have as a sample its values

$$\mathbf{x}_{*i} = (x_{i0}, x_{i1}, ... x_{ip})',$$

$i = 1, 2, ..., n$. We suppose that X_0 is the dependent variable and $X_1, X_2, ..., X_p$ are the arguments – the independent variables. In such a situation it is not possible to choose the set of arguments, fix them and then measure the value of dependent variable. If we once get a matrix of arguments, we can never repeat the measurements with the same values

---

*Kitsos, C.P., and Müller, W.G., Eds., *Proceedings of MODA4*, Physica Verlag, Heidelberg, 1995

of arguments. But we can yet define the linear regression function and estimate it. That problem is called the statistical regression analysis, see C.R.Rao (1965), p.220.

Is the variance matrix (3) in the new situation still valid? Can we find a better estimation for variability of estimates linear regression coefficients if we really have a problem of statistical regression analysis? How to define the linear regression function for given random vector $\mathbf{X}_*$?

We shall answer these questions in the following paper.

2 The Linear Regression Function

Let us observe the random vector $\mathbf{X}_*$. We are looking for the best linear function of arguments

$$l(\alpha_*, \mathbf{X}) = \alpha_0 + \alpha_1 X_1 + ... + \alpha_p X_p$$

to describe the dependent variable X_0. As usual, for determining that linear function it is suitable to use the least-square (LS) condition

$$E(X_0 - l(\alpha_*, \mathbf{X}))^2 = \min_{a_*} E(X_0 - l(\mathbf{a}_*, \mathbf{X}))^2. \qquad (4)$$

The linear function $l(\alpha_*, \mathbf{X})$ fulfilling that condition is called linear regression function (LRF).

It is well known that LRF exists if vector $\mathbf{X}_*$ has all the second-order moments. Let us denote $E\mathbf{X}_* = \mu_*$,

$$\mu_* = [\, \mu_0 \;\vdots\; \mu' \,]'$$

and $D X_* = \Sigma_*$,

$$\Sigma_* = \begin{bmatrix} \sigma_{00} & \vdots & \sigma_0' \\ \cdots\cdots\cdots \\ \sigma_0 & \vdots & \Sigma \end{bmatrix}.$$

From LS condition (4) follows that the coefficients of LRF have to satisfy the system

$$\begin{cases} \alpha_0 + \alpha'\mu &= \mu_0 \\ \Sigma\alpha &= \sigma_0 \end{cases}$$

and hence, if Σ has full rank,

$$\begin{cases} \alpha_0 &= \mu_0 - \alpha'\mu \\ \alpha &= \Sigma^{-1}\sigma_0 \end{cases} \qquad (5)$$

The coefficient of determination ρ^2 is given by formula

$$\rho^2 = \frac{\sigma_0'\Sigma^{-1}\sigma_0}{\sigma_{00}}. \qquad (6)$$

In real problems the mean vector μ_* and variance matrix Σ_* are unknown. We have to use their sample estimations

$$\bar{x}_* = \frac{1}{n}\sum_{i=1}^n \mathbf{x}_{*i}$$

and

$$S_* = \frac{1}{n}\sum_{i=1}^n \mathbf{x}_{*i}\mathbf{x}_{*i}' - \bar{\mathbf{x}}_*\bar{\mathbf{x}}_*',$$

which give the consistent estimations of LRF coefficients

$$
\begin{cases}
a_0 & = \quad \bar{x}_0 - \mathbf{a}'\mathbf{x} \\
\mathbf{a} & = \quad \mathbf{S}^{-1}\mathbf{s}_0.
\end{cases}
$$

The estimations are the same as in fixed-argument regression analysis.

3 The Distribution of LS Estimates

It is self-evident that the distribution of LS estimates must depend on the distribution of observed vector $\mathbf{X}_*$. It is also understandable that the exact distribution of them is very complicated and probably it is not possible to get here general results. But there is more hope to get some asymptotic results which may work well in case of large samples.

One quite rough asymptotic approximation is available by approximate linearization (see, for example, Barndorff-Nielsen, Cox (1989), p.33-34).

Let $\mathbf{V_n}$, a random vector, converging in probability to a constant vector (with same dimension) $\mathbf{c}$,

$$
\mathbf{V_n} \xrightarrow{p} \mathbf{c}
$$

and

$$
\sqrt{n}(\mathbf{V_n} - \mathbf{c}) \xrightarrow{l} P,
$$

where P is a distribution and $\xrightarrow{l}$ marks converging in distribution.

Denoting by $\mathbf{V}_*$ a random vector with distribution P, $\mathbf{V}_* \sim P$, we have

$$
\sqrt{n}(\mathbf{V_n} - \mathbf{c}) \sim \mathbf{V}_* + o(1),
$$

where $o(1) \xrightarrow{p} 0$. Let $\mathbf{h}(\mathbf{V})$ be a twice differentiable vector-function with nonvanishing first derivative at $\mathbf{c}$, $\frac{\partial \mathbf{h}}{\partial \mathbf{V}}(\mathbf{c}) \neq 0$, and with bounded second derivative. Consider the sequence $\{\mathbf{h}(\mathbf{V}_n)\}$ and denote

$$
\mathbf{H_n^*} = \sqrt{n}(\mathbf{h}(\mathbf{V}_n) - \mathbf{h}(\mathbf{c})).
$$

From Taylor's formula follows that

$$
\mathbf{H_n^*} = \frac{\partial \mathbf{h}}{\partial \mathbf{V}}(\mathbf{c})\mathbf{V}_* + o(1).
$$

Denoting with Ξ the variance matrix of vector $\mathbf{V}_*$ we get

$$
D\mathbf{H_n^*} \to \xi = \frac{\partial \mathbf{h}}{\partial \mathbf{V}}(\mathbf{c})\Xi\frac{\partial \mathbf{h}}{\partial \mathbf{V}}(\mathbf{c})'.
$$

This matrix ξ is named asymptotic variance matrix of $\sqrt{n}\mathbf{h}(\mathbf{V_n})$.

As we are looking for asymptotic variance matrix of LRF coefficients estimates, it is obvious that as $\mathbf{V}_n$ we have to use the vector

$$
[\ \bar{\mathbf{X}}'_* \ \vdots \ vec_*\mathbf{L}'_2\ ],
$$

where $_*\mathbf{L}_2$ is the matrix of sample second-order moments

$$
_*\mathbf{L}_2 = \frac{1}{n}\sum_{i=1}^{n}\mathbf{x}_{*i}\mathbf{x}'_{*i}
$$

and the *vec* operator tansforms a matrix into a vector by stacking the columns of the matrix one underneath the other, see Magnus, Neudecker (1988) p. 30.

Let us denote the matrix of k-oder moments of vector $\mathbf{X}_*$ as $_*\mathbf{M}_k$. Hence

$$[\,\bar{\mathbf{X}}'_* \;\vdots\; vec_*\mathbf{L}'_2\,] \xrightarrow{p} [\,\mu'_* \;\vdots\; vec_*\mathbf{M}'_2\,]$$

it is obvious that as $\mathbf{c}$ we have to use vector $[\,\mu'_* \;\vdots\; vec_*\mathbf{M}'_2\,]$. In such a choise

$$\mathbf{V}_* \sim N(\mathbf{0}, \Xi),$$

where matrix Ξ [see Parring (1979a)]

$$\Xi = \begin{bmatrix} _*\mathbf{M}_2 - \mu_*\mu'_* & \vdots & _*\mathbf{M}'_3 - \mu(vec_*\mathbf{M}_2)' \\ \cdots\cdots\cdots\cdots\cdots\cdots\cdots\cdots\cdots\cdots\cdots\cdots\cdots\cdots\cdots\cdots \\ _*\mathbf{M}_3 - vec_*\mathbf{M}_2\mu' & \vdots & _*\mathbf{M}_4 - vec_*\mathbf{M}_2(vec_*\mathbf{M}_2)' \end{bmatrix},$$

has blocks which are functions of moments up to 4th order of observed vector $\mathbf{X}_*$:

$$\begin{aligned} _*\mathbf{M}_2 &= E(\mathbf{X}_*\mathbf{X}_*'), \\ _*\mathbf{M}_3 &= E(\mathbf{X}_*\mathbf{X}_*'\mathbf{X}_*), \\ _*\mathbf{M}_4 &= E(\mathbf{X}_*\mathbf{X}_*'\mathbf{X}_*\mathbf{X}_*'). \end{aligned}$$

As the vector-function $\mathbf{h}$ we have to use the function defined by system (5)

$$\alpha_* = [\,\mu_0 - \sigma'_0\Sigma^{-1}\mu \;\vdots\; \sigma'_0\Sigma^{-1}\,]'.$$

Using equation $\Sigma = _*\mathbf{M}_2 - \mu\mu'$ it is possible to calculate the derivative $\frac{\partial\alpha_*}{\partial\mathbf{v}}[\mu'_* \;\vdots\; vec_*\mathbf{M}'_2]'$ (see Parring 1979b) we get the asymptotic variance matrix of LRF estimates

$$\xi = \begin{bmatrix} \sigma_{00}(1-\rho^2) - 2\mu'\Sigma^{-1}\mathbf{f} + \mu'\Pi\mu & \vdots & \mathbf{f}'\Sigma^{-1} - \mu'\Pi \\ \cdots\cdots\cdots\cdots\cdots\cdots\cdots\cdots\cdots\cdots\cdots\cdots\cdots\cdots\cdots\cdots \\ \Sigma^{-1}\mathbf{f} - \Pi\mu & \vdots & \Pi \end{bmatrix}, \tag{7}$$

where

$$\begin{aligned} \Pi &= \Sigma^{-1}C\Sigma^{-1}, \\ C &= (c_{ij}), \qquad i,j = 1,2,\ldots,p, \\ c_{ij} &= \gamma'(_*\bar{\mathbf{M}}_4)_{ij}\gamma, \\ \gamma &= [\,1 \;\vdots\; -\alpha'\,]', \\ (_*\bar{\mathbf{M}}_4)_{ij} &= E(X_i - \mu_i)(X_j - \mu_j)(\mathbf{X}_* - \mu_*)(\mathbf{X}_* - \mu_*)', \\ f &= (f_1, f_2, \ldots, f_p), \\ f_i &= \gamma'(_*\bar{\mathbf{M}}_3)_i\gamma, \\ (_*\bar{\mathbf{M}}_3)_i &= E(X_i - \mu_i)(\mathbf{X}_* - \mu_*)(\mathbf{X}_* - \mu_*)'. \end{aligned}$$

Remembering formulae (5) and (6) it is possible to calculate the matrix ξ if we know all the central moments up to 4th order. In the most real problems we do not know them, but then we may calculate the sample estimations of these moments.

4   The Example

Let us examine a simple example which shows that the asymptotic variance matrices of LRF coefficients for different distributions may be quite different.

Let us at first assume that observed random vector $\mathbf{X}_*$ has multivariate normal distribution, $\mathbf{X}_* \sim N(\mu, \Sigma)$. Then

$$_*\bar{\mathbf{M}}_3 = 0$$

and

$$_*\bar{\mathbf{M}}_4 = (\mathbf{I}_{(p+1)^2} + \mathbf{I}_{p+1,p+1})\Sigma \otimes \Sigma + vec\Sigma(vec\Sigma)',$$

see for example Kollo (1989). Here $\mathbf{I}_{(p+1)^2}$ is $(p+1)^2 \times (p+1)^2$-dimensional unity matrix and $\mathbf{I}_{p+1,p+1}$ is $(p+1)^2 \times (p+1)^2$ matrix of permutations. We get the estimations of LRF coefficients by formula (5), the coefficient of determination by formula (6) and the asymptotic variance matrix by formula (7)

$$\xi = \sigma_{00}(1 - \rho^2) \begin{bmatrix} 1 + \mu'\Sigma^{-1}\mu & \vdots & -\mu'\Sigma^{-1} \\ \cdots\cdots\cdots\cdots\cdots\cdots\cdots \\ -\Sigma^{-1}\mu & \vdots & \Sigma^{-1} \end{bmatrix}. \tag{8}$$

Let us have another random vector $\mathbf{Y}_*$, the distribution of which is the mixture of multivariate normal distributions

$$(1 - \beta)N(\mu_*, \Gamma) + \beta N(\mu_*, \Gamma + \sigma^2 \mathbf{I}).$$

We can choose parameters of that mixture Γ, β and σ so that the variance matrix of that mixture

$$D\mathbf{Y}_* = \Gamma + \beta\sigma^2 \mathbf{I}$$

is equal to the variance matrix of $\mathbf{X}_*$, $D\mathbf{Y}_* = D\mathbf{X}_* = \Sigma_*$. In that case the LRF coefficients for $\mathbf{Y}_*$ are the same as for $\mathbf{X}_*$. It is easy to see that for mixture

$$_*\bar{\mathbf{M}}_3 = 0$$

and

$$_*\bar{\mathbf{M}}_4 = (\mathbf{I}_{p^2} + \mathbf{I}_{p,p})\Sigma_* \otimes \Sigma_* + vec\Sigma_*(vec\Sigma_*)' + \sigma^4\beta(1-\beta)(\mathbf{I}_{p+1^2} + \mathbf{I}_{p+1,p+1} + vec\mathbf{I}_{p+1}(vec\mathbf{I}_{p+1})').$$

So we get the asymptotic variance matrix

$$\zeta = \sigma_{00}(1-\rho^2) \begin{bmatrix} 1 + \mu'\Sigma^{-1}\mu & \vdots & -\mu'\Sigma^{-1} \\ \cdots\cdots\cdots\cdots\cdots\cdots\cdots \\ -\Sigma^{-1}\mu & \vdots & \Sigma^{-1} \end{bmatrix} + \sigma^4(1-\beta)\beta \begin{bmatrix} \mu'\Sigma^{-1}\mathbf{U}\Sigma^{-1}\mu & \vdots & -\mu'\Sigma^{-1}\mathbf{U}\Sigma^{-1} \\ \cdots\cdots\cdots\cdots\cdots\cdots\cdots\cdots \\ -\Sigma^{-1}\mathbf{U}\Sigma^{-1}\mu & \vdots & \Sigma^{-1}\mathbf{U}\Sigma^{-1} \end{bmatrix}, \tag{9}$$

where

$$\mathbf{U} = \gamma'\gamma\mathbf{I}_p + 2\alpha\alpha'.$$

5   The Simulation Experiment

The difference of asymptotic variance matrices (8) and (9) is obvious. But there arise some important practical questions:

- how large must be a sample to make the asymptotic variance matrix useful?

- how large must be a sample to make the sample estimation of asymptotic variance matrix useful?

We try to answer these questions in a special case by simulation.

Let us fix p=1, so the observed vector is two-dimensional. Following the preceding example we suppose that $\mathbf{X}_*$ has normal distribution with mean

$$\mu_* = (10 \quad 10)'$$

and variance matrix

$$\Sigma_* = \begin{pmatrix} 8 & 5 \\ 5 & 5 \end{pmatrix}.$$

From formula (5) we get the vector of coefficients of LRF

$$\alpha_* = (0 \quad 1)',$$

from formula (6) the coefficient of determination

$$\rho_2 = \frac{5}{8} = 0.625$$

and from formula (8) the asymptotic variance matrix of sample estimates of LRF coefficients

$$\xi_X = \begin{pmatrix} 63 & -6 \\ -6 & 0.6 \end{pmatrix}.$$

We suppose also that the distribution $\mathbf{Y}_*$ is the mixture of normal distributions. The mixture of normal distributions may occur quite often in real problems – for example when we examine a population of birds or fishes and the variation of measured variable depend on the age of individual what is quite difficult to find out.

Let us choose $\sigma^2 = 4.75$, $\beta = \frac{1}{4.75} \approx 0.21$ and

$$\Gamma = \begin{pmatrix} 7 & 5 \\ 5 & 4 \end{pmatrix}.$$

Then for vector $\mathbf{Y}_*$ the variance matrix is

$$D\mathbf{Y}_* = \begin{pmatrix} 8 & 5 \\ 5 & 5 \end{pmatrix},$$

coefficients of LRF are

$$\alpha_* = (0 \quad 1)',$$

coefficient of determination is

$$\rho = \frac{5}{8} = 0.625$$

and the asymptotic variance matrix of sample estimates of LRF coefficients is

$$\xi_Y = \begin{pmatrix} 63 & -6 \\ -6 & 0.6 \end{pmatrix} + \begin{pmatrix} 60 & -6 \\ -6 & 0.6 \end{pmatrix} = \begin{pmatrix} 123 & -12 \\ -12 & 1.2 \end{pmatrix}.$$

To attain in our special case answer to the first question was generated 2000 samples with given sample size and calculated the LRF coefficients for each sample. So was built up a sample of LRF coefficients. The variance matrix for that sample is given in Table 1, it refered as "sample estimate of variance matrix".

The theoretical estimation of the variance matrix is $\frac{1}{n}\xi$. It refered in Table 1 as "true asymptotic variance matrix". The results are given in the following Table 1.

Table 1

True asymptotic variance matrix and its sample estimation

Number of cases	:	Normal distribution		:	Mixture of normal distributions	
	:	Sample estimate of variance matrix	True asymptotical variance matrix	:	Sample estimate of variance matrix	True asymptotical variance matrix
20	:	$\begin{pmatrix} 3.6612 & -.3495 \\ -.3495 & .0348 \end{pmatrix}$	$\begin{pmatrix} 3.1500 & -.3000 \\ -.3000 & .0300 \end{pmatrix}$	:	$\begin{pmatrix} 6.0188 & -.5893 \\ -.5893 & .0592 \end{pmatrix}$	$\begin{pmatrix} 6.1500 & -.6000 \\ -.6000 & .0600 \end{pmatrix}$
50	:	$\begin{pmatrix} 1.3714 & -.1305 \\ -.1305 & .0130 \end{pmatrix}$	$\begin{pmatrix} 1.2600 & -.1200 \\ -.1200 & .0120 \end{pmatrix}$	:	$\begin{pmatrix} 2.3635 & -.2332 \\ -.2332 & .0236 \end{pmatrix}$	$\begin{pmatrix} 2.4600 & -.2400 \\ -.2400 & .0240 \end{pmatrix}$
100	:	$\begin{pmatrix} 0.6405 & -.0609 \\ -.0609 & .0061 \end{pmatrix}$	$\begin{pmatrix} 0.6300 & -.0600 \\ -.0600 & .0060 \end{pmatrix}$	:	$\begin{pmatrix} 1.2422 & -.1209 \\ -.1209 & .0120 \end{pmatrix}$	$\begin{pmatrix} 1.2300 & -.1200 \\ -.1200 & .0120 \end{pmatrix}$
200	:	$\begin{pmatrix} 0.3061 & -.0291 \\ -.0291 & .0029 \end{pmatrix}$	$\begin{pmatrix} 0.3150 & -.0300 \\ -.0300 & .0030 \end{pmatrix}$	:	$\begin{pmatrix} 0.5857 & -.0570 \\ -.0570 & .0057 \end{pmatrix}$	$\begin{pmatrix} 0.6150 & -.0600 \\ -.0600 & .0060 \end{pmatrix}$

It is easy to see that on our special case the true asymptotic variance matix is quite good approximation if the sample size from near by 100.

To attain to on our special case answer to the second question only the mixture of normal distributions was considered. There were generated 1000 samples with given sample size and tested for each sample hypothesis $H_0 : \alpha_1 = 1$ (which is true for given parameters) using traditional t-statistic $t = \frac{a_1}{s_{a_1}}$. The standard deviation s_{a_1} was calculated in two different ways: following formula (8) and following formula (9). The significance level $\alpha = 0.05$ was used and the relative frequency of rejecting the null-hypothesis was registred. The results are given in Table 2.

212

Table 2
The relative frequency of Type I error

Number of cases	Significance level	Relative frequency of rejecting H_0 estimation(8)	Relative frequency of rejecting H_0 estimation (9)
20	0.05	0.181	0.189
50	0.05	0.167	0.124
100	0.05	0.170	0.099
200	0.05	0.189	0.083
500	0.05	0.161	0.060

It is easy to understand that the classical estimation of variance matrix (8) underestimates on our case the variance of regression coefficient. Hence the real significance level is quite different of choosed significant level. The difference do not change with increasing the sample size.

Results for sample estimate of asymptotic variance (9) are different. For small samples the real siginificants level is also quite different from choosed one. But with increasing the sample size the difference will decrease and for sample size 500 the difference is not significant.

The programs for simulation were written in IML language of SAS.

6 Conclusion

The distribution of LRF coefficients depends from distribution of observed random vector. It is possible to calculated the asymptotic variance matrix for LRF coefficients which depends only from central moments of observed vector up to 4th order. For different distributions the asymptotic variance matrices may be significantly different (see also Parring (1992)). For big samples using the sample estimate of asymptotic variance may be useful.

References

[1] Barndorff-Nielsen O.E., Cox D.R. (1989) Asymptotic Techniques for Use in Statistics. Chapman and Hall

[2] Kollo T. (1991) Matrix derivative in multivariate statistics. Tartu (in Russian)

[3] Magnus J.R., Neudecker H. (1988) Matrix differential calculus with applications in statistics and econometrics, John Wiley& Sons

[4] Parring A.-M. (1979a) The asymptotic characteristics of some sample functions. Acta et Commentationes Universitatis Tartuensis , 492, 86-90 (in Russian)

[5] Parring A.-M. (1979b) The estimating of linear regression and its asymptotic behavior. Acta et Commentationes Universitatis Tartuensis , 492, 91-99 (in Russian)

[6] Parring A.-M. (1992) The asymptotic variance of regression coefficients. Acta et Commentationes Universitatis Tartuensis , 942, 76-85

[7] Rao C.R. (1965) Linear Statistical Inference and Its Applications, John Wiley& Sons

Some Simulation Results on Cross-Validation and Competitors for Model Choice

BERND DROGE

Institute of Mathematics, Humboldt University, Unter den Linden 6, 10099 Berlin, Germany.

Abstract:

The behaviour of model selection procedures based on different criteria such as cross-validation is investigated in a simulation study. Emphasis is on the relationship to the problem of estimating the prediction quality of a model.

1 Introduction

We consider a regression situation where the information available on the problem under consideration does not favour a specific model, so that we have to choose a good one from those being tentatively proposed. For this selection process, many data-driven procedures have been discussed in the literature (see e.g. the surveys by Hocking, 1976, and Thompson, 1978), and those based on the cross-validation criterion of Stone (1974) and the generalized cross-validation criterion of Craven and Wahba (1979) belong to the most popular of them. The use of both criteria may be motivated by being estimates of the mean squared error of prediction (MSEP), which describes the prediction performance of a model. Other model selection criteria may also be interpreted as estimates of the MSEP, and therefore a comparison of their properties as estimates of the MSEP may provide some justification for the choice of a specific one.

One advantage of cross-validation and generalized cross-validation over some other selection criteria such as Mallows' (1973) C_p is that they do not require estimation of the error variance. However, both criteria have also some disadvantages compared with other MSEP estimates such as C_p or some variant of the bootstrap, see e.g. Bunke and Droge (1984). Additionally, cross-validation may fail in some nonlinear regression situations, as it has been illustrated in Bunke et al. (1993). To overcome this failure, Droge (1994) has introduced a new variant of cross-validation, called full cross-validation, which outperforms the traditional cross-validation as estimate of the MSEP.

---

*Kitsos, C.P., and Müller, W.G., Eds., *Proceedings of MODA4*, Physica Verlag, Heidelberg, 1995

The aim of this paper is to investigate the properties of model selection procedures based on the different cross-validation approaches. We are particularly interested in learning whether the hope is justified, that a better estimate of the MSEP provides a more appropriate model selection criterion in the sense of leading to a model with smaller MSEP. Since it appears hard to establish analytically finite-sample properties of model selection procedures (for some results see e.g. Droge, 1993, and Droge and Georg, 1995), we resort on simulation. For the sake of completeness we include some other model selection criteria like Akaike's (1970) final prediction error (FPE) in our investigations.

2 Model Selection Criteria

We assume to have observations of a response variable Y at nonrandom design points $x_1, \ldots, x_n$, which follow the model

$$y_i = f(x_i) + \varepsilon_i, \quad i = 1, \ldots, n. \tag{1}$$

Here the ε_i are zero mean, uncorrelated random errors having some common variance σ^2, and f is an unknown real-valued regression function that we wish to estimate.

In this setting there exists a variety of parametric and nonparametric approaches for estimating the regression function f. In this paper we focus on the problem of selecting an appropriate linear model. That is, we assume that there are p known functions of the explanatory variables, say $g_1, \ldots, g_p$, associated with the response variable, and the aim is to approximate the regression function by an appropriate linear combination of some of these functions. Each such linear combination is characterized by the subset of indices of the included functions, say $\lambda \subset \{1, \ldots, p\}$. Possibly not all linear combinations are allowed, so that the class of competing models is characterized by a subset Λ of the power set of $\{1, \ldots, p\}$. Using the least squares approach for fitting the models to the data gives, for each $\lambda \in \Lambda$, the following estimator of $f(x)$

$$\hat{f}_\lambda(x) = \sum_{i \in \lambda} \hat{\beta}_i g_i(x),$$

where the coefficients $\hat{\beta}_i$ are the minimizers of

$$\sum_{j=1}^{n} (y_j - \sum_{i \in \lambda} \beta_i g_i(x_j))^2$$

with respect to β_i ($i \in \lambda$). Future values of the response variable at the design point x_i will usually be predicted by $\hat{y}_i(\lambda) = \hat{f}_\lambda(x_i)$ ($i = 1, \ldots, n$). Then, with $\hat{y}(\lambda) = (\hat{y}_1(\lambda), \ldots, \hat{y}_n(\lambda))^T$, we may write

$$\hat{y}(\lambda) = H(\lambda)y,$$

where the "hat matrix" $H(\lambda) := ((h_{ij}(\lambda)))_{i,j=1,\ldots,n}$ is just the projection onto the column space of the $n \times |\lambda|$-matrix $G(\lambda) = ((g_j(x_i)))_{i=1,\ldots,n}^{j \in \lambda}$ ($|\lambda|$ denotes the number of elements in λ).

The prediction performance of the model indexed by λ will be described by a mean squared error (MSEP),

$$MSEP(\lambda) = \sigma^2 + MSE(\lambda), \tag{2}$$

where

$$MSE(\lambda) = \tfrac{1}{n} \sum_{i=1}^{n} \mathrm{E}(f(x_i) - \hat{y}_i(\lambda))^2$$

is the mean squared error for estimating the regression function. Therefore, the prediction problem is closely related to the problem of estimating f. Ideally one would select a model by choosing that value of λ, which minimizes the MSEP defined in (2). Unfortunately, the MSEP is unkown. Therefore, in practice, one has to use an estimate of the MSEP, which is then minimized with respect to λ.

One of the most popular MSEP estimates in applications is the *cross-validation* (CV) criterion of Stone (1974) defined by

$$CV(\lambda) = \tfrac{1}{n} \sum_{i=1}^{n} (y_i - \hat{y}_{-i}(\lambda))^2,$$

where $\hat{y}_{-i}(\lambda)$ is the prediction at x_i leaving out the i-th data point. It is well-known that CV may be expressed in terms of the ordinary residuals $y_i - \hat{y}_i(\lambda)$, provided that $h_{ii}(\lambda) < 1$ for $i = 1, \ldots, n$:

$$CV(\lambda) = \tfrac{1}{n} \|y - \hat{y}(\lambda)\|_{C(\lambda)}^2, \tag{3}$$

where $\|z\|_C = z^T C z$ for $z \in I\!\!R^n$ and some $n \times n$-matrix C, and

$$C(\lambda) = \mathrm{diag}[(1 - h_{11}(\lambda))^{-2}, \ldots, (1 - h_{nn}(\lambda))^{-2}].$$

Cross-validation works well in many applications. However, to avoid the introductory mentioned difficulties with CV in nonlinear regression, we proposed the so-called *full cross-validation* (FCV) criterion (see Bunke et al., 1993, and Droge, 1994):

$$FCV(\lambda) = \tfrac{1}{n} \sum_{i=1}^{n} (y_i - \bar{y}_i(\lambda))^2, \tag{4}$$

where $\bar{y}_i(\lambda)$ is the least squares prediction at x_i with substituting y_i by $\hat{y}_i(\lambda)$ instead of deleting it. As shown by Droge (1994), (4) may be rewritten as

$$FCV(\lambda) = \tfrac{1}{n} \|y - \hat{y}(\lambda)\|_{D(\lambda)}^2,$$

where

$$D(\lambda) = \mathrm{diag}[(1 + h_{11}(\lambda))^2, \ldots, (1 + h_{nn}(\lambda))^2].$$

Craven and Wahba (1979) have proposed *generalized cross-validation* (GCV) as another useful method for selecting the parameter λ of linear estimates $\hat{f}_\lambda$. Assuming that $\mathrm{tr}[H(\lambda)] < n$, the GCV criterion is defined by

$$GCV(\lambda) = \tfrac{1}{n} \|y - \hat{y}(\lambda)\|^2 / (1 - \tfrac{1}{n} \mathrm{tr}[H(\lambda)])^2.$$

Obviously, GCV weights the ordinary residuals $(y_i - \hat{y}_i)$ by the average of the weights used for the CV criterion (cf. (3)). Applying this idea to FCV, Droge (1994) has introduced the following *generalized full cross-validation* (GFCV) criterion

$$GFCV(\lambda) = \tfrac{1}{n} \|y - \hat{y}(\lambda)\|^2 (1 + \tfrac{1}{n} \mathrm{tr} H[(\lambda)])^2.$$

Note that $\mathrm{tr}[H(\lambda)] = |\lambda|$ for our model selection problem if $G(\lambda)$ has full rank.

Droge (1994) has compared the different cross-validation criteria as estimates of the MSEP, concluding that FCV and GFCV outperform their traditional counterparts. More precisely, for the problem of selecting linear regression models it has been shown that the absolute values of the biases of FCV and GFCV are smaller than those of CV and GCV,

216

respectively. Moreover, under the assumption of normally distributed errors in (1) we have Var($GFCV$) < Var(GCV), and FCV has a smaller variance than CV at least in a minimax sense.

In the next section we investigate by simulation the behaviour of the model selection procedures associated with the above introduced criteria. For the sake of completeness we include some alternative criteria which are functions of the residual sum of squares,

$$RSS(\lambda) = \tfrac{1}{n}\|y - \hat{y}(\lambda)\|^2,$$

and which are widely discussed in the literature:

$$FPE(\lambda) = \tfrac{n+|\lambda|}{n-|\lambda|}RSS(\lambda) \quad \text{(Final prediction error, see Akaike, 1970)}$$

$$AIC(\lambda) = log(RSS(\lambda)) + 2|\lambda|/n \quad \text{(Information criterion A of Akaike, 1974)}$$

$$BIC(\lambda) = log(RSS(\lambda)) + (log(n))|\lambda|/n \quad \text{(Asymptotic Bayesian modification of AIC, see Schwarz, 1978)}$$

$$C_\lambda = RSS(\lambda) + 2|\lambda|\hat{\sigma}^2/n \quad (C_p\text{-criterion of Mallows, 1973)}$$

$$SH(\lambda) = \tfrac{n+2|\lambda|}{n}RSS(\lambda) \quad \text{(Asymptotically optimal criterion of Shibata, 1981).}$$

We remark that the given expressions for AIC and BIC are specialized for the case of normally distributed observations.

3 Simulations

The behaviour of the model selection procedures based on the different criteria of the preceding section was investigated in a simulation study. We considered the problem of selecting an appropriate order of a polynomial model to estimate the regression function depending on one explanatory variable. Consequently, we had $g_i(x) = x^{i-1}$ for $i = 1,\ldots,p$, and, with $\lambda_k := \{1,\ldots,k\}$, $\Lambda = \{\lambda_k|k = 1,\ldots,p\}$, so that each model ($\lambda_k$) could be characterized by the order of the polynomial (k). To be precise, we orthonormalized the polynomials with respect to the data as it is provided by the function *poly* in Splus (cf. Chambers and Hastie, 1992), in which our simulations were performed. For practical reasons we chose $p = n/3$, so that we had, roughly speaking, at least three observations per parameter. From the definitions of the criteria in Section 2 it is obvious that a variety of them would not be applicable when the class of competing models contains all possible polynomials, since they would select then the largest model.

As true regression function we chose two alternatives with different shapes:

    Situation 1 (logistic curve): $f(x) = 1/\{1 + \exp(1 - 6x)\}$,
and
    Situation 2 (sinus curve): $f(x) = \sin(2\pi x)$.

The observations y_i were taken at $n = 30$ equispaced design points $x_i = (i - 1/2)/n$ with $\varepsilon_i \sim N(0,\sigma^2)$ ($i = 1,\ldots,n$) and with two different values of σ^2 (0.1 and 0.01).

Figure 1 illustrates the situation when $\sigma^2 = 0.01$. It shows for each of the considered cases, the observations for one of the data sets together with the true regression function. Additionally, the projection of the true regression function onto the model with minimal MSEP as well as the least squares fit of this model to the presented data are included.

The MSEP-values for the competing models, normalized by the variances, are given in Table 1. They show that in situation 1, for both values of the error variance, the

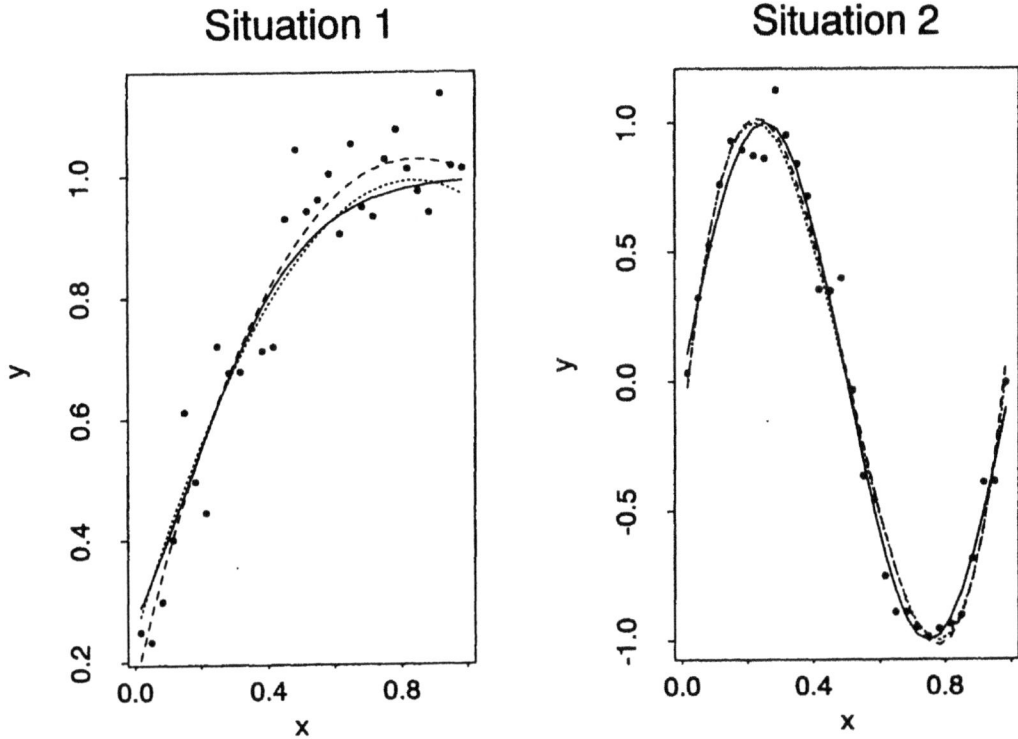

Figure 1: *For both situations: One simulated sample in case of $\sigma^2 = 0.01$, the true regression function (solid line), its projection onto the model with minimal MSEP (dotted line) and the least squares fit of this model to the data (dashed line).*

polynomial model of order 3 (i.e. of degree 2) was that with minimal MSEP. On the average, all considered criteria had also their minima for this model.

Table 1: *Values of MSEP, normalized by the error variance, for the considered situations and orders of polynomial.*

Situation	σ^2	Order of polynomial									
		1	2	3	4	5	6	7	8	9	10
1	0.1	1.53	1.13	1.10	1.13	1.17	1.20	1.23	1.27	1.30	1.33
1	0.01	5.99	1.70	1.11	1.14	1.17	1.20	1.23	1.27	1.30	1.33
2	0.1	6.03	3.01	3.05	1.18	1.21	1.20	1.23	1.27	1.30	1.33
2	0.01	51.03	20.5	20.6	1.55	1.58	1.20	1.23	1.27	1.30	1.33

In situation 2, the MSEP-optimal orders of polynomial were 4 and 6 for $\sigma^2 = 0.1$ and $\sigma^2 = 0.01$, respectively. Here, only the averaged values of CV, GCV, C_p, FPE and BIC attained their global minima at the optimal orders for both values of σ^2. When $\sigma^2 = 0.01$ this was also the case for GFCV and AIC. In the remaining cases, the averaged values of the criteria attained only local minima at the optimal order, except for FCV which was nearly constant for $6 \leq |\lambda| \leq 8$ and then monotonously decreasing when $\sigma^2 = 0.01$. The difficulties with some criteria in situation 2 seem to be connected with the rather slight increase of the MSEP after its optimum compared with the decrease before that point.

Figure 2 depicts the behaviour of some of the criteria and of the MSEP in the case $\sigma^2 = 0.1$. The values for AIC and BIC are not included since they may only be regarded as MSEP estimates after some transformation. FPE behaves similar to C_p, and SH is not far from FCV or GFCV. Note that C_p was always calculated with an error variance estimation using the polynomial model of order 10.

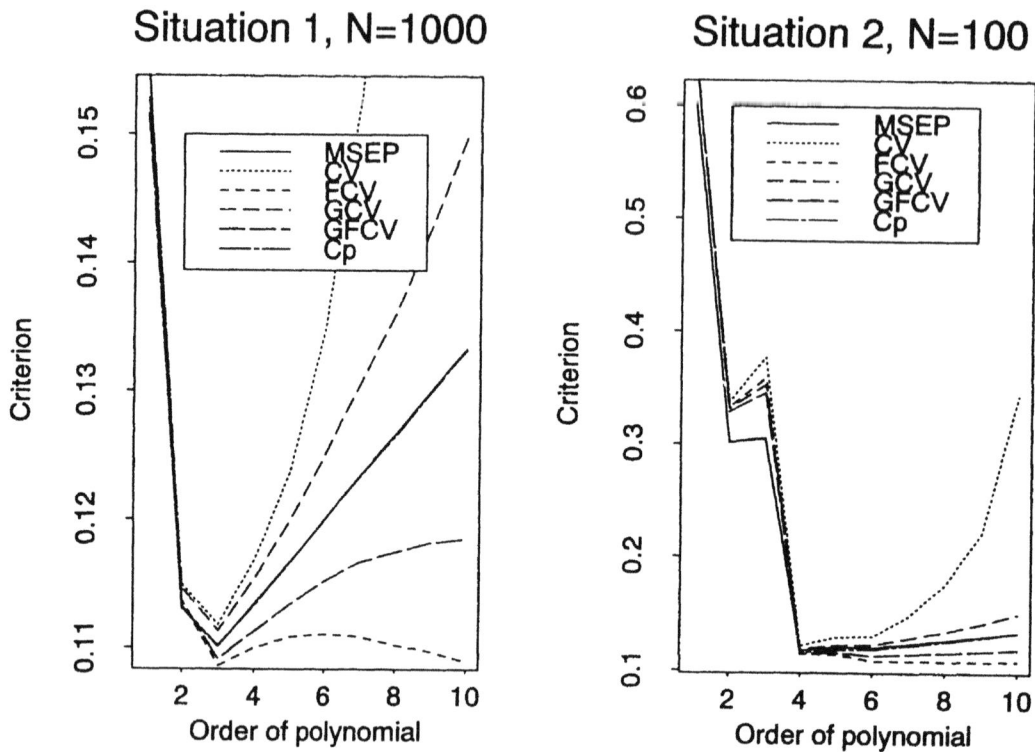

Figure 2: *For both situations: Interpolated values of MSEP and of some criteria averaged over the simulated samples as function of the polynomial order when $\sigma^2 = 0.1$; N denotes the simulation size.*

The efficiency of a model selection procedure may be measured by the percentile deviation of the average mean squared error of chosen model from the mean squared error of the optimal model. For each choice of polynomial order, given the true regression function and σ^2, the mean squared error can be calculated from Table 1 using (2). The corresponding results of our simulation study are summarized in Table 2.

Two different types of selection procedures have been applied. The first and usual one, denoted by procedure A, selects the minimizer of the considered criterion over the set of admitted polynomial orders. Alternatively, from a practical point of view, a model with fewer parameters than the global criterion-minimizer could also be a good candidate for selection provided the value of the criterion increases only slightly, say at most by 40 per cent. Such a rule of thumb appears plausible if one takes into account the errors in estimating the prediction quality of the models by the criteria. For a discussion of this aspect we refer e.g. to Bunke et al. (1993). Our second approach (procedure B) is a slight modification of the above. It selects that local minimizer of the considered criterion which involves as few as possible parameters and fulfills a rule of thumb of the discussed type.

One feature of the results is that procedure B provides a considerable improvement upon procedure A for all considered criteria and settings. Moreover, the most acceptable

Table 2: *Percentile deviation of the average mean squared error of chosen model from the mean squared error of the best model. Procedures A and B select the global and first reasonable local minimizer, respectively, of the criteria.*

| Criterion | Situation 1 | | | |
| | 1000 simulations, $\sigma^2 = 0.1$ | | 100 simulations, $\sigma^2 = 0.01$ | |
	Procedure A	Procedure B	Procedure A	Procedure B
CV	36.8	20.2	25.6	6.3
FCV	98.1	19.1	82.2	12.6
GCV	39.5	20.0	18.8	8.2
GFCV	72.2	18.6	63.7	11.9
FPE	50.8	19.8	36.5	10.3
AIC	53.9	19.8	40.8	11.1
BIC	38.6	33.9	17.4	7.9
C_p	42.7	19.9	24.1	8.4
SH	87.9	19.0	73.9	13.8

| Criterion | Situation 2 | | | |
| | 100 simulations, $\sigma^2 = 0.1$ | | 100 simulations, $\sigma^2 = 0.01$ | |
	Procedure A	Procedure B	Procedure A	Procedure B
CV	13.8	4.1	16.4	7.3
FCV	36.2	9.1	32.6	8.7
GCV	10.8	4.0	14.9	3.4
GFCV	26.5	7.2	26.8	6.7
FPE	16.3	4.9	19.8	4.6
AIC	19.1	5.0	21.5	4.8
BIC	9.2	3.5	11.8	2.6
C_p	11.1	4.0	13.8	5.1
SH	32.2	7.8	31.6	8.8

model selection criteria seem to be BIC (except for one setting), GCV, C_p and CV. Note that the minimax regret procedure of Droge and Georg (1995) showed nearly the same behaviour as that based on C_p. In case of selecting the global minimizer of a criterion (procedure A) it is not advisable to use FCV, GFCV and SH.

In contrast to situation 2, a smaller error variance improves the performance of the considered model selection procedures in situation 1. The reason for this is that the true regression function in the second situation is harder to approximate by a polynomial than in situation 1 (the order of the best possible polynomial changes from 4 to 6 when the variance decreases from 0.1 to 0.01, whereas this optimal order was 3 for both variances in situation 1, see also Table 1). Consequently, in situation 1 the linear model was rather often selected by each of the criteria (and sometimes even the constant model) when $\sigma^2 = 0.1$, whereas this never occured when $\sigma^2 = 0.01$. In situation 2, the optimal order of the polynomial was in general not underestimated by the considered procedures, except for very few cases when $\sigma^2 = 0.01$.

Now, returning to the problem of estimating the MSEP, Figure 3 shows the associated MSE's of the different criteria for situation 1 and $\sigma^2 = 0.1$. Obviously, FCV and GFCV outperform in this sense their traditional counterparts CV and GCV, respectively,

220

confirming the theoretical results of Droge (1994). Hence, our simulation study indicates that a better estimation of the MSEP does not necessarily lead to a better performance of the associated model selection procedure. This effect is mainly due to the increasing underestimation of the MSEP by FCV and GFCV for increasing parameter dimensions, see Figure 2. Unbiased MSEP estimates such as C_p are preferable as well as criteria like CV and BIC, which penalize increasing model dimensions even stronger than C_p.

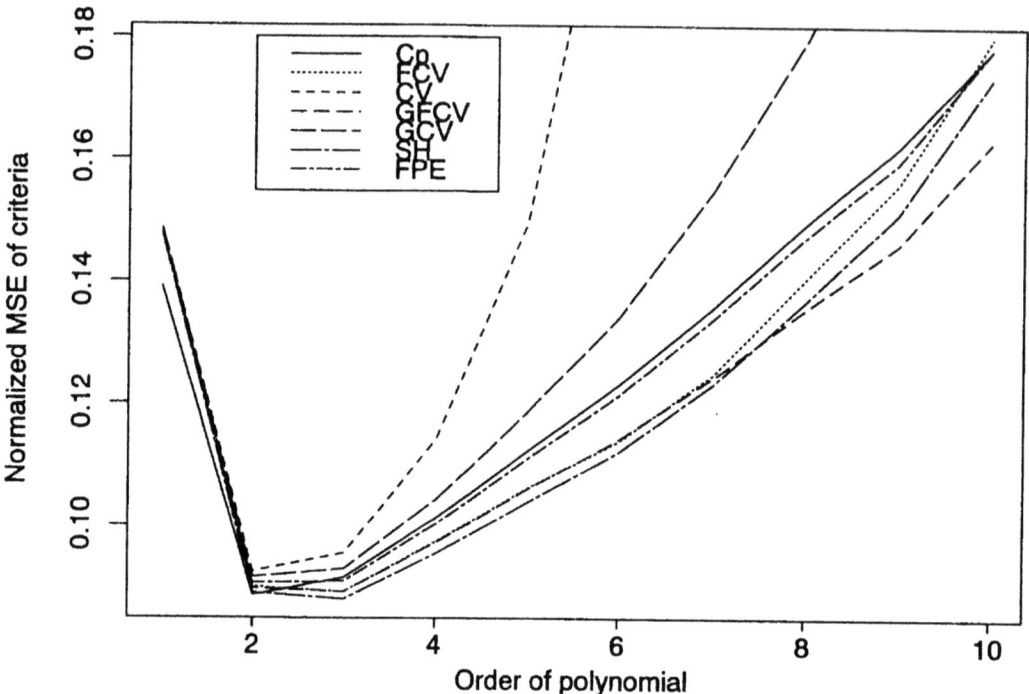

Figure 3: *Interpolated values of the MSE's of the different criteria as estimates of the MSEP, divided by σ^4, in situation 1 when $\sigma^2 = 0.1$; 1000 simulations.*

FCV and GFCV may be expected to behave well as model selection criteria if the number of parameters of the competing models is comparatively small and does not vary too much, as it was the case for the nonlinear models considered by Bunke et al. (1993).

The low efficiency of SH for model selection has similar reasons as that of GFCV. Recalling the definitions in Section 2, it is easy to see that $SH(\lambda)$ can be rewritten in the same way as $C_p(\lambda)$, with the only exception that $\hat{\sigma}^2$ has to be replaced by $RSS(\lambda)$, which depends on the model and has not the correct normalizing factor for being an unbiased estimate of σ^2. $FPE(\lambda)$ may be represented analogously, but again with a model-dependent variance estimation, $n(n - |\lambda|)^{-1} RSS(\lambda)$, instead of $\hat{\sigma}^2$, which tends to be disadvantageous.

Finally we remark that the problem of order selection in polynomial regression has been discussed in numerous articles, see, for example, Ellerton et al. (1986). Some finite-sample properties of model selection procedures have been derived (as already introductory mentioned) by Droge (1993) and Droge and Georg (1995), whereas we refer to Eubank and Jayasuriya (1993) and Müller (1993) (as well as to the references quoted therein) concerning asymptotic results.

Acknowledgement

I am grateful to Dr. J. Polzehl for helpful discussions on Splus and to the referee for his comments on the first draft of the manuscript. This work was supported by the Deutsche Forschungsgemeinschaft, Sonderforschungsbereich 373 "Quantifikation und Simulation ökonomischer Prozesse", Humboldt-Universität zu Berlin, Berlin, Germany.

References

- AKAIKE, H. (1970). Statistical predictor identification. *Ann. Inst. Statist. Math.* **22**, 203-217.

- AKAIKE, H. (1974). A new look at the statistical model identification. *I.E.E.E. Trans. Auto. Control* **19**, 716-723.

- BUNKE, O. and DROGE, B. (1984). Bootstrap and cross-validation estimates of the prediction error for linear regression models. *Ann. Statist.* **12**, 1400-1424.

- BUNKE, O., DROGE, B. and POLZEHL, J. (1993). Model selection and variable transformations in nonlinear regression. CORE Discussion Paper No. 9327, C.O.R.E., UCL, Belgium.

- CHAMBERS, J.M. and HASTIE, T. (1992). *Statistical Models in S.* Wadsworth & Brooks/Cole Computer Science Series, Pacific Grove.

- CRAVEN, P. and WAHBA, G. (1979). Smoothing noisy data with spline functions: estimating the correct degree of smoothing by the method of generalized cross-validation. *Numer. Math.* **31** 377-403.

- DROGE, B. (1993). On finite-sample properties of adaptive least squares regression estimates. *Statistics* **24**, 181-203.

- DROGE, B. (1994). Some comments on cross-validation. Discussion Paper No. 7, Sonderforschungsbereich 373, Humboldt-Universität Berlin.

- DROGE, B. and GEORG, T. (1995). On selecting the smoothing parameter of least squares regression estimates using the minimax regret approach. *Statistics & Decisions* **13**, 1-20.

- ELLERTON, R.R., KITSOS, C.P. and RINCO, S. (1986). Choosing the optimal order of a response polynomial-structural approach with minimax criterion. *Comm. Statist. Theory Methods* **15**, 129-136.

- EUBANK, R.L. and JAYASURIYA, B.R. (1993). The asymptotic average squared error for polynomial regression. *Statistics* **24**, 311-319.

- HOCKING, R.R. (1976). The analysis and selection of variables in linear regression. *Biometrics* **32**, 1-49.

- MALLOWS, C.L. (1973). Some comments on C_p. *Technometrics* **15**, 661-675.

- MÜLLER, M. (1993). Asymptotische Eigenschaften von Modellwahlverfahren in der Regressionsanalyse. Doctoral Thesis, Department of Mathematics, Humboldt University, Berlin (in German).

- SCHWARZ, G. (1978). Estimating the dimension of a model. *Ann. Statist.* **6**, 461-464.

- SHIBATA, R. (1981). An optimal selection of regression variables. *Biometrika* **68**, 45-54.

- STONE, M. (1974). Cross-validatory choice and assessment of statistical predictions. *J. Roy. Statist. Soc.* B **36**, 111-147.

- THOMPSON, M.L. (1978). Selection of variables in multiple regression: Part I. A review and evaluation. *Internat. Statist. Rev.* **46**, 1-49. Part II. Chosen procedures, computations and examples. *Ibid.* **46**, 129-146.

Robust Estimation of Non-linear Aspects

CHRISTOS P. KITSOS and CHRISTINE H. MÜLLER

Department of Statistics, University of Business and Economics, Athens 10434, Greece
and
1st Mathematical Institute, Free University, Arnimallee 2-6, 1000 Berlin 33, Germany.

Abstract:

As a first step for dealing with efficient robust estimation in non-linear models, we regard the problem of efficient robust estimation of non-linear aspects (functions) $\varphi(\beta)$ of the unknown parameter β of a linear model. For robust estimation of a general non-linear aspect we propose estimators which are based on one-step-M-estimators and derive their asymptotic behaviour at the contaminated linear model, where the errors have contaminated normal distributions. The asymptotic behaviour provides criteria for robustness and optimality of the estimators and the corresponding designs. Because it is impossible to find globally optimal robust estimators and designs locally optimal solutions are used for efficiency comparisons. Simple formulas for the efficiency rates are given for the general case. Using these results the efficiency rates for estimating robustly the relative variation of a circadian rhythm are calculated. These efficiency rates are very similar to those for non-robust estimation although on principle there is an important difference.

1 Introduction

A general linear model

$$Y_{nN} = x(t_{nN})'\beta + Z_{nN}, \qquad n = 1, ..., N$$

is considered where Y_{nN} are observations, $t_{nN} \in T$ are experimental conditions, $x : T \to I\!\!R^p$ is a known "regression" function, $\beta \in B \subset I\!\!R^p$ is an unknown parameter vector, Z_{nN} are error variables. In this model a general non-linear aspect $\varphi(\beta)$ of β shall be estimated, where $\varphi : B \to I\!\!R^r$ is a given non-linear function.

In classical linear models it is assumed that the error variables $Z_{1N}, ..., Z_{NN}$ are independent and identically distributed and often it is assumed that they are normally

---

*Kitsos, C.P., and Müller, W.G., Eds., *Proceedings of MODA4*, Physica Verlag, Heidelberg, 1995

distributed with mean 0 and known or unknown variance σ^2, i.e.

$$\frac{1}{\sigma}Z_{nN} \sim P := n_{(0,1)},$$

where $n_{(0,1)}$ denotes the standard normal distribution. But if some outlying observations (gross errors) may appear the normal distribution is not correct and the classical estimators based on least squares estimators may be biased very much.

For estimating the whole unknown parameter vector several approaches were developed to derive outlier robust estimators (see Hampel et al. (1986) and Rousseeuw and Leroy (1987)). One often used approach is the approach of Hampel (1978) and Krasker (1980) who derived the robustness criterion by calculating the influence function of the estimator, where mainly M-estimators were regarded. A similar approach which leads to the same robustness criterion is the more model oriented and more general approach of Bickel (1981, 1984) and Rieder (1985, 1987). They studied the asymptotic behaviour of general asymptotically linear estimators at contaminated linear models and in particular at conditionally contaminated linear models where the errors Z_{nN} can have different contaminated normal distributions for different experimental conditions. The class of asymptotically linear estimators in particular includes one-step-M-estimators which generalize M-estimators. This approach is especially appropiate for designed experiments and for estimation general linear aspects $C\beta$, $C \in \mathbb{R}^{r \times p}$, of β as Kurotschka and Müller (1992) showed. Therefore we here also use this model oriented approach to study the robustness properties and efficiency properties of estimators for general non-linear aspects $\varphi(\beta)$ of β. Because some problems in non-linear models can be transformed in a linear model with an interesting non-linear aspect (see also the example in Section 4 of this paper) the approach of this paper may also a first step for regarding the efficiency of robust estimators in non-linear models.

In Section 2 we describe the conditionally contaminated linear model and introduce the class of regarded estimators for the non-linear aspect which are based on one-step-M-estimators. In this section we also derive the asymptotic behaviour of the estimators at the contaminated linear model which provides the robustness and optimality criterion. As for classical non-robust estimation (see for example Silvey (1980), Buonaccorsi et al. (1986), Kitsos (1986, 1992), Kitsos et al. (1988), McDonald and Studden (1990)) it is impossible to find globally optimal designs for robust estimation. Moreover, in opposite to non-robust estimation, it is also impossible to find globally optimal robust estimators. Only locally optimal designs and robust estimators can be derived and used for efficiency comparisons as is shown in Section 3. In this section it is also shown that the locally optimal solutions have a simple form so that general formulas for the efficiency rates can be derived. In Section 4 the efficiency rates are investigated for the problem of estimating the relative variation of a circadian rhythm. The efficiency comparisons show that on principle there is an important difference to those for classical non-robust estimation which were given by Kitsos et al. (1988). But nevertheless the calculated efficiency rates are very similar to those for classical non-robust estimation. This may imply, as we discuss in Section 5, that strategies which were developed for deriving optimal designs for non-robust non-linear estimation problems can be transferred to robust non-linear estimation problems.

2 Robust Estimators for Non-linear Aspects and their Asymptotic Behaviour

For robust estimation of a non-linear aspect $\varphi(\beta)$ we consider estimators of the form $\widehat{\varphi}_N = \varphi(\widehat{\beta}_N)$ where $\widehat{\beta}_N$ is an estimator for β. An appropriate class of estimators for β which includes the least squares estimator as well as many robust estimators is the class of one-step-M-estimators as were regarded in Bickel (1975) and Müller (1992). These one-step-M-estimators are given by

$$\widehat{\beta}_N(y_N, d_N) = \widehat{\beta}_N^0(y_N, d_N) + \frac{1}{N} \sum_{n=1}^{N} \psi\left(\frac{y_{nN} - x(t_{nN})'\widehat{\beta}_N^0(y_N, d_N)}{\widehat{\sigma}_N}, t_{nN} \right) \widehat{\sigma}_N$$

where $d_N := (t_{1N}, ..., t_{NN})'$, $y_N := (y_{1N}, ..., y_{NN})'$ and $\psi : \mathbb{R} \times T \to \mathbb{R}^p$ is called the score function of the estimator. The initial estimator $\widehat{\beta}_N^0$ and the variance estimator $\widehat{\sigma}_N$ can be robust or non-robust estimators, for example the initial estimator can be the least squares estimator or some M-estimator, and the variance estimator can be the mean squared residuals or Huber's Proposal 2 (Huber (1964)).

For these estimators we now derive their asymptotic behaviour at a conditionally contaminated linear model which was also regarded by Bickel (1981, 1984), Rieder (1987) and Müller (1992, 1994). In such a model it is assumed that the error variables $Z_{1N}, ..., Z_{NN}$ of the linear model are also independent but distributed according to a contaminated normal distribution where the contamination may be different for different experimental conditions. I.e.

$$\frac{1}{\sigma} Z_{nN} \sim Q_{nN}(dz) := (1 - N^{-1/2} R \, \epsilon(t_{nN})) \, n_{(0,1)}(dz) \tag{1}$$
$$+ N^{-1/2} R \, \epsilon(t_{nN}) \, g(z, t_{nN}) \, n_{(0,1)}(dz)$$

with $R > 0$, $\frac{1}{N} \sum_{n=1}^{N} \epsilon(t_{nN}) \leq 1$, $\int g(z, t) \, n_{(0,1)}(dz) = 1$, $g(z, t) \geq 0$ for all $z \in \mathbb{R}$, $t \in T$. Thereby the markov kernel $g(\cdot, t) n_{(0,1)}$ models the form and $\epsilon(t) \geq 0$ the proportion of contamination. The set $\mathcal{P}_R$ of all sequences $(Q^N := \bigotimes_{n=1}^{N} Q_{nN})_{N \in \mathbb{N}}$ defines a conditionally contamination neighbourhood around the classical model $(P^N)_{N \in \mathbb{N}}$. If $(Z_{1N}, ..., Z_{NN})$ is distributed according to Q^N and β is the true parameter vector then Q_β^N will denote the distribution of $(Y_{1N}, ..., Y_{NN})$, in particular P_β^N the distribution of $(Y_{1N}, ..., Y_{NN})$ for the classical model.

To derive the asymptotic behaviour of the estimators we assume that a finite set T_0 and an asymptotic design measure δ exist with $\delta(T_0) = 1$, $\{t_{1N}, ..., t_{NN}\} \subset T_0$ for all $N \in \mathbb{N}$, $\lim_{N \to \infty} \frac{1}{N} \sum_{n=1}^{N} e_{t_{nN}}(\{t\}) = \delta(\{t\})$ for all $t \in T_0$ and

$$\int \psi(z, t) x(t)' z P(dz) \delta(dt) = E_{p \times p}, \tag{2}$$

where e_t denotes the Dirac measure on t and $E_{p \times p}$ the identity matrix. Then under some regularity conditions on the initial and the variance estimator an one-step-M-estimator $\widehat{\beta}_N$ for β is asymptotically normally distributed in the conditionally contaminated linear model (see Müller (1992)). If $\varphi(\beta)$ is continuously differentiable on an open set $B \subset \mathbb{R}^p$, where

$$\dot{\varphi}_\beta := \frac{\partial}{\partial \beta} \varphi(\beta) \in \mathbb{R}^{r \times p}$$

denotes the matrix of partial derivatives of φ with respect to the elements of β, then also $\varphi(\widehat{\beta}_N)$ is asymptotically normally distributed in the conditionally contaminated linear model. I.e. for all $\beta \in B$ and all $(Q^N)_{N \in \mathbb{N}} \in \mathcal{P}_R$ we have

$$\mathcal{L}(\sqrt{N}(\varphi(\widehat{\beta}_N) - \varphi(\beta))|Q_\beta^N) \stackrel{N \to \infty}{\longrightarrow} \mathcal{N}(\dot{\varphi}_\beta \, b(\psi, (Q^N)_{N \in \mathbb{N}}), \sigma^2 V(\psi, \delta, \beta)) \tag{3}$$

with

$$V(\psi, \delta, \beta) := \dot{\varphi}_\beta \int \psi \psi' \, d(P \otimes \delta) \, \dot{\varphi}_\beta'$$

and maximum asymptotic bias

$$\sup\{|\dot{\varphi}_\beta \, b(\psi, (Q^N)_{N \in \mathbb{N}})|; \, (Q^N)_{N \in \mathbb{N}} \in \mathcal{P}_R\} = R \, \sigma \operatorname*{ess\,sup}_{P \otimes \delta} |\dot{\varphi}_\beta \, \psi| \, .$$

If the score function of $\widehat{\beta}_N$ satisfies $\operatorname{ess\,sup}_{P \otimes \delta} |\psi| < \infty$, i.e. $\widehat{\beta}_N$ is robust, then also ess $\sup_{P \otimes \delta} |\dot{\varphi}_\beta \, \psi| < \infty$ so that $\varphi(\widehat{\beta}_N)$ is robust. Moreover we can set without loss of generality $R = 1 = \sigma$.

3 Efficiency of Robust Estimators and Designs

In (3) it is shown that the maximum asymptotic bias $\operatorname{ess\,sup}_{P \otimes \delta} |\dot{\varphi}_\beta \, \psi|$ and the asymptotic covariance matrix $V(\psi, \delta, \beta)$ of an estimator $\varphi(\widehat{\beta}_N)$ based on an one-step-M-estimator $\widehat{\beta}_N$ for β with score function ψ depend on β. If there is no bias bound for the maximum asymptotic bias then the matrix $\int \psi \psi' \, d(P \otimes \delta)$ is minimized uniformly by $\psi_\infty(z, t) := I(\delta)^{-1} x(t) z$ within all score functions ψ fulfilling condition (2), where we have $I(\delta) := \int x(t) x(t)' \, \delta(dt)$. I.e. the covariance matrix $V(\psi_\infty, \delta, \beta)$ is the minimum for all β so that the estimator based on the least squares estimator is asymptotically optimal. (Note that the least squares estimator behaves asymptotically like an one-step-M-estimator with score function ψ_∞.) But if the maximum asymptotic bias should be bounded by some bias bound b, in principle it is impossible to minimize $\int \psi \psi' \, d(P \otimes \delta)$ uniformly so that an optimal robust estimator cannot be found independently of β. Also optimal designs δ cannot be found independently of β, and this problem already occurs for classical non-robust estimation. All classical non-linear optimal experimental design theory suffer on this β-dependence so that strategies were invented to overcome this problem. See for an overview Ford et al. (1989).

One basic strategy is to compare a given estimator and design with the locally optimal estimators and designs. This provides the efficiency rate of the estimator and the design and this efficiency rate shall be investigated in this section.

For deriving the locally optimal estimators and designs for robust estimation we regard the local behaviour of an estimator $\varphi(\widehat{\beta}_N)$ based on an robust one-step-M-estimator $\widehat{\beta}_N$ for β with score function ψ. Locally, i.e. for given β, this estimator behaves asymptotically like

$$\dot{\varphi}_\beta \widehat{\beta}_N = \dot{\varphi}_\beta \widehat{\beta}_N^0 + \frac{1}{N} \sum_{n=1}^{N} \dot{\varphi}_\beta \psi \left(\frac{y_{nN} - x(t_{nN})' \widehat{\beta}_N^0 (y_N, d_N)}{\widehat{\sigma}_N}, t_{nN} \right) \widehat{\sigma}_N,$$

i.e. like an one-step-M-estimator for the linear aspect $\tilde{\varphi}(\tilde{\beta}) = \dot{\varphi}_\beta \tilde{\beta}$ with score function $\dot{\varphi}_\beta \psi$ (see Müller (1992)). Now we can try to find locally optimal robust estimators within all these robust one-step-M-estimators for the linear aspect $\tilde{\varphi}(\tilde{\beta}) = \dot{\varphi}_\beta \tilde{\beta}$. I.e., given a bias bound b for estimating this linear aspect, we have to find a score function $\psi_{\beta,b,\delta}$ which solves

$$\psi_{\beta,b,\delta} = \arg\min\{\operatorname{tr} V(\psi,\delta,\beta); \; \dot\varphi_\beta \psi \in \Psi_\delta(\dot\varphi_\beta) \text{ with } \operatorname*{ess\,sup}_{P\otimes\delta} |\dot\varphi_\beta \psi| \le b\},$$

where $\Psi_\delta(\dot\varphi_\beta)$ is the class of all score functions for estimating the linear aspect $\bar\varphi(\bar\beta) = \dot\varphi_\beta \bar\beta$ and in particular satisfying $\int \psi(z,t) x(t)' z P(dz)\delta(dt) = \dot\varphi_\beta$.

Similarly locally optimal designs for robust estimation with bias bound b can be defined as solutions $\delta_{\beta,b}$ of

$$\delta_{\beta,b} = \arg\min\{\operatorname{tr} V(\psi_{\beta,b,\delta},\delta,\beta); \; \delta \in \Delta\}$$

where Δ is the set of the regarded design measures.

For estimating a general linear aspect $C\beta$ the solutions concerning optimal robust estimators are characterized in Kurotschka and Müller (1992) and the solutions concerning optimals designs for robust estimation are characterized in Müller (1994). In particular in Müller (1994) it is shown that the classical A-optimal designs, which minimize $\operatorname{tr} C\, I(\delta)^- C'$ (where $I(\delta)^-$ denotes the g-inverse of $I(\delta)$), are also optimal for robust estimation and this holds independently of the bias bound b. Therefore the locally optimal designs for estimation without bias bound are also locally optimal for robust estimation with bias bound b, i.e.

$$\delta_{\beta,b} = \delta_\beta := \arg\min\{\operatorname{tr} \dot\varphi_\beta I(\delta)^- \dot\varphi_\beta'; \; \delta \in \Delta\}.$$

Moreover, the score functions

$$\psi_{\beta,b} := \psi_{\beta,b,\delta_\beta}$$

of the locally optimal robust estimators with bias bound b at the locally optimal designs δ_β have a very simple form, namely (see Müller (1994))

$$\psi_{\beta,b}(z,t) = \begin{cases} I(\delta_\beta)^- x(t)\operatorname{sgn}(z)\sqrt{\frac{\pi}{2}} & \text{for } b = \sqrt{\operatorname{tr} \dot\varphi_\beta I(\delta_\beta)^- \dot\varphi_\beta' \frac{\pi}{2}} \\ I(\delta_\beta)^- x(t)\operatorname{sgn}(z)\dfrac{\min\{|z|, b\, y_{b,\beta}\}}{y_{b,\beta}\sqrt{\operatorname{tr} \dot\varphi_\beta I(\delta_\beta)^- \dot\varphi_\beta'}} & \text{for } b > \sqrt{\operatorname{tr} \dot\varphi_\beta I(\delta_\beta)^- \dot\varphi_\beta' \frac{\pi}{2}} \end{cases} . \quad (4)$$

Thereby, $y_{b,\beta}$ is the positive solution of $y\sqrt{\operatorname{tr} \dot\varphi_\beta I(\delta_\beta)^- \dot\varphi_\beta'} = 2\,\Phi(b\,y) - 1$ while Φ denotes the distribution function of the normal distribution $n_{(0,1)}$. Hence, locally optimal robust estimators at locally optimal designs are easy to compute and can be applied for efficiency comparisons.

Now, a given design δ and a given estimator $\varphi(\widehat\beta_N)$ for $\varphi(\beta)$, where $\widehat\beta_N$ is a robust one-step-M-estimator for β with bias bound b and score function ψ, can be compared with the locally optimal robust estimator at the locally optimal design. Because, according to (3), the maximum asymptotic bias of $\varphi(\widehat\beta_N)$ is equal to

$$b(\beta) := b(\beta,\psi,\delta) := \operatorname*{ess\,sup}_{P\otimes\delta} |\dot\varphi_\beta \psi|,$$

i.e. depends on β, the estimator $\varphi(\widehat\beta_N)$ at the design δ should be only compared with locally optimal robust estimators with the same local bias bound $b(\beta)$ at the locally optimal design δ_β, to make fair comparisons. Hence the asymptotic efficiency measure is defined as

$$E_\beta(\psi, \delta) := \frac{\operatorname{tr} V(\psi_{\beta, b(\beta)}, \delta_\beta, \beta)}{\operatorname{tr} V(\psi, \delta, \beta)} \in [0, 1]$$

where $\psi_{\beta, b(\beta)}$ is the score function of the locally optimal one-step-M-estimator with local bias bound $b(\beta)$.

If δ is an A-optimal design for estimating β then the score function ψ_b of the optimal robust one-step-M-estimator for β at δ with bias bound b has the form (see Müller (1994))

$$\psi_b(z, t) = \begin{cases} I(\delta)^{-1} x(t) \operatorname{sgn}(z) \sqrt{\frac{\pi}{2}} & \text{for } b = \sqrt{\operatorname{tr} I(\delta)^{-1} \frac{\pi}{2}} \\ I(\delta)^{-1} x(t) \operatorname{sgn}(z) \frac{\min\{|z|, b\, y_b\}}{y_b \sqrt{\operatorname{tr} I(\delta)^{-1}}} & \text{for } b > \sqrt{\operatorname{tr} I(\delta)^{-1} \frac{\pi}{2}} \end{cases}, \tag{5}$$

where y_b is the positive solution of $y\sqrt{\operatorname{tr} I(\delta)^{-1}} = 2\,\Phi(b\,y) - 1$. Hence, in this case with the equivalence theorem for linear optimality (see Federov (1972), p. 125), we get for the local bias bounds

$$b(\beta) = \frac{b}{\sqrt{\operatorname{tr} I(\delta)^{-1}}} \operatorname{ess\,sup}_\delta |\dot\varphi_\beta I(\delta)^{-1} x(t)| = \operatorname{ess\,sup}_\delta \frac{|\dot\varphi_\beta I(\delta)^{-1} x(t)|}{|I(\delta)^{-1} x(t)|} b, \tag{6}$$

where $b(\beta) \geq \sqrt{\operatorname{tr} \dot\varphi_\beta I(\delta_\beta)^{-} \dot\varphi'_\beta \frac{\pi}{2}}$ for all $b \geq \sqrt{\operatorname{tr} I(\delta)^{-1} \frac{\pi}{2}}$. Then the efficiency measure is

$$E_\beta(\psi_b, \delta) = \begin{cases} \frac{g(b(\beta)\, y_{b,\beta})}{y_{b,\beta}^2 \operatorname{tr} \dot\varphi_\beta I(\delta)^{-1} \dot\varphi'_\beta} \frac{2}{\pi} & \text{for } b = \sqrt{\operatorname{tr} I(\delta)^{-1} \frac{\pi}{2}} \\ \frac{g(b(\beta)\, y_{b,\beta})}{g(b\, y_b)} \cdot \frac{y_b^2}{y_{b,\beta}^2} \cdot \frac{\operatorname{tr} I(\delta)^{-1}}{\operatorname{tr} \dot\varphi_\beta I(\delta)^{-1} \dot\varphi'_\beta} & \text{for } b > \sqrt{\operatorname{tr} I(\delta)^{-1} \frac{\pi}{2}} \end{cases}.$$

Because of $\lim_{v \to \infty} g(v) = 1$, $\lim_{b \to \infty} y_b = \frac{1}{\sqrt{\operatorname{tr} I(\delta)^{-1}}}$ and $\lim_{b \to \infty} y_{b,\beta} = \frac{1}{\sqrt{\operatorname{tr} \dot\varphi_\beta I(\delta_\beta)^{-} \dot\varphi'_\beta}}$ it is easy to see that for $b \to \infty$ the efficiency $E_\beta(\psi_b, \delta)$ tends to

$$\frac{\operatorname{tr} \dot\varphi_\beta I(\delta_\beta)^{-} \dot\varphi'_\beta}{\operatorname{tr} \dot\varphi_\beta I(\delta)^{-1} \dot\varphi'_\beta}, \tag{7}$$

i.e. to the design efficiency of δ relative to δ_β as appeared in classical non-robust estimation (see Kitsos et al. (1988)).

4 Example: The Rhythmometry Problem

Consider the cosinor model of a circadian rhythms which was regarded in Kitsos et al. (1988). In such a model the observations are given by the non-linear model

$$Y_{nN} = \theta_0 + \theta_1 \cos(2\pi t_{nN} + \theta_2) + Z_{nN}$$

with $t_{nN} \in T := [0, 1)$ and θ_0 being the mean, θ_1 the amplitude and θ_2 the phase. In Kitsos et al. (1988) the interest was focussed on estimating the ratio

$$g = g(\theta_0, \theta_1) = \frac{\theta_1}{\theta_0}$$

as it gives a potentially useful measure of relative variation, falling usually in practice within the range $0 < g < 1$.

By reparameterization one gets a linear model

$$
\begin{aligned}
Y_{nN} &= \beta_0 + \beta_1 \cos(2\pi t_{nN}) + \beta_2 \sin(2\pi t_{nN}) + Z_{nN} \\
&= x(t_{nN})'\beta + Z_{nN}
\end{aligned}
$$

where $\beta_0 = \theta_0$, $\beta_1 = \theta_1 \cos(\theta_2)$, $\beta_2 = -\theta_1 \sin(\theta_2)$, $x(t) = (1, \cos(2\pi t), \sin(2\pi t))'$ and $\beta = (\beta_0, \beta_1, \beta_2)'$. Therefore the aspect $g(\theta_0, \theta_1)$ of the non-linear cosinor model is a non-linear aspect

$$
\varphi(\beta) := \frac{\sqrt{\beta_1^2 + \beta_2^2}}{\beta_0} = g(\theta_0, \theta_1)
$$

of a linear model. Note that $\dot{\varphi}_\beta = \frac{1}{\theta_0}(-\frac{\theta_1}{\theta_0}, \frac{\beta_1}{\theta_1}, \frac{\beta_2}{\theta_1})$.

In Kitsos et al. (1988) is was assumed that the errors Z_{nN} are normally distributed with mean 0 and constant variance σ^2. But there may exist some outlying observations so that the errors have a conditionally contaminated normal distribution of form (1). Without loss of generality we can here assume that the variance σ^2 as well as the radius R of the contamination neighbourhood $\mathcal{P}_R$ are equal to 1.

In a conditionally contaminated cosinor model the non-linear aspect $\varphi(\beta)$ should be estimated with a robust estimator, for example with an estimator of the form $\varphi(\widehat{\beta}_N)$, where $\widehat{\beta}_N$ is an optimal robust one-step-M-estimator for β. If the design

$$
\delta = \frac{1}{3}(e_0 + e_{\frac{1}{3}} + e_{\frac{2}{3}})
$$

is used then this design is A- and D-optimal for estimating β (see Kitsos et al. (1988)) and the score function of the optimal robust one-step-M-estimator for β with bias bound $b > \sqrt{5\frac{\pi}{2}}$ results according to (5) in

$$
\psi_b(z, t) = (1, 2\cos(2\pi t), 2\sin(2\pi t))' \frac{\operatorname{sgn}(z) \min\{|z|, b\, y_b\}}{\sqrt{5} y_b}.
$$

Thereby y_b is the positive solution of $y\sqrt{5} = 2\Phi(b\, y) - 1$, i.e. $y_b = 0.216$ for $b = 3$ and $y_b = 0.444$ for $b = 6$, and $\sqrt{5\frac{\pi}{2}} = \sqrt{\operatorname{tr} I(\delta)^{-1}\frac{\pi}{2}}$ is the minimum possible bias bound. Then, according to (6) the maximum asymptotic bias of the estimator $\varphi(\widehat{\beta}_N)$ is

$$
b(\beta) = \frac{b}{\sqrt{5}\theta_0} \max\left\{ \left| \frac{-\theta_1}{\theta_0} + 2\cos(2\pi t + \theta_2) \right| ; \ t \in \{0, \tfrac{1}{3}, \tfrac{2}{3}\} \right\}.
$$

These estimator should be compared with the locally optimal robust one-step-M-estimators for $\varphi(\beta)$ with local bias bounds $b(\beta)$ at the locally optimal designs.

Locally optimal designs, i.e. A-optimal designs for estimating $\dot{\varphi}_\beta \beta$, are (see Kitsos et al. (1988))

$$
\delta_\beta = \frac{1}{2}\left(\left(1 - \frac{\theta_1}{\theta_0}\right) e_{\frac{-\theta_2}{2\pi}} + \left(1 + \frac{\theta_1}{\theta_0}\right) e_{\frac{\pi - \theta_2}{2\pi}} \right) \tag{8}
$$

if $0 < \frac{\theta_1}{\theta_0} < 1$. According to (4) the score function of the locally optimal robust one-step-M-estimator for $\varphi(\beta)$ with bias bound $b(\beta)$ at δ_β has the form

$$
\psi_{\beta, b(\beta)}(z, t) = I(\delta_\beta)^- \begin{pmatrix} 1 \\ \cos(2\pi t) \\ \sin(2\pi t) \end{pmatrix} \frac{\operatorname{sgn}(z) \min\{|z|, b(\beta)\, y_{b,\beta}\} \theta_0}{y_{b,\beta}}, \tag{9}
$$

where $y_{b,\beta}$ is the positive solution of $\frac{y}{\theta_0} = 2\Phi(b(\beta)\,y) - 1$. Always this solution will exist because $b(\beta) > \frac{1}{\theta_0}\sqrt{\frac{\pi}{2}}$ if $b > \sqrt{5\frac{\pi}{2}}$, where $\frac{1}{\theta_0}\sqrt{\frac{\pi}{2}} = \sqrt{\operatorname{tr}\dot\varphi_\beta I(\delta_\beta)^-\dot\varphi'_\beta\frac{\pi}{2}}$ is the locally minimum possible bias bound.

Table 1: Relative efficiencies $E_\beta(\psi_b,\delta)$ and, in brackets, local bias bounds $b(\beta)$ for the rhythmometry problem with $\theta_0 = 1$

	$b = 3$					$b = 6$		$b = \infty$
θ_1	$\theta_2 = 0$	$\theta_2 = \frac{\pi}{12}$	$\theta_2 = \frac{\pi}{6}$	$\theta_2 = \frac{\pi}{4}$	$\theta_2 = \frac{\pi}{3}$	$\theta_2 = 0$	$\theta_2 = \frac{\pi}{3}$	
0.1	0.414	0.414	0.414	0.414	0.413	0.497		0.498
	(2.549)	(2.458)	(2.458)	(2.726)	(2.817)	(5.098)	(5.635)	
0.2	0.408	0.409	0.408	0.407	0.407	0.490		0.490
	(2.415)	(2.324)	(2.592)	(2.860)	(2.952)	(4.830)	(5.903)	
0.3	0.399	0.399	0.398	0.397	0.397	0.478		0.478
	(2.281)	(2.300)	(2.726)	(2.994)	(3.086)	(4.562)	(6.172)	
0.4	0.388	0.386	0.385	0.385	0.384	0.462		0.463
	(2.147)	(2.434)	(2.860)	(3.129)	(3.220)	(4.293)	(6.440)	
0.5	0.374	0.370	0.369	0.369	0.369	0.444		0.444
	(2.012)	(2.568)	(2.995)	(3.263)	(3.354)	(4.025)	(6.708)	
0.6	0.355	0.352	0.352	0.352	0.352	0.423		0.424
	(2.147)	(2.702)	(3.129)	(3.397)	(3.488)	(4.293)	(6.977)	
0.7	0.335	0.334	0.334	0.333	0.333	0.401		0.402
	(2.281)	(2.837)	(3.263)	(3.531)	(3.622)	(4.562)	(7.245)	
0.8	0.316	0.315	0.315	0.315	0.315	0.378		0.379
	(2.415)	(2.971)	(3.397)	(3.665)	(3.757)	(4.830)	(7.513)	
0.9	0.296	0.296	0.296	0.296	0.296	0.355		0.356
	(2.549)	(3.105)	(3.531)	(3.799)	(3.891)	(5.098)	(7.782)	

Table 1 shows the efficiencies $E_\beta(\psi_b,\delta)$ of the estimator $\varphi(\widehat\beta_N)$ for nine different values for θ_1, five different values for θ_2 and three different values for b including the case $b = \infty$, i.e. the classical non-robust case. For θ_0 only the value $\theta_0 = 1$ was used because the efficiencies are independent of θ_0. This is due to the fact that

$$
\begin{aligned}
b(\beta)\,y_{b,\beta} &= \frac{b}{\sqrt{5}}\max\left\{\left|\frac{-\theta_1}{\theta_0} + 2\cos(2\pi t + \theta_2)\right|; \; t \in \{0,\tfrac{1}{3},\tfrac{2}{3}\}\right\}\frac{y_{b,\beta}}{\theta_0} \\
&= \frac{b}{\sqrt{5}}\max\left\{\left|\frac{-\theta_1}{\theta_0} + 2\cos(2\pi t + \theta_2)\right|; \; t \in \{0,\tfrac{1}{3},\tfrac{2}{3}\}\right\}(2\Phi(b(\beta)\,y_{b,\beta}) - 1)
\end{aligned}
$$

and

$$
\begin{aligned}
y_{b,\beta}^2 \operatorname{tr}\dot\varphi_\beta I(\delta)^{-1}\dot\varphi'_\beta &= \operatorname{tr}\dot\varphi_\beta I(\delta_\beta)^-\dot\varphi'_\beta \quad y_{b,\beta}^2\,\frac{\operatorname{tr}\dot\varphi_\beta I(\delta)^{-1}\dot\varphi'_\beta}{\operatorname{tr}\dot\varphi_\beta I(\delta_\beta)^-\dot\varphi'_\beta} \\
&= \frac{y_{b,\beta}^2}{\theta_0^2}\left(\frac{\theta_1^2}{\theta_0^2} + 2\right) \\
&= (2\Phi(b(\beta)\,y_{b,\beta}) - 1)^2\left(\frac{\theta_1^2}{\theta_0^2} + 2\right)
\end{aligned}
$$

depend only on the ratio $\frac{\theta_1}{\theta_0}$ and the phase θ_2. But for non-robust estimation, i.e. for $b = \infty$, the efficiencies (given by (7)) depend only on the ratio $\frac{\theta_1}{\theta_0}$ and not on the mean θ_0 and the phase θ_2 as already Kitsos et al. (1988) have shown. Hence the main difference between robust and non-robust estimation is the additional dependence on the phase θ_2 for robust estimation. For $b < \infty$ Table 1 provides also the values for the local bias bounds $b(\beta)$.

5 Discussion

As Table 1 for the rhythmometry problem shows, the efficiencies for robust estimation ($b < \infty$) are very similar to the efficiencies for classical non-robust estimation ($b = \infty$), in particular for the bias bound $b = 6$ but also for the bias bound $b = 3$ which is near to the minimum possible bias bound of $\sqrt{5\frac{\pi}{2}} = 2.802$. The main difference to non-robust estimation is that the efficiency rates depend additionally on the phase θ_2. But the influence of the phase θ_2 is very low, although also those values θ_2 were used which produce the largest ($\theta_2 = \frac{\pi}{3}$) and the smallest ($\theta_2 = 0$, at least for $\frac{\theta_1}{\theta_0} \geq 0.5$) values for the local bias bounds $b(\beta)$. Even at $b = 6$ the efficiencies were identical at $\theta_2 = 0$ and $\theta_2 = \frac{\pi}{3}$, and this coincides with the behaviour of the efficiencies for non-robust estimation.

From the results for the rhythmometry problem one can expect that for general non-linear problems the efficiencies for robust estimation are very similar to the efficiencies for non-robust estimation. Hence the strategies which were investigated for deriving optimal designs for non-robust non-linear estimation problems as maximin efficient designs, fully sequential designs, batch sequential designs and two-stage designs (see Silvey (1980), Ford et al. (1989), Kitsos (1986, 1989, 1992)) may be also used for robust non-linear estimation problems. These strategies may also provide strategies for deriving particular robust estimators. For example if we have a two-stage design with N_1 observations at the first stage and N_2 observations at the second stage, then we can use in the second stage the locally optimal design $\delta_{\widehat{\beta}^1}$ and the locally optimal robust estimator with score function $\psi_{\widehat{\beta}^1, b}$, where the estimator $\widehat{\beta}^1$ for β is obtained in the first stage.

For the rhythmometry problem a very practicable two-stage design is the following design: In the first stage a four point design of the form $d_4 := (\tau, \tau + \frac{1}{4}, \tau + \frac{1}{2}, \tau + \frac{3}{4})'$ with $\tau \in [0, \frac{1}{4})$ is used. Because of $\cos(2\pi t)^2 + \sin(2\pi t)^2 = 1$ this design is D-optimal for non-robust estimation of β for all $\tau \in [0, \frac{1}{4})$ (see Federov (1972), p.75). Moreover this design is approximately locally optimal for all relevant β and for $\tau = 0$ the least squares estimator $\widehat{\beta}_4^*$, say, for estimating β has a very simple form, namely

$$\widehat{\beta}_4^*((y_1, y_2, y_3, y_4)', d_4) = \begin{pmatrix} \frac{1}{4}(y_1 + y_2 + y_3 + y_4) \\ \frac{1}{2}(y_1 - y_3) \\ \frac{1}{2}(y_2 - y_4) \end{pmatrix}.$$

Basing on this estimator in the second stage the locally optimal design $\delta_{\widehat{\beta}_4^*}$ given by (8) and the locally optimal robust one-step-M-estimator with score function $\psi_{\widehat{\beta}_4^*, b}$ given by (9) can be used. Note that at the locally optimal design $\delta_{\widehat{\beta}_4^*}$ the whole parameter vector β is not anymore estimable. But this design has besides its greater efficiency the advantage that only at two different experimental conditions the observations are made, which is more practicable than a three or four point design, in particular for medical investigations of a circadian rhythm.

232

References

[1] BICKEL, P.J. (1975). One-step Huber estimates in the linear model. *Journal of the American Statistical Association* **70**, 428-434.

[2] BICKEL, P.J. (1981). Quelque aspects de la statistique robuste. In École d'Été de Probabilités de St. Flour. *Springer Lecture Notes in Mathematics* **876**, 1-72.

[3] BICKEL, P.J. (1984). Robust regression based on infinitesimal neighbourhoods. *The Annals of Statistics* **12**, 1349-1368.

[4] BUONACCORSI, J.P. and IYER, H.K. (1986). Optimal designs for ratios of linear combinations in the general linear model. *Journal of Statistical Planning and Inference* **13**, 345-356.

[5] FEDEROV, V.V. (1972). *Theory of Optimal Experiments.* Academic Press, New York.

[6] FORD, I., KITSOS, C.P. and TITTERINGTON, D.M. (1989). Recent advances in nonlinear experimental design. *Technometrics* **13**, 49-60.

[7] HAMPEL, F.R. (1978). Optimally bounding the gross-error-sensitivity and the influence of position in factor space. *Proceedings of the ASA Statistical Computing Section*, ASA, Washington, D.C., 59-64.

[8] HAMPEL, F.R., RONCHETTI, E.M., ROUSSEEUW, P.J. and STAHEL, W.A. (1986). *Robust Statistics - The Approach Based on Influence Functions.* John Wiley, New York.

[9] HUBER, P.J. (1964). Robust estimation of a location parameter. *Annals of Mathematical Statistics* **35**, 73-101.

[10] KITSOS, C.P. (1986). Design and inference for nonlinear problems. *Ph.D. thesis*, University of Glasgow, U.K.

[11] KITSOS, C.P. (1989). Fully-sequential procedures in nonlinear design problems. *Comp. Statist. & Data Analysis* **8**, 13-19.

[12] KITSOS, C.P. (1992). Quasi-sequential procedures for the calibration problem. In *COMPSTAT 1992, Vol. 2*, Y. Dodge and J. Whittaker (eds.), 227-231, Physica-Verlag.

[13] KITSOS, C.P., TITTERINGTON, D.M. and TORSNEY, B. (1988). An optimal design problem in rhythmometry. *Biometrics* **44**, 657-671.

[14] KRASKER, W.S. (1980). Estimation in linear regression models with disparate data points. *Econometrica* **48**, 1333-1346.

[15] KUROTSCHKA, V. and MÜLLER, Ch.H. (1992). Optimum robust estimation of linear aspects in conditionally contaminated linear models. *The Annals of Statistics* **20**, 331-350.

[16] McDONALD, G.C. and STUDDEN, W.J. (1990). Design aspects of regression-based ratio estimation. *Technometrics* **32**, 417-424.

[17] MÜLLER, Ch.H. (1992). One-step-M-estimators in conditionally contaminated linear models. *Preprint No. A-92-11*, Freie Universität Berlin, Fachbereich Mathematik. *To appear in Statistics & Decisions*

[18] MÜLLER, Ch.H. (1994). Optimal designs for robust estimation in conditionally contaminated linear models. *Journal of Statistical Planning and Inference* **38**, 125-140.

[19] RIEDER, H. (1987). Robust regression estimators and their least favorable contamination curves. *Statistics & Decisions* **5**, 307-336.

[20] ROUSSEEUW, P.J. and LEROY, A.M. (1987). *Robust Regression and Outlier Detection*. John Wiley, New York.

[21] SILVEY, S.D. (1980). *Optimal design*. Chapman and Hall, London.

Robust Minimax Adaptive M-Estimators of Regression Parameters

GEORGIY L. SHEVLYAKOV and NIKITA O. VIL'CHEVSKIY

Department of Applied Mathematics, State Technical University,
Polytechnicheskaya ul., 29, St.Petersburg, 195251, Russia.

Abstract:

The minimax robust M-estimators of regression parameters designed over the classes with a bounded variance of a distribution are obtained. The properties of these new estimators and their adaptive versions are studied in asymptotics and in a finite sample size case.

1 Introduction

One of the basic approaches to the synthesis of robust estimation procedures is the minimax principle. In this case, in a given class of densities the least favorable one which minimizes the Fisher information is determined. The unknown parameters of a regression model are then estimated by means of the maximum likelihood method for this density [1].

Such an approach makes it possible to construct robust statistical procedures which are stable with respect to deviations from apriori assumptions about the distribution in the case of prior uncertainty of the probability models being used.

The robust minimax procedures provide a guaranteed level of the estimator's accuracy (measured by the supremum of an asymptotic variance) for any density distribution of a given class.

The form of the solution obtained by the minimax approach essentially depends upon the characteristics of a distribution class. Thus, it seems rather important for applications to consider the distribution classes with apriori or aposteriori available characteristics. What is more that is performing the required computational and adaptive properties to robust procedures.

The paper describes the robust minimax M-estimators of a location parameter and regression parameters designed over the classes with a bounded variance (in practice the information on this characteristic is usually available apriori or while data processing). The

---

*Kitsos, C.P., and Müller, W.G., Eds., *Proceedings of MODA4*, Physica Verlag, Heidelberg, 1995

properties of these new estimators and their adaptive versions are studied in asymptotics and in a finite sample size case.

2 Robust Minimax M-Estimators

Consider the following linear regression model

$$X = \Phi\Theta + E. \tag{1}$$

In this model, $X = (x_1, ..., x_n)^T$ is a vector of observed values; $\Theta = (\theta_1, ..., \theta_m)^T$ is a vector of unknown parameters; $\Phi = (\phi_{ij})_{n,m}$ is a given model matrix and $E = (e_1, ..., e_n)^T$ is a vector of unobservable random errors with a density f belonging to a certain class $\mathcal{F}$.

The M - estimate of a regression parameter Θ is defined as [1]

$$\widehat{\Theta}_n = \arg\min_{\Theta} \sum_{i=1}^n \rho(x_i - \sum_{j=1}^m \theta_j \phi_{ij}), \tag{2}$$

or

$$\sum_{i=1}^n \psi(x_i - \sum_{j=1}^m \widehat{\theta}_j \phi_{ij})\phi_{ik} = 0, \quad k = 1, ..., m, \tag{3}$$

where $\rho(u)$ is a loss function, $\psi(u) = \rho'(u)$ with ψ belonging to a certain class Ψ.

Then the minimax approach implies the determination of the least favorable density f for the class $\mathcal{F}$ minimizing the Fisher information $I(f)$ for a location parameter

$$f = \arg\min_f I(f), \quad I(f) = \int_{-\infty}^{\infty} (f'/f)^2 f dx, \tag{4}$$

followed by designing the optimum maximum likelihood estimate (MLE) with a loss function and its derivative of the form

$$\rho^*(u) = -\log f^*(u), \quad \psi^*(u) = -(f^*(u))'/f^*(u). \tag{5}$$

Under some regularity and convexity conditions put upon the classes $\mathcal{F}$ and Ψ [2], the asymptotic covariance matrix $V(\psi, f)$ has the saddle point (ψ^*, f^*) with the corresponding minimax property

$$V(\psi^*, f) \leq V(\psi^*, f^*) \leq V(\psi, f^*).$$

For all classes of distributions the following conditions are common:

$$f(x) \geq 0, \quad f(-x) = f(x), \quad \int_{-\infty}^{\infty} f(x)dx = 1. \tag{6}$$

The symmetry of distributions is a rather restrictive condition but the minimax property for M-estimates of regression parameters holds only in this case [1].

Depending on additional restrictions on the class $\mathcal{F}$, different forms of the density f^* and the appropriate loss functions ρ^* may result.

It follows from [1] that, for the class of nonsingular densities

$$\mathcal{F}_1 = (f : f(0) \geq 1/(2a) > 0)$$

the least favorable density is the Laplace one

$$f_1^*(x) = \frac{1}{2a} \exp(-\frac{|x|}{a})$$

with the optimum loss function $\rho^*(u) = |u|$ and the least modules method (LMM) estimator;

for the class of distributions with a bounded variance

$$\mathcal{F}_2 = (f : \sigma^2(f) = \int_{-\infty}^{\infty} x^2 f(x) dx \leq \overline{\sigma}^2)$$

the least favorable density is the Gaussian

$$f_2^*(x) = \frac{1}{\sqrt{2\pi}} \, exp(-\frac{x^2}{2\overline{\sigma}^2})$$

with the optimum loss function $\rho^*(u) = u^2$, and the least squares method (LSM) estimator.

For the class of ε - contaminated distributions

$$\mathcal{F}_3 = (f : f(x) \geq (1 - \varepsilon)p(x), \ 0 < \varepsilon < 1)$$

the least favorable density has the exponential "tails":

$$f_3^*(x) = \begin{cases} (1 - \varepsilon)p(x), & |x| \leq \Delta, \\ A f_1^*(Bx), & |x| > \Delta, \end{cases}$$

where $f(x)$ is a given density; ε is a parameter characterising the degree of apriori uncertainty; the constants A, B, Δ are chosen to satisfy the conditions of normalization and the sewing smoothness at the point $x = \Delta$. If $p(x)$ is Gaussian then the optimum estimator is a compromise between LSM and LMM estimators with a linear bounded derivative of a loss function [1] :

$$\psi_3^*(u) = \max(-k, \min(u, k)).$$

3 The Least Favorable Density for the Class $\mathcal{F}_{12}$

Consider the intersection of the classes $\mathcal{F}_1$ and $\mathcal{F}_2$ with the inequality constraint upon the variance: $\sigma^2(f) \leq \overline{\sigma}^2$.

Theorem. The solution of the problem (4) for the distribution class

$$\mathcal{F}_{12} = (f : f(0) \geq 1/(2a) > 0, \ \sigma^2(f) \leq \overline{\sigma}^2) \tag{7}$$

is of the following form:

$$f_{12}^*(x) = \begin{cases} f_2^*(x), & \overline{\sigma}^2/a^2 \leq 2/\pi, \\ f(x; \nu, \overline{\sigma}), & 2/\pi < \overline{\sigma}^2/a^2 \leq 2, \\ f_1^*(x), & \overline{\sigma}^2/a^2 > 2, \end{cases} \tag{8}$$

where

$$f(x; \nu, \overline{\sigma}) = \frac{\Gamma(-\nu)\sqrt{2\nu + 1 + 1/S(\nu)}}{\sqrt{2\pi} \, \overline{\sigma} \, S(\nu)} \mathcal{D}_\nu{}^2(\frac{|x|}{\overline{\sigma}}\sqrt{2\nu + 1 + 1/S(\nu)} \,); \tag{9}$$

the parameter ν $(-\infty < \nu \leq 0)$ is determined from the equation

$$\frac{\overline{\sigma}}{a} = \frac{\sqrt{2\nu + 1 + 1/S(\nu)} \, \Gamma^2(-\nu/2)}{\sqrt{2\pi} \, 2^{\nu+1} \, S(\nu) \, \Gamma(-\nu)}.$$

Here $\mathcal{D}_\nu(\cdot)$ are Weber-Hermite functions or functions of parabolic cylinder :

$$S(\nu) = [\psi(1/2 - \nu/2) - \psi(-\nu/2)]/2, \quad \psi(x) = d\ln\Gamma(x)/dx.$$

Various ramifications of the solution (8) are connected with the degree in which the constraints are taken into account: in the first zone ($\bar{\sigma}^2 \leq 2a^2/\pi$) it is only the variance that matters, in the third zone ($\bar{\sigma}^2 > 2a^2$) only the density constraint is essential, in the intermediate zone both constraints are used. From (5) and (8) it follows that, with relatively small variances (in the first zone), LSM is optimum, with large variances (in the third zone) LMM is optimum, in the middle zone compromise between LSM and LMM algorithms are the best ones.

The similar results can be obtained for the intersection of the classes $\mathcal{F}_2$ and $\mathcal{F}_3$: the "tails" of the least favorable density f_{23}^* are exponential only with sufficiently large variances: $\sigma^2(f) > \sigma^2(f_3^*)$; if $\sigma^2(f) < \sigma^2(f_3^*)$ then the density is of the Weber-Hermite type (9).

4 Robust Adaptive M-estimators

Considering the problems of designing robust estimators we have supposed the availability of apriori information on the characteristics of a distribution class. However, in practical problems of estimation these characteristics are usually unknown and can be determined while data processing.

In applications the approximate value of the upper bound of a distribution variance could be obtained from the restrictions of a physical, technical or any other data measurement procedure. A statistician may estimate this value analysing data extreme values or using the upper confident bounds of a distribution variance. We suggest another way in the adaptive procedure.

As additional information is coming in (observation data come in successively), it is feasible to develop estimators which are capable of adapting to the ever increasing volume of data and correcting the characteristics of a distribution class for improving accuracy. With not very large sample sizes, such an approach is heuristic and the simplest for the examination by Monte Carlo technique.

Consider the adaptive algorithm of robust estimation of regression parameters (2) ,called ARLI - regression:

(i) Choose initial LMM - estimates for Θ : $\widehat{\Theta}_{LMM}$.

(ii) Evaluate the errors estimates

$$\widehat{\varepsilon}_i = x_i - \sum_{j=1}^m \widehat{\theta}_{j\,LMM}\,\phi_{ij}, \quad i = 1, ..., n.$$

(iii) Evaluate the estimates of the characteristics of the class $\mathcal{F}_{12}$

$$\widehat{\bar{\sigma}^2} = (n - m)^{-1}\sum_{i=1}^n \widehat{\varepsilon}_i^2, \quad \widehat{a} = (n + 1)[\widehat{\varepsilon}_{(h+1)} - \widehat{\varepsilon}_{(h)}], \quad n = 2h, n = 2h + 1.$$

(iv) Use the robust minimax algorith (2) with the loss function ρ_{12}^* .

The behavior of the ARLI-regression algorithm with respect to finite size samples (n = 20 - 100 with step 10 and 100 - 1000 with step 100) was studied by Monte Carlo. The following conclusions can be made on the results of simulation.

The ARLI - estimator has proved to be better than the LSM - and the LMM - estimators both in asymptotics (see Table 1) and in a finite sample size case (see Table 2

) . Its adaptive properties are obvious in the case of a mixture of the Gaussian and the Laplace densities :

$$f(x) = \frac{1-\varepsilon}{\sqrt{2\pi}} \exp(-\frac{x^2}{2}) + \frac{\varepsilon}{\sqrt{2}} \exp(-\frac{|x|}{\sqrt{2}}), \ 0 \le \varepsilon \le 1.$$

ε	$ARLI$	LSM	LMM
.0	1.0	1.0	.637
.5	.92	.65	.71
1.0	1.0	.5	1.0

Table 1: The asymptotic relative efficiency of the ARLI, LSM and LMM- estimates of a location parameter

ε	$ARLI$	LSM	LMM
.0	.73	1.0	.64
.5	.84	.67	.79
1.0	.98	.53	1.0

Table 2: The relative efficiency of the ARLI, LSM and LMM - estimates of a location parameter (n = 20)

For large sample sizes ($n > 100 - 200$), the ARLI - estimates fit well with the type of a distribution law involved.

For many distribution laws, including the Gaussian law, and with small sample sizes, the ARLI - estimators are close to the LMM - estimator. This effect can be explained by the fact that, with small sample sizes and at the sufficiently high level of significance experimental data may be explained by rather arbitrary disrtibution laws. For large sample sizes an unique distribution law comes into effect and determines the further data processing.

References

1. Huber, P.J. (1981) Robust Statistics. Wiley, New York.
2. Huber, P.J. (1967) The behaviour of maximum likelihood estimates under nonstandart conditions. - In: Proc. of the 5-th Berkeley Symp. on Math. Stat. and Prob., V.1, Berkeley Univ. California Press, 221-223.

Modeling Heterogeneity and Extraneous Variation Using Weighted Distributions

DIPAK K. DEY , FENGCHUN PENG and DANIEL LAROSE

Department of Statistics, University of Connecticut, Storrs, CT 06269-3120, USA,
Department of Biostatistics, Medical Center, University of Rochester, NY 14623, USA
and
Department of Statistics, University of Connecticut, Storrs, CT 06269-3120, USA.

Abstract:

In one-parameter exponential families, the variance is a function of the mean. One powerful method of modeling heterogeneity and overdispersion in an exponential family is to use a parametrized weighted distribution. In this paper we interpret such a weighted distribution model as an overdispersed generalized linear model by introducing covariates and forming a very general class of models. Here, such models are fit from a Bayesian perspective, using non-informative priors in order to let the data (likelihood) drive the inference. Bayesian calculations are carried out using a Metropolis-within-Gibbs sampling algorithm. An illustrative example using a previously analyzed data set is presented with emphasis on model comparison.

1 Introduction

Model determination is one of the fundamental problems in statistics. One powerful method of modeling heterogeneity and overdispersion in an exponential family is to use a parametrized weighted distribution. In this paper we interpret such a weighted distribution model as an overdispersed generalized linear model by introducing covariates and forming a very general class of models.

The method of weighted distributions takes into account the method of ascertainment by adjusting the probabilities of actual occurrence of events to arrive at a specification of the probabilities of the events as observed and recorded. For example, if one wanted to investigate the population of criminals in the country, it would be highly expensive (and perhaps futile) to attempt to obtain a random sample from the entire population of

---

*Kitsos, C.P., and Müller, W.G., Eds., *Proceedings of MODA4*, Physica Verlag, Heidelberg, 1995

criminals. Bayarri and DeGroot (1992) suggested that it may be preferable to study the population of criminals already in jail, which represents a weighted distribution $f^w(y|\theta)$ of the original $f(y|\theta)$ according to some weighted function $w(y)$, the probability of y number of criminals being caught.

Formally, suppose the random variable (or random vector) Y is distributed over some population of interest with density $f(y|\theta)$ where θ is some underlying parameter and that it is desired to make inference about θ. The usual statistical methodology assumes a random sample $y_1, y_2, \cdots, y_n$ from $f(y|\theta)$. There are many situations, however, when it is impossible or too expensive to obtain a random sample. Even if a random sample can be obtained, the experimenter may decide not to do so, since a bias sample, carefully chosen, may be more informative by associating with the original likelihood a weighted function $w(y)$ which may itself depend on a parameter τ, where τ may be unknown. Thus the observed data is a random sample from the following weighted function

$$f^w(y \mid \theta, \tau) = \frac{w(y, \tau) f(y|\theta)}{E_f[w(y, \tau)]} \qquad (1)$$

where the expectation in the denominator of (1) is the normalizing constant.

There are several ways of choosing the weight function. Rao (1985) observed that the "methods of ascertainment", for example how observations are recorded, must be accounted for in identifying a parent population of a sample. He introduced size bias, where $w(y) = y$, and remarks that the resulting weighted distribution $f^w(y|\theta)$ belongs to the same family as $f(y|\theta)$ for many standard distributions.

While examining a problem relating to clumped sampling, Diaconis and Efron (1985) and Efron (1986), found there was more dispersion in the data sets than the existing models could accommodate. Consequently, they introduced a model called the double exponential family, which allows the data analyst to model overdispersion while carrying out the usual regression analysis for the mean as a function of the predictors. Such overdispersion may be due to one or more possible causes, such as heterogeneity, selection bias, clumped sampling etc.

There are different ways of creating a larger class of models incorporating overdispersion. Historically, the most frequently used approach for creating a larger class has been through mixture models. For instance, the one parameter exponential family defining the Generalized Linear Model (GLM) can be mixed with a two parameter exponential family for the canonical parameter θ, equivalently mean parameter μ, resulting in a two parameter marginal mixture family for the data. Shaked (1980) showed that such mixing necessarily inflates the model variance. However, since the likelihood depends upon sample size while the mixture distribution does not, the relative overdispersion of the resulting mixture family to the original exponential family tends to infinity as sample size does. In other words, taking additional observations within a population does not increase our knowledge regarding heterogeneity across populations. (See Gelfand and Dalal, 1990 in this regard.) We also note that the resulting overdispersed family of mixture models will generally be awkward to work with since it will no longer be an exponential family (e.g. Beta-binomial, Poisson-gamma). Efron (1986) presented an alternative approach through so-called double exponential families. Such families are derived as the saddle point approximation to the density of an average of n^* random variables from a one parameter exponential family for large n^*. The parameter n^* written suggestively by Efron as $n\rho$ for actual sample size n introduces ρ as a second parameter to the model along with the canonical parameter θ.

Adding a dispersion parameter , say ϕ, to the usual one parameter exponential family results in an exponential dispersion model, EDM (Jorgensen, 1987). Because ϕ enters as a scale parameter, associated inference is usually handled differently from that for θ or μ. A one parameter exponential family in θ arises for each given ϕ but as a two parameter model in (θ, ϕ) we no longer have an exponential family . Recently Ganio and Schafer (1992) circumvent this problem in an approximate fashion by viewing the EDM as embedded within Efron's double exponential family and appealing to the associated asymptotic inference. Unlike in the mixture case, the asymptotics here result in overdispersion relative to the original exponential family which tends to a constant as $n \to \infty$. (See Efron, 1986 and Gelfand & Dalal, 1990).

Gelfand & Dalal (1990) argued that an appeal to asymptotics was not necessary to justify such models. More generally, for a given one parameter exponential family they introduced a two parameter exponential family of models. This family includes Efron's model as a special case and also includes a family discussed in Lindsay (1986). Retaining the exponential family structure simplifies inference (as we shall detail in the subsequent sections). Relative overdispersion behaves as in Efron's models.

Another approach for handling heterogeneity in populations is through the use of random effects. (See e.g., Breslow and Clayton, 1993 for a discussion and review of the literature). Here a standard GLM is employed but for each individual or population in the sample a random effect is added to the fixed covariate term in the definition of the mean structure. Such linear mixture models retain the familiar GLM formulation but are nonregular in that, as sample size tends to infinity so does the number of model parameters. Customary approaches for model adequacy and model choice fail since the usual asymptotic distribution theory is no longer valid.

The approach proposed in this paper is to view an overdispersed model as a perturbation of the original model so that the original model can be formed as a weighted distribution of the form (1). Our models are regular and parsimonious compared to a GLM but still offer the possibility of capturing overdispersion within an exponential family framework.

We adopt a Bayesian perspective in fitting these models since we are drawn to the unifying use of inference summaries based upon the posterior implicit therein. However, since we are primarily concerned with the modeling incorporated in the likelihood, we take an automatic or objective Bayesian stance employing appropriate noninformative prior specifications. For large sample sizes our inference will be close to that arising from maximum likelihood; for smaller samples our estimates of variability should be more appropriate than asymptotic ones arising from maximum likelihood. We illustrate our approach through the well known toxoplasmosis data examined by Efron (1986) and by Ganio and Schaffer (1992). Required Bayesian computation is handled through a Metropolis-within-Gibbs Markov chain Monte Carlo approach resulting in samples essentially from the joint posterior distribution which may be summarized to provide any desired inference. Such samples may also be used as the starting point for sampling from predictive distributions to investigate questions of model adequacy and model choice.

2 Overdispersed Models for Clumped Data

Consider the one parameter exponential family of the form

$$f(y \mid \theta) = b(y) \exp\{\theta y - \chi(\theta)\} \qquad (2)$$

where, if y is continuous, f is assumed to be a density with respect to Lebesgue measure, while if y is discrete, f is assumed to be a density with respect to counting measure. Now consider a function $T(y)$ of $y \epsilon \mathcal{Y}$, where $\mathcal{Y}$ is the sample space and an overdispersion parameter $\tau(> 0)$. Finally define a weight function as $\exp\{\tau T(y)\}$. Then it follows from (1) that the corresponding weighted model has the form

$$
\begin{aligned}
f^w(y \mid \theta, \tau) &= \frac{\exp\{\tau T(y)\}}{E[\exp\{\tau T(y)\}]} f(y|\theta) \\
&= b(y) \exp\{\theta y + \tau T(y) - \chi(\theta)\}/E_\theta[\exp[\tau T(y)]].
\end{aligned}
\tag{3}
$$

Defining $\exp\{\rho(\theta, \tau)\} = E_\theta[\exp\{\theta y + \tau T(y)\}]$ it follows that (3) can be expressed as

$$
f^w(y \mid \theta, \tau) = b(y) \exp\{\theta y + \tau T(y) - \rho(\theta, \tau)\}.
\tag{4}
$$

The form (3) is the so called two-parameter exponential family of models as proposed by Gelfand and Dalal (1990). They show that if $T(y)$ is convex, for a common mean, the $var(y)$ increases in τ, which is the most desirable property of modeling overdispersion.

Expression (2) is the customary one parameter exponential family from which a GLM is developed. In particular $\mu \equiv E(y) = \chi'(\theta)$, $var(y) = \chi''(\theta) = V(\mu)$. Here $\chi'(\theta)$ is strictly increasing in θ so that μ and θ are one-to-one ($\theta = (\chi')^{-1}(\mu)$). $V(\mu)$ is called the variance function. A GLM is defined through a link function g, a strictly increasing differentiable transformation from μ to $\eta \epsilon R^1$, i.e. $g(\mu) = \eta = x^T\beta$, where x and β are, respectively, a known vector of explanatory variables and an unknown vector of model parameters. If a scale parameter is incorporated (2) becomes

$$
f(y \mid \theta, \phi) = b(y, \phi) \exp\{(\theta y - \chi(\theta))/a(\phi)\}.
\tag{5}
$$

Now $var(y) = \frac{V(\mu)}{a(\phi)}$. If y is viewed as an average of n random variables then a usual form for $a(\phi)$ is $(n\phi)^{-1}$. In this form $\phi^* = n\phi$ is referred to as the dispersion parameter and (4) is called an exponential dispersion model (EDM) (Jorgensen, 1987). Ganio and Schafer (1992) extended the GLM based on (2) by defining in the EDM, $\phi = h(z^T\alpha)$ with z a known vector and α an unknown parameter vector. They assume $z^T\alpha$ includes an intercept. Note that the two parameter family in (4) differs from that in (3) in the sense that (3) is a customary two parameter exponential family, whereas (4) is a customary one parameter family for each fixed ϕ. Efron (1986) defined the double exponential family of models through the density

$$
\bar{f}(y \mid \theta, \rho, n) = c(\theta, \rho, n)\rho^{\frac{1}{2}} \exp\{n\rho(\theta y - \chi(\theta)) + n(1 - \rho)(\theta(y)y - \chi(\theta(y)))\}
\tag{6}
$$

where $\theta(y) = (\chi')^{-1}(y)$. Here y is viewed as an average of n i.i.d. random variables, θ is the canonical parameter and ρ is a dispersion parameter. In regression problems he assumes a GLM in θ (canonical link) i.e. $\theta = x^T\beta$ and for ρ, a model $\rho = h(z^T\alpha)$. Using various expansions, Efron shows that (6) permits attractive approximation as n grows large. Most notably $\bar{f}$ behaves like (4) with $a(\phi) = (n\rho)^{-1}$. Hence Ganio and Schafer(1992) treat their extended GLM based upon (4) as an example of Efron's double exponential family and carry out their model fitting following his examples.

We note that, regardless of n, (6) is of the form (3) with $T(y) = \theta(y)y - \chi(\theta(y))$, $\tau = n(1 - \rho)$ and $\theta = n\rho\theta$. Straightforward calculation shows this $T(y)$ is convex so that, in fact (6) is a special case of (3). As Gelfand and Dalal (1990) show, other choices of

$T(y)$ may be more appropriate and in any event it seems preferable to work with the exact form (3) rather than with approximation to (6).

Returning to (3), under usual regularity conditions, we have the following properties. If $\rho^{(r,s)} \equiv \frac{\partial \rho^{r+s}}{\partial \theta^r \partial \tau^s}$ then $\rho^{(1,0)} = E(y|\theta,\tau) \equiv \mu$, $\rho^{(2,0)} = var(y|\theta,\tau)$, $\rho^{(0,1)} = E(T(y)|\theta,\tau)$, etc. It is sometimes convenient to consider (3) through a mean parametrization

$$f(y \mid \mu, \tau) = b(y)e^{(y-\mu)\psi^{(1,0)}(\mu,\tau)+\tau T(y)+\psi(\mu,\tau)}. \tag{7}$$

In (7), we employ the same sort of notation for ψ as for ρ. By comparison with (4) we have $\theta = \psi^{(1,0)}(\mu,\tau)$, $\rho(\theta,\tau) = -\psi(\mu,\tau) + \mu\psi^{(1,0)}(\mu,\tau)$. Using (6) and straightforward calculation, we can show that $E\left(\frac{\partial^2 \log f}{\partial \tau \partial \mu}\right) = 0$, i.e, that μ and τ are orthogonal parameters in the sense of Barndorff-Nielsen (1978, p.184) and Cox and Reid (1987). The only practical drawback to working with (2) is that $\rho(\theta,\tau)$ is not available explicitly. While $\chi(\theta)$ in (3) is usually an explicit function of θ, the best we can generally do with ρ is to note that $\rho(\theta,\tau) = \log \int b(y)e^{\theta y+\tau T(y)}dy$ and approximate ρ through a univariate numerical integration or summation. In the examples we have investigated thus far this has not presented a problem.

To incorporate covariates in our model of the form (3) we consider independent response y_i with associated covariates $x_i, p \times 1$ and $z_i, q \times 1$, i=1,2,..,n. The components of x and z need not be exclusive. We define $\theta_i = g(x_i^T \beta), \tau_i = h(z_i^T \alpha)$ resulting in the likelihood, up to proportionality,

$$L(\beta,\alpha;Y) = \prod_{i=1}^{n} e^{\theta_i y_i + \tau_i T(y_i) - \rho(\theta_i,\tau_i)}. \tag{8}$$

3 Prior Specification for Overdispersed Models

As noted in section 1, we take a Bayesian perspective in "fitting" the model in (8). Furthermore we propose noninformative prior specification in order to let the data (likelihood) drive the inference. One such choice is the flat prior $f(\beta,\alpha) = 1$. A second possibility is Jeffreys's prior, a commonly used reference prior for Bayesian analysis. This prior is defined to be the square root of the determinant of the Fisher's information matrix associated with (4). It has been the focus of renewed attention in recent work of Kass (1989, with discussion, and 1990). Ibrahim and Laud (1991) calculated Jeffreys's prior in the case of a GLM developed from (5) assuming ϕ is known. We now extend this calculation to (4).

Straightforwardly we may show that

$$E\left(\frac{\partial^2 \log L(\beta,\alpha,y)}{\partial \beta_i \partial \beta_j}\right) = -\sum_i \rho^{(2,0)}(\theta_i,\tau_i)x_{ij}x_{ik}(g'(x_i^T\beta))^2$$

$$E\left(\frac{\partial^2 \log L(\beta,\alpha,y)}{\partial \alpha_i \partial \alpha_j}\right) = -\sum_i \rho^{(0,2)}(\theta_i,\tau_i)z_{ij}z_{ik}(h'(z_i^T\alpha))^2$$

$$E\left(\frac{\partial^2 \log L(\beta,\alpha,y)}{\partial \beta_j \partial \alpha_k}\right) = -\sum_i \rho^{(1,1)}(\theta_i,\tau_i)x_{ij}z_{ik}(g'(x_i^T\beta))(h'(z_i^T\alpha)).$$

Let X denote the nxp design matrix arising from the x_i's, Z the nxq design matrix arising from the z_i's, M_θ an nxn diagonal matrix with $(M_\theta)_{ii} = \rho^{(2,0)}(\theta_i,\tau_i)(g'(x_i^T\beta))^2$, M_τ an nxn diagonal matrix with $(M_\tau)_{ii} = \rho^{(0,2)}(\theta_i,\tau_i)(h'(z_i^T\alpha))^2$

246

and $M_{\theta,\tau}$ an nxn diagonal matrix with $(M_{\theta,\tau})_{ii} = \rho^{(1,1)}(\theta_i, \tau_i)(g'(x_i^T\beta))(h'(z_i^T\alpha))$. Then

$$I(\beta, \alpha) = \begin{pmatrix} X^T M_\theta X & X^T M_{\theta,\tau} Z \\ Z^T M_{\theta,\tau} X & Z^T M_\tau Z \end{pmatrix} \qquad (9)$$

and Jeffreys's prior is $|I(\beta, \alpha)|^{\frac{1}{2}}$.

To work with Jeffreys's prior requires calculation of ρ, $\rho^{(2,0)}$, $\rho^{(0,2)}$ and $\rho^{(1,1)}$. This in turn, requires calculation of six integrals of the form $\int y^c T^d(y) b(y) e^{\theta y + \tau T(y)} dy$ for the set of (c,d) from $\{(0,0)\ (1,0)\ (2,0)\ (0,1)\ (0,2)\ (1,1)\}$. Numerical integration or summation for such integrals is generally routine.

Suppose instead we define an extension of the GLM using the mean parametrization (7) setting $\mu = g(x^T\beta)$ with again $\tau = h(z^T\alpha)$. In the case of (2), i.e., $\tau = 0$ this is, in fact, the more usual way of formulating a GLM. Paralleling (4) let us now define M_μ as an nxn diagonal matrix with $(M_\mu)_{ii} = \psi^{(2,0)}(\mu_i, \tau_i)(g'(x_i^T\beta))^2$ and M_τ an nxn diagonal matrix with $(M_\tau)_{ii} = \psi^{(0,2)}(\mu_i, \tau_i)(h'(z_i^T\alpha))^2$. The orthogonality of μ and τ results in

$$I(\beta, \alpha) = \begin{pmatrix} X^T M_\mu X & 0 \\ 0 & Z^T M_\tau Z \end{pmatrix}. \qquad (10)$$

Hence Jeffreys's prior is $|I(\beta, \alpha)|^{\frac{1}{2}} = |X^T M_\mu X|^{\frac{1}{2}} |Z^T M_\tau Z|^{\frac{1}{2}}$. In the example of section 4, we work with (9) rather than (10) in order to more easily reconcile our analysis with that of Efron (1986) and of Ganio and Schafer (1992).

An important question to ask at this point is whether in combination with the likelihood in (8), either a flat prior for (β, α) or Jeffreys's prior as in (9) results in a proper posterior for (β, α). This problem is addressed in Dey et. al. (1993).

4 An Overdispersed Binomial Model

We now apply the Bayesian model developed from (4) to the toxoplasmosis data, using Jeffreys's prior. We consider a data set involving the disease toxoplasmosis. The data consists of the number of subjects y_j in a sample (not a genuine random sample!) of n_j testing positive for toxoplasmosis in $j = 1, 2, ..., 34$ cities in El Salvador. A supplied covariate is x_j, the annual rainfall for city j. The data appear in Efron (1986) and so are not reproduced here. We do note that the n_j range from 1 to 82 with nearly half less than or equal to 10, four greater than 50. Efron (1978) fit an ordinary logistic regression to these data and found a highly significant cubic term in x_j. Nevertheless the cubic logistic regression model failed a customary chi-squared goodness-of-fit test. Efron (1986) revisits this example arguing that the reason for this failure is that the binomial model of variation implicit in the logistic regression is inadequate. The observed deviations from the fitted logistic regression are too large to be explained by ordinary binomial variation. Stated in a different way, the large n_j have a large influence on the fit because they force variation in $\frac{y_j}{n_j}$ which is proportionately smaller in n_j. We shall return to this point later.

Using his model, (6), Efron assumes a cubic form for θ_j in standardized rainfall, $\tilde{x}_j = (x_j - \bar{x})/S_x$. He uses a somewhat contrived form for ρ_j, $\rho_j = M(1 + e^{-z^T\alpha})^{-1}$ with $M = 1.25$ and $z_j^T\alpha$ a quadratic in standardized sample size $\tilde{n}_j = (n_j - \bar{n})/S_n$. Necessarily, with a larger family of models, improved fitting occurs. Ganio and Schafer (1992) also examined the data set utilizing their EDM based on (5) with various models for $\phi = h(z^T\alpha)$. They find (p. 800) "little reason to eschew the simple generalized linear model with ...[ϕ a constant]". Efron appears to concur (p: 718) obtaining SD's large enough so that the components of α are not at all significantly different from 0. He adds the caveat that "[i]n general it seems more difficult to estimate ρ_j than θ_j so [the regression model for ρ_j] should be kept simple."

In the context of this example, (4) becomes an overdispersed binomial model. That is, at $\tau = 0$ we obtain a Binomial(n, p) density. The canonical parameter θ is the logit, $\log(\frac{p}{1-p})$. Any choice of convex T(y) provides a proper density for y since its domain is a finite set. We tried $T(y) = (\frac{y}{n})\log(\frac{y}{n}) + (1 - (\frac{y}{n}))\log(1 - (\frac{y}{n}))$, taking $T(0) = T(n) = 0$, which corresponds to Efron's double exponential family (see the discussion following (6)). We also used $T(y) = y^2$, which for small τ approximates an arbitrary mixture of (2) provided the mixing distribution has finite second moment (see Cox, 1983, p. 272). Regardless of T(y), exact calculation of $\rho(\theta, \tau)$ only requires evaluation of a finite sum, i.e., at a given n_j, $\rho(\theta, \tau) = \log \sum_{y=0}^{n_j} \binom{n_j}{y} e^{\theta y + \tau T(y)}$.

Paralleling the earlier work of Efron (1986) and of Ganio and Schafer (1992) we set θ_j to be a cubic in $\tilde{x}_j$. We combine this with three choices for τ_j : model 1, $\tau_j = \alpha_0 + \alpha_1 \tilde{n}_j$; model 2, $\tau_j = \alpha_0$; model 3, $\tau_j = 0$; We ask two questions. Are these models adequate? Amongst those that are, which one would we choose? The Bayesian approach to answering these questions is based upon predictive distributions. Under an improper prior, the *prior* predictive distribution $\int \int L(\beta, \alpha; y)|I(\beta, \alpha)|^{\frac{1}{2}} d\beta d\alpha$ is improper hence difficult to calibrate. As an alternative we adopt a cross-validation approach, in particular, considering the proper densities $f(y_r \mid y_{(r)})$, r = 1, 2, ..., 34, where $y_{(r)}$ denotes y with y_r removed. We, in fact, condition on the actual observations $y_{(r),obs}$ creating the predictive distribution for y_r under the model and all the data except y_r. For model determination we would then compare, in some fashion, $f(y_r \mid y_{(r),obs})$ with the observation $y_{r,obs}$. Such cross validation is discussed in Gelfand, Dey and Chang (1992) and in further references provided there. We consider diagnostic developed from the set of $(f(y_r \mid y_{(r),obs}), y_{(r),obs})$ which is useful for model adequacy and choice.

A natural approach for model adequacy is to draw, for each r, a sample from $f(y_r \mid y_{(r),obs})$, and compare this sample with $y_{r,obs}$. In particular using this sample we might obtain the .025 and .975 quantiles of the $f(y_r \mid y_{(r),obs})$ say $\underline{y}_r$ and $\overline{y}_r$ and see how many of the 34 $y_{r,obs} \in [\underline{y}_r, \overline{y}_r]$. For the toxoplasmosis data all three models seem adequate. Under each model at least 32 of the 34 intervals contained the corresponding $y_{r,obs}$. Suppose instead we obtain the lower and upper quartiles of $f(y_r \mid y_{(r),obs})$ and see how many $y_{r,obs}$ fall in their interquartile ranges. We find 26 for model 1, 25 for model 2 and 20 for model 3. Under the true model we would expect half, i.e., 17. Hence model 3 performs closest to expectation while model 1 and model 2 may be overfitting. Recall our earlier remark regarding the influence of larger n_r on the fit. Table 1 considers the four $n_r > 50$.

Table 1 : For the Toxoplasmosis data a comparison of predictive intervals for the large samples

Data			Model 1			Model 2			Model 3		
r	n_r	y_r	$\underline{y}_r$	$\overline{y}_r$	L_r	$\underline{y}_r$	$\overline{y}_r$	L_r	$\underline{y}_r$	$\overline{y}_r$	L_r
18	54	33	17	42	25	22	37	15	20	37	17
24	77	41	26	58	32	32	51	19	33	51	18
27	82	46	15	49	34	24	43	19	25	45	20
30	75	53	26	59	33	31	50	19	32	51	19

248

Note that only model 1 permits enough dispersion to contain $y_{27,obs}$ and $y_{30,obs}$.

We conclude with a brief discussion regarding the computation of the $f(y_r \mid y_{(r)})$ and the sampling from them in the present context. See Gelfand, Dey and Chang (1992) and Gelfand and Dey (1994) for a more general discussion. For each model we used a Metropolis-within-Gibbs Markov chain Monte Carlo algorithm (Müller, 1993) to develop samples from the posterior beginning with multiple starts in the vicinity of the maximum likelihood estimate. Evaluation of the likelihood required repeated calculation of the function $\rho(\theta_j, \tau_j)$. Evaluation of Jeffreys's prior required repeated evaluation of certain $\rho^{(r,s)}(\theta_j, \tau_j)$ as discussed after (9). As with $\rho(\theta, \tau)$ these are finite sums.

Samples from $f(y_r \mid y_{(r)})$ are drawn in two stages. Given samples (β_j^*, α_j^*), j = 1, 2, ..., m from $f(\beta, \alpha \mid y)$ we convert these to samples from $f(\beta, \alpha \mid y_{(r)})$ by resampling with weights $q_j = \frac{(f(y_{r,obs} \mid \beta_j^*, \alpha_j^*)^{-1}}{\sum_{j=1}^m (f(y_{r,obs} \mid \beta_j^*, \alpha_j^*))^{-1}}$. See Smith and Gelfand (1992) in this regard.

Since $f(y_r \mid y_{(r)}) = \int f(y_r \mid \beta, \alpha) f(\beta, \alpha \mid y_{(r)}) d\beta d\alpha$, if β', α' is a draw from $f(\beta, \alpha \mid y_r)$ then if $y_r' \sim f(y_r \mid \beta', \alpha')$, the marginal distribution of y'_r is $f(y_r \mid y_{(r)})$. To draw y_r' from $f(y_r \mid \beta', \alpha')$ is easy in the present case since it is a density over the finite set $\{0, 1, 2, ..., n_r\}$.

References

Bayarri, M.J. and DeGroot, M.H. (1992), "A "BAD" View of Weighted Distributions and Selection Models," In *Bayesian Statistics* 4, (J. Bernardo et al. eds.), Oxford University Press, Oxford, 17-33,(with discussion).

Barndorff-Nielsen, O.E. (1978), *Information and Exponential Families in Statistical Theory*, John Wiley & Sons, New York.

Breslow, N. and Clayton, D. (1993), "Approximate Inference in Generalized Linear Mixed Models," *Journal of the American Statistical Association*, 88, 9-25.

Cox, D.R. (1983), "Some Remarks on Overdispersion," *Biometrika* 70, 269-74.

Cox, D.R. and Reid, N. (1987), "Parameter Orthogonality and Approximate Conditional "Inference" (with discussion)," *Journal of the Royal Statistical Society*, Ser. B, 49, 1-39.

Dellaportas, P. and Smith, A.F.M. (1993). "Bayesian Inference for Generalized Linear Models via Gibbs Sampling," *Applied Statistics*.

Dey, D.K., Gelfand, A.E. and Peng, F. (1993), "Overdispersed Generalized Linear Models." *Unpublished Manuscript*.

Diaconis, P. and Efron, B.(1985), "Testing the Independence of a Two- Way Table: New Interpretations of the Chi-square Statistic (with discussion)". *Annals of Statistics*, 13, 845-913.

Efron, B. (1978), "The Geometry of Exponential Families," *The Annals of Statistics*, 6, 362-376.

Efron, B. (1986), "Double Exponential Families and Their Use in Generalized Linear Regression," *Journal of the American Statistical Association*, 81, 709-21.

Ganio, L.M. and Schafer, D.W. (1992). "Diagnostics for Overdispersion," *Journal of the American Statistical Association*, 87, 795-804.

Gelfand, A.E. and Dalal, S.R. (1990), "A Note on Overdispersed Exponential Families," *Biometrika*, 77, 55-64.

Gelfand, A.E. and Dey, D.K. (1994), "Bayesian Model Choice: Asymptotics and Exact Calculations," *Journal of the Royal Statistical Society*, Ser. B, 56, 501-514.

Gelfand, A.E., Dey, D.K. and Chang H. (1992), "Model Determination Using Predictive Distributions with Implementation Via Sampling-Based Methods," In *Bayesian Statistics 4*, (J. Bernardo et al. eds.), Oxford University Press, Oxford, 147-167.

Ibrahim, J.G., and Laud, P.W. (1991). "On Bayesian Analysis of General Linear Models Using Jeffreys's Prior", *Journal of the American Statistical Association*, 86, 981-986.

Iyengar, S. and Greenhouse, J.B. (1988). "Selection Models and the File Drawer Problem", *Statistical Science* 3, 109-135.

Jeffreys, H. (1961), *Theory of Probability*, Oxford University Press, London.

Jorgensen, B. (1987). "Exponential Dispersion Models (with discussion)," *Journal of the Royal Statistical Society*, Ser. B, 49, 150.

Kass, R.E. (1989), "The Geometry of Asymptotic Inference (with discussion)," *Statistical Science*, 4, 188-234.

Kass, R.E. (1990), "Data Translated Likelihood and Jeffreys's Rules," *Biometrika*, 77, 107-114.

Lindsay, B. (1986), "Exponential Family Mixture Models (with least squares estimators)," *The Annals of Statistics*, 14,124-37.

McCullagh, P., and Nelder, J.A. (1989), *Generalized Linear Models*, Chapman & Hall, London.

Müller P. (1993), "A Generic Approach to Posterior Integration and Gibbs Sampling," *Journal of the American Statistical Association*, To appear.

Patil, G.P. (1984), "Studies in Statistical Ecology Involving Weighted Distribution," *In Statistics: Applications and New Directions*, 475-503. Calcutta: Indian Statistical Institute.

Rao, C.R. (1985), "Weighted Distribution Arising out of Methods of Ascertainment: What Population Does a Sample Represent?" *In a Celebration of Statistics: The ISI Centenary Volume (A.G. Atkinson and S.E. Fienberg, eds.), 543-569. New York: Springer-Verlag.*

Shaked, M. (1980), "On Mixtures from Exponential Families," *Journal of the Royal Statistical Society*, Ser. B, 42, 192-198.

Smith, A.F.M. and Gelfand, A.E. (1992), "Bayesian Statistics Without Tears: a Sampling-Resampling Perspective, " *The American Statistician*, 46, 2, 84-88.

Wedderburn, R. (1976), "On the Existence and Uniqueness of the Maximum Likelihood Estimates for Certain Generalized Linear Models," *Biometrika*, 63, 27-32.

Gibbs Sampling for ARCH Models in Finance

Wolfgang Polasek and Peter Müller

Institut für Statistik und Ökonometrie, University of Basel, Petersgraben 51,
4051 Basel, Switzerland
and
Institute of Statistics and Decision Sciences, Duke University, Durham,
NC 27708-0251, USA.

Abstract:

The paper develops a simple estimation procedure for Bayesian ARCH models: The Gibbs-importance algorithm (also called independence chain) is applied for the simulation step involving the ARCH parameters. We demonstrate this approach to model the volatility between the Dollar, the DM and the Yen. An extension of the model to multivariate VAR-VARCH models is proposed.

1 Introduction

Autoregressive conditional heteroskedasdicity (ARCH) models have become an important tool for modeling volatility in finance. Simple estimation procedures do not obey the parameter constraints and their finite sample distribution is not known. Recent advances in Bayesian inferences allow an exact derivation of the complete posterior distribution by stochastic simulation. Based on Gibbs sampling (see Gelfand and Smith 1990 or Smith and Gelfand 1992) and Markov Chain Monte Carlo methods (MCMC) we will show how the posterior distribution for a Bayesian ARCH model can be derived. Bayesian ARCH models have been applied increasingly for financial time series (see e.g. Müller 1991 or Jacquier etal. 1992), because the forecasting properties of so-called Bayesian vector autoregressive or B-VAR models have been found to outperform the corresponding classical models. A further advantage for the Gibbs sampler is that one can impose almost any additional complication to a basic model. The example discusses the exchange rates of daily time series between the Deutschmark, Yen and the Dollar. In a multivariate model the volatility between countries is analyzed.

---

*Kitsos, C.P., and Müller, W.G., Eds., *Proceedings of MODA4*, Physica Verlag, Heidelberg, 1995

1.1 ARCH Models in Finance

ARCH models have been developed by Engle (1982) and Bollerslev et al. (1992) for financial time series and have enjoyed since then a variety of developments. Many financial time series are found to be random walk processes in the levels which are transformed to zero mean processes by taking the first differences. Regression models in the first differences have usually found no or eventually not too much structure (e.g. like from time to time in exchange rate models), but the covariance structure seem to produce so-called volatility clusters. There are long times with rather small changes, but if the variances start to bounce up and down like seismic data, it takes some time for them to settle down again on a normal level. Those volatility clusters tend to spread in the market and form some kind of transmission patterns. The marginal data of such conditional heteroskedastic time series are then more leptokurtic than normal distributions. Unfortunately, not all of the leptokurtosis can be captured by ARCH models and therefore more and richer classes of volatility models are developed and tested for their usefulness. This line of research is also pursued in this paper. We concentrate on a model class which might be called in analogy to the usual transfer function models ARCH transfer function models. For technical reasons we are trying to concentrate on ARCH models with 'exogenous' inputs and we don't treat generalized ARCH (GARCH) or ARMACH models. We start with the univariate model in section 2 and we analyze in section 4 the multivariate VAR-VARCH model. The example is dicussed in section 3.

2 The Univariate ARCH Model

We consider the following ARCH type model:

$$
\begin{aligned}
y_t &= \mathbf{x}_t'\beta + \epsilon_t, \\
\epsilon_t &\sim N(0, h_t), \\
h_t &= \mathbf{z}_t'\gamma, \quad h_t > 0,
\end{aligned}
\tag{1}
$$

where the observations y_t are the daily exchange rates for USD/DM, recorded over the period July 1988 to December 1992. The rates are as recorded at 12 midnight. The observations are transformed to log differences. The covariates $\mathbf{x}_t$ in the mean equation are the lagged values y_{t-1} and the exchange rates for USD/Sfr and USD/Yen from the preceding period $t - 1$. The equation for the volatility contains covariates in the vector $\mathbf{z}_t = (\epsilon_{t-1}^2, Sfr_{t-1}^2, Yen_{t-1}^2)$.

Together with a prior distribution $p(\beta, \gamma, \epsilon_0)$ on the ARCH parameters and intitial conditions, the model defines a coherent probability distribution on observables and parameters:

$$
p(\beta, \gamma, \epsilon_0, y_1, \ldots, y_T) = p(\beta, \gamma, \epsilon_0) \prod_{t=1}^{T} p(y_t | y_{t-1}, \beta, \gamma).
$$

In particular notice that the previous residuals ϵ_{t-1} which enters as covariate in the conditional distribution $p(y_t | y_{t-1}, \beta, \gamma)$ causes no concern as $\epsilon_{t-1} = y_{t-1} - x_{t-1}'\beta$ is defined only in terms of parameters and observations which are being conditioned upon. As prior we choose a constant noninformative prior for $p(\beta, \gamma)$, and for ϵ_0 we use a point mass at zero.

2.1 Estimation

To estimate the model parameters and to forecast future volatility we implemented a Markov chain Monte Carlo scheme. The chosen algorithm is similar to the Markov chain Monte Carlo used in Jaquier, Polson and Rossi (1993). We construct a Markov chain Monte Carlo algorithm of the type suggested as "independence chain" in Tierney (1991). Specification of an independence chain requires for each parameter θ_i the specification of a "probing distribution" $g_i(\theta_i|\theta_{-i})$, where θ_{-i} is the full parameter vector without θ_i. The "probing distribution" should be chosen to mimick the full conditional posterior $p(\theta_i|\theta_{-i}, \mathbf{y})$. Iterative sampling from these probing distributions, followed by an appropriate rejection step, defines a Markov chain which can be shown to converge to the desired posterior distribution $p(\theta|\mathbf{y})$ (c.f. Tierney 1991).

Rather than iteratively sampling the individual parameters, we implement the chain by successive substituion of the subvectors β and γ. Diebold and Robert (1990) argue that sampling in blocks is preferable wherever possible.

The joint distribution of model (1)-(3) is given by

$$p(\beta, \gamma, \mathbf{y}) = \prod_{t=1}^{T} |\mathbf{z}_t'\gamma|^{-\frac{1}{2}} \exp\left\{ -\frac{1}{2}(y_t - \mathbf{x}_t'\beta)^2 / \mathbf{z}_t'\gamma \right\}.$$

2.2 The Full Conditional and Probing Distributions

The choice of the probing distributions is motivated by approximating regression problems. For $p_\beta(\beta|\gamma)$ consider the following regression problem:

$$y_t = \mathbf{x}_t'\beta + \epsilon_t, \tag{2}$$

$$Var(\epsilon_t) = h_t = \mathbf{z}_t'\gamma, \tag{3}$$

where the residuals which are required in $\mathbf{z}_t$ are computed using the values of β from the previous iteration of the Markov chain. The posterior distribution on β in this model provides the probing distribution $g(\beta|\gamma)$.

$$g(\beta|\gamma, \mathbf{z}) = N[\hat{\mathbf{b}}, \hat{\mathbf{H}}] \tag{4}$$
$$\hat{\mathbf{H}} = (\mathbf{X}'\mathbf{D}_h'\mathbf{X})^{-1} \quad \text{with} \quad \mathbf{D}_h = diag(h_1, \ldots, h_T),$$
$$\hat{\mathbf{b}} = \hat{\mathbf{H}}^{-1}\mathbf{X}'\mathbf{D}_h^{-1}\mathbf{y}.$$

The choice of the probing distribution for γ is motivated by an approximate regression problem. Thus, we derive the probing distribution $g(\gamma|\beta)$ from the regression model in the squared residuals:

$$\epsilon_t^2 = \mathbf{z}_t'\gamma + \omega_t, \qquad \omega_t \sim N(0, 2h_t), \tag{5}$$

Now the heteroskedastic error term $h_t = \mathbf{z}_t'\gamma$ is calculated new in each iteration, where the γ values from the previous iterations are substituted. The doubling of the variance corresponds to a χ^2 approximation of a squared normal error term. Thus, the probing distribution with the diffuse prior is

$$p(\gamma|\beta, \mathbf{y}) = N[\hat{\gamma}, \hat{\mathbf{G}}] \tag{6}$$

$$\hat{\mathbf{G}} = \left(\mathbf{Z}'\mathbf{D}_h^{-1}\mathbf{Z}\right)^{-1} \quad \text{with} \quad \begin{pmatrix} \mathbf{z}_1' \\ \vdots \\ \mathbf{z}_T' \end{pmatrix}$$

where $\mathbf{D}_h$ is given in (4) and $\mathbf{e}' = (\epsilon_1^2, \ldots, \epsilon_T^2)$ with $\epsilon_t = y_t - \mathbf{x}_t'\beta$ is the vector of squared error terms.

2.3 The Independence Chain - Hastings Algorithm

This section describes briefly the Metropolis-Hastings algorithm which was used in the Markov chain simulation. More details can be found in Tierney (1991).

[1] Generate candidates z from a probing distribution $g(z)$ for the posterior distribution $\pi(x)$.

[2] Form the "importance weight function"

$$w(x) = \frac{\pi(x)}{g(x)}.$$

[3] Accept the candidate z with the probability

$$\alpha = \min\left\{\frac{w(z)}{w(x)}, 1\right\}$$

where x is the value from the last iteration.

It is important to note that the Markov chain defined by an independence chain has exactly the desired posterior distribution as limiting distribution, i.e. the approximation which is being used to define the probing distribution is only in the motivating intuition, not in the final inference.Using a combination of straightfoward Gibbs sampler steps and independence chain steps outlined above, we implemented the following Markov chain Monte Carlo algorithm:

Step 1: We start the Markov chain by setting all parameters equal to some ad-hoc initial guesses β^0, γ^0. The ad-hoc estimate β^0 is derived by simple least squares fits

Step 2: Replace γ^0 by γ^1 generated using a Metropolis-Hastings step as described in 2.3 with the probing distribution given in (4).

Step 3: Replace β^0 by a draw from the multivariate normal conditional posterior distribution $\beta^1 \sim p(\beta|\gamma^1, y)$.

Step 4: Repeat steps 2 and 3 until the Markov chain is judged to have practically converged.

Step 5: Evaluate desired posterior integrals (posterior means, standard deviations and marginal distributions for the paramters) by appropriate ergodic averages over the simultion output.

3 Example: The Lag1 US/DM Volatility Model (Univariate Estimation Results)

In our application we estimate first the mean and then the variance effects of the univariate daily exchange rate equation for y_t, the USD/DM exchange rate, as in equation in (1):

$$y_t = \beta_0 + \beta_1 y_{t-1} + \beta_2 Sfr_{t-1} + \beta_3 Yen_{t-1} + \epsilon_t.$$

Figure 1 shows a rather symmetric distributions for the coefficients close to 0. The USD/DM- and the USD/SFr effects are slightly positive but 'insignificant', since the mean of the posterior distribution is close to zero, while the Yen has a negative effect of -0.12 (in the mean) on the USD/DM exchange rate. This means that a devaluation trend in e.g. the DM has an appreciation effect in the Yen and vice versa. Figure 2 shows the marginal posterior distribution for $\gamma_1, \gamma_2, \gamma_3$ of the volatility equation

$$h_t = \gamma_0 + \gamma_1 \epsilon_{t-1}^2 + \gamma_2 S f r_{t-1}^2 + \gamma_3 Y e n_{t-1}^2$$

where ϵ_t is the residual in (1) which can be interpreted as a residual DM effect. The marginal distributions show the expected right skewed effect, and again, the Yen effect is slightly higher than the others.

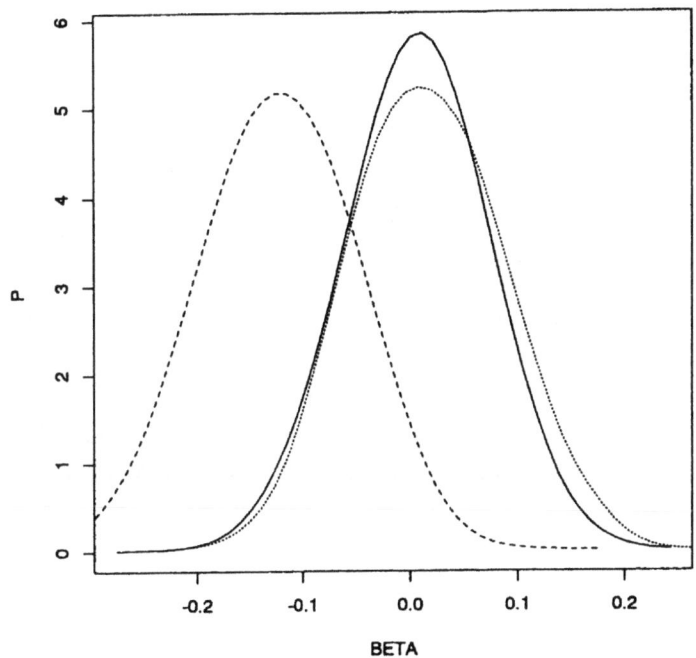

Figure 1: Marginal posterior densities $p(\beta_j|y)$.

Also, the Yen volatility effect seems to be less skewed with a side peak. Further estimation with e.g. a mixture model (Diebold and Roberts 1990) will show if this peak can be interpreted as a mixture effect (mixture volatility model) of two types of volatility effects, possibly a trading time delay effect.

Convergence diagnostics of the Gibbs sampler are given in Figure 3. The parallel runs of length 5000 did converge pretty fast. Higher order volatility models are currently investigated and exhibit more technical problems. The stationarity condition for ARCH models become more stringent and affects the convergence behavior. Also, the correlation structure between effects becomes more visible and the interpretation becomes less obvious.

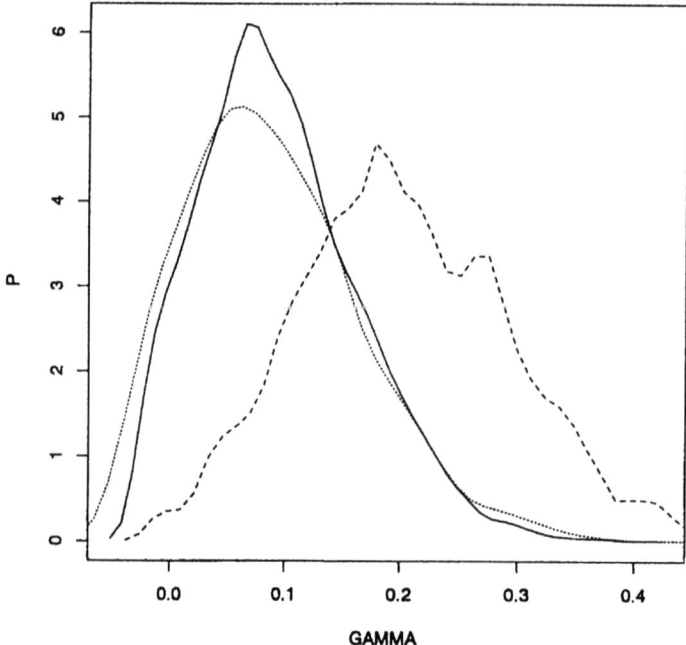

Figure 2: Marginal posterior densities $p(\gamma_j|y)$.

4 The Multivariate ARCH Model

A multivariate (VAR-VARCH) model has to take into account all simultaneous time series effects of an exchange rate model. Let $z_t = (y_{1t}, y_{2t}, y_{3t})' = (DM_t, Sfr_t, Yen_t)'$.

$$
\begin{aligned}
DM_t &= y_{t1} = \beta_{10} + \beta_{11}y_{1,t-1} + \beta_{12}Sfr_{t-1} + \beta_{13}Yen_{t-1} + \epsilon_{1t}, \\
Sfr_t &= y_{2t} = \beta_{20} + \beta_{21}y_{1,t-1} + \beta_{22}DM_{t-1} + \beta_{23}Yen_{t-1} + \epsilon_{2t}, \\
Yen_t &= y_{3t} = \beta_{30} + \beta_{31}y_{1,t-1} + \beta_{32}Sfr_{t-1} + \beta_{33}DM_{t-1} + \epsilon_{3t}.
\end{aligned}
$$

The residuals are distributed normally as $\epsilon_t = (\epsilon_{1t}, \epsilon_{2t}, \epsilon_{3t})' \sim N[\mathbf{0}, \mathbf{H}_t]$ with covariance matrix

$$
\mathbf{H}_t = \begin{pmatrix} h_{11t} & h_{12t} & h_{13t} \\ h_{21t} & h_{22t} & h_{13t} \\ h_{31t} & h_{32t} & h_{33t} \end{pmatrix}.
$$

The volatility equations are

$$
\begin{aligned}
h_{11t} &= \gamma_{10} + \gamma_{11}\epsilon_{1,t-1}^2 + \gamma_{12}Sfr_{t-1}^2 + \gamma_{13}Yen_{t-1}^2, \\
h_{22t} &= \gamma_{20} + \gamma_{21}\epsilon_{2,t-1}^2 + \gamma_{22}DM_{t-1}^2 + \gamma_{23}Yen_{t-1}^2, \\
h_{33t} &= \gamma_{30} + \gamma_{31}\epsilon_{3,t-1}^2 + \gamma_{32}Sfr_{t-1}^2 + \gamma_{33}DM_{t-1}^2.
\end{aligned}
$$

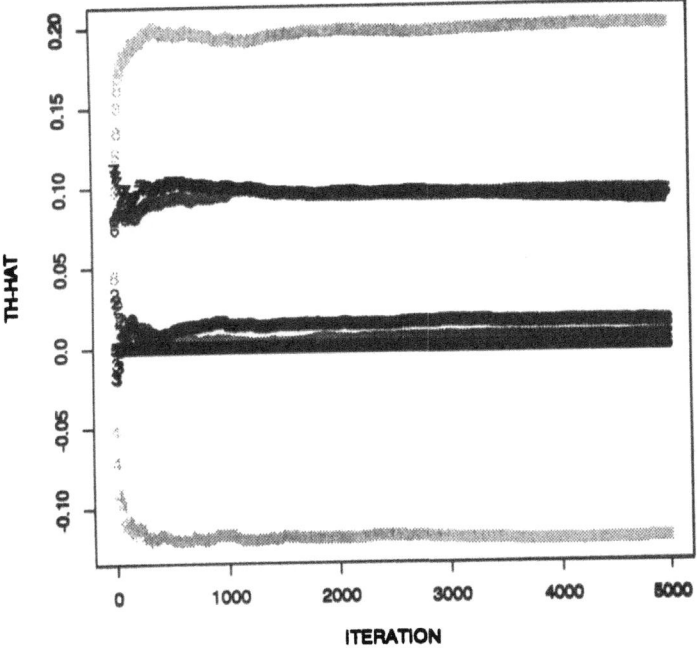

Figure 3: Trajectories of the ergodic averages for the parameters.

For the probing distribution we suggest taking the normal distribution with a double variance for the diagonal elements of **H** as before in (4) and (5), and for the off-diagonal elements a Wishart distribution of the residuals:

$$\mathbf{H}_t \sim N[\mathbf{E}'\mathbf{E}, T - 3].$$

Note that the M-dimensional VAR system can be written as

$$\mathbf{y}_1 = \mathbf{X}_1 \mathbf{b}_1 + \epsilon_1,$$
$$\ldots$$
$$\mathbf{y}_M = \mathbf{X}_M \mathbf{b}_M + \epsilon_M,$$

or compactly as

$$\mathrm{vec}\,\mathbf{Y} = \mathrm{diag}\,(\mathbf{X}_1, \ldots, \mathbf{X}_M)\,\mathrm{vec}\mathbf{B} + \mathrm{vec}\mathbf{E}$$

with the $T \times M$ data matrix $\mathbf{Y} = (\mathbf{y}_1 : \ldots : \mathbf{y}_M)$, and the $T \times M$ residual matrix $\mathbf{E} = (\epsilon_1 : \ldots : \epsilon_M)$.

4.1 The Vector ARCH Matrix Probing Distribution

In order to work out the full conditonal and probing dristribution for the VAR-VARCH model more easily, we suggest a slight reordering of the equations in the vectorized VAR model. The multivariate ARCH time series model can be reformulated as

$$\mathrm{vec}\mathbf{Y}' \sim \mathrm{N}_{M \times T}\,[\mathbf{X}\mathrm{vec}\mathbf{B}, \mathbf{D}_H]$$

with $\mathbf{Y}$ a $T \times M$ matrix of dependent variables

$$\mathbf{X} = \begin{pmatrix} \mathbf{X}_1 \\ \vdots \\ \mathbf{X}_T \end{pmatrix}$$

a $TM \times KM$ stacked matrix of regressors, $\mathbf{M}_t = \text{diag}(\mathbf{x}'_{1t}, \ldots, \mathbf{x}'_{Mt})$ a $M \times KM$ diagonal matrix for each time point t, $\mathbf{B} = (\mathbf{b}_1, \ldots, \mathbf{b}_M)$ a $KM \times M$ matrix of regression coefficients, and $\mathbf{D}_H = \text{diag}(\mathbf{H}_1, \ldots, \mathbf{H}_T)$ a $TM \times TM$ block diagonal covariance matrix.

Let $\mathbf{y}(t)$ be the tth row of $\mathbf{Y}$ and $\epsilon(t)$ the tth of $\mathbf{E}$, then each equation of the system can be written as

$$\begin{aligned} \mathbf{y}(t)' &= \text{diag}(\mathbf{x}'_{1t}, \ldots, \mathbf{x}'_{Mt})\ \text{vec}\mathbf{B} + \epsilon(t) \\ &= \mathbf{X}_t \text{vec}\mathbf{B} + \epsilon(t), \end{aligned}$$

where

$$\epsilon(t)' = \begin{pmatrix} \epsilon_{t1} \\ \vdots \\ \epsilon_{tM} \end{pmatrix} \sim \mathrm{N}\left[\mathbf{0}, \mathbf{H}_t = \begin{pmatrix} h_1^2 & & h_{12} & \ldots & h_{M-1,M} \\ \ldots & & \ldots & \ldots & \ldots \\ h_{M-1,M} & \ldots & \ldots & & h_M^2 \end{pmatrix}_t \right],$$

and the model for the $J = M(M+1)/2$ distinct values of the covariance matrix $\mathbf{H}_t$ is given by the multivariate regression system

$$\text{vech}\mathbf{H}_t = \mathbf{z}'_t\Gamma, \qquad t = 1, \ldots, T.$$

For an auxiliary regression we have to approximate the matrix $\mathbf{H}_t$ by the residuals of the last iteration (indexed by i-1), i.e. $\text{vech}\mathbf{H}_t = (h^2_{1t}, \ldots, h^2_{Mt}, \ldots, h_{12,t}, \ldots, h_{M-1,M,t})'_{i-1}$ and where Γ is a $2M \times J$ ARCH coefficient matrix and $\text{vech}\mathbf{H}_t$ means the vectorisation of the lower triangular part of the symmetric matrix $\mathbf{H}_t$ and $\mathbf{z}_t$ contains the covariates of the volatility model, e.g. squared past errors of time series values. The covariates of the mixed model of order 1 looks like

$$\mathbf{z}'_t = \left(\epsilon^2_{1,t-1}, \ldots, \epsilon^2_{M,t-1}, y^2_{1,t-1}, \ldots, y^2_{M,t-1}\right).$$

Now we have to stack the components of the multivariate regression system into the $2M \times T$ regressor matrix $\mathbf{Z}' = (\mathbf{z}_1, \ldots, \mathbf{z}_T)$ and the $J \times T$ matrix of dependent variables $\mathbf{E}_h$ given by

$$\mathbf{E}'_h = (\text{vech}\mathbf{H}_1; \ldots, \text{vech}\mathbf{H}_T);$$

Thus we can write the system as

$$\begin{aligned} \mathbf{E}_h &= \mathbf{Z}\Gamma + \mathbf{U}, \qquad\qquad \mathbf{U} \sim \mathrm{N}_{T \times J}[\mathbf{0}, \Omega \otimes \mathbf{I}_T], \\ (T \times J) &= (T \times 2M)(2M \times J) + (T \times J) \end{aligned}$$

where Γ is the $2M \times J$ coefficient matrix as before, and Ω is a $J \times J$ covariance matrix between the ARCH components. The multivariate OLS estimate is given by

$$\widehat{\Gamma} = (\mathbf{Z}'\mathbf{Z})^{-1}\mathbf{Z}'\mathbf{E}_h,$$

and the distribution of the estimator is

$$\widehat{\Gamma} \sim N_{2M \times J} \left[\widehat{\Gamma}, \widehat{\Omega} \otimes (\mathbf{Z}'\mathbf{Z})^{-1} \right],$$

with $\widehat{\Omega} = \widehat{\mathbf{U}'\mathbf{U}}/T$, $\mathbf{U} = \widehat{(\mathbf{E}_h - \mathbf{Z}\widehat{\Gamma})}_{i-1}$, where the index $i-1$ stands for the previous iteration of the Gibbs sampler. Note that $H_{T,i-1} = \text{Var}(\epsilon(t)_{i-1})$ and the last residual is given by $\epsilon(t)'_{i-1} = y(t)' - \mathbf{X}_t \text{vec} \mathbf{B} + \epsilon(t)$.

The joint distribution of the VARCH model for the data $\mathbf{Y}$ and the parameter $\theta = (\mathbf{B}, \Gamma)$ is

$$p(\theta, \mathbf{Y}) = \prod_{t=1}^{T} |\mathbf{H}_t|^{-\frac{1}{2}} \exp \left\{ -\frac{1}{2} (\mathbf{y}(t)' - \mathbf{X}_t \text{vec} \mathbf{B})' \mathbf{H}_t^{-1} (\mathbf{y}(t)' - \mathbf{X}_t \text{vec} \mathbf{B} \right\},$$

with $\mathbf{H}_t = \text{vech}^{-1}(\mathbf{z}'_t)$ where vech^{-1} stands for a 'covariance constructing transformation'. Only the lower triangular coefficients of a symmetric matrix are stored in a vector and used to build up the entire matrix.

The full conditional distribution are given as following:

[1] The full conditional for $\mathbf{B}$ is

$$p(\mathbf{B}|\Gamma, \mathbf{Y}) = N_{KM \times M} [\mathbf{B}_{**}, \mathbf{H}_{**}],$$

with the parameters

$$\mathbf{H}_{**}^{-1} = \mathbf{X}' \mathbf{D}_H^{-1} \mathbf{X}, \qquad \mathbf{D}_H = \text{diag}(\mathbf{H}_1, \ldots, \mathbf{H}_T),$$
$$\text{vec} \mathbf{B}_{**} = \mathbf{H}_{**} \mathbf{X}' \mathbf{D}_H^{-1} \text{vec} \mathbf{Y}'.$$

[2] For the probing distribution for Γ we take the restricted sampling distribution of the multivariate OLS estimator:

$$g(\Gamma|\mathbf{B}, \mathbf{Y} = N \left[\widehat{\Gamma}, \widehat{\Omega} \otimes (\mathbf{Z}'\mathbf{Z}) \right] I_{(\mathbf{H}_t > 0)},$$

where $I_{(\mathbf{H}_t > 0)}$ is the indicator function that the sampled $\widehat{\Gamma}$ coefficients have a positive definite covariance matrix $\mathbf{H}_t$, which itself is constructed as

$$\mathbf{H}_t = \text{vech}^{-1} \left(\mathbf{z}'_t \Gamma \right), \qquad t = 1, \ldots, T,$$

the inverse transformation of the 'half-vectorization' of a symmetric matrix $\mathbf{H}_t$. Note that this has to be checked simultaneously for all time points t; this procedure might turn out to be quite time consuming.

5 Conclusions

The approach in this paper has demonstrated that the Gibbs sampler solves the estimation problem of simple univariate ARCH models quite satisfactory. For the parameters of the variance equation a Metropolis step was proposed and we have used as probing distribution a normal with parameter values of the previous iteration. This approach was also extended to a VAR-VARCH approach. The example estimated the volatility of a first order model of exchange rates for the US dollar. We find that the Yen explains more of the volatility of the US dollar than the European currencies.

In a subsequent paper it will be shown that ARCH models can be quite easily extended to cope with further aspects of volatility, like outliers, mixture models or models which imposes some kinds of constraints, like in form of tightness constraints on the lag distribution (see e.g. Polasek 1994 or Polasek and Jin 1994). Also, MCMC posterior distribution allows the simulation of the predictive distribution which can be used for forecasting and model selection.

References

Bollerslev T., Chou R.Y., and Kroner K.F. (1992) *ARCH modeling in finance*, J. of Econometrics 52, 5-59.

Diebold J. und Robert C. (1990) *Bayesian estimation of finite mixture distributions (I): Technical aspects*, Tech. report 110, University Paris.

Gelfand A.E., and Smith A.F.M. (1990) *Sampling based approaches to calculating marginal densities*, JASA 85, 398-409.

Gelfand A.E., and Dey D.K. (1992) *Bayesian model choice: Asymp- totics and exact calculation*, University of Connecticut, mimeo.

Gelfand A.E., Smith A.F.M., and Lee T-M. (1992) *Bayesian analysis of constrained parameter and truncated data problems*, JASA 87, 523-532.

Gelfand A.E., Hills, S.E., Racine-Poon A., and Smith A.F.M., (1990) *Illustrations of Bayesian inferences in normal data models using Gibbs sampling*, JASA 85 (#412), 972-985.

Jacquier E., Polson N.G. and Rossi P.E. (1992) *Bayesian analysis of stochastic volatility models*, Working paper 92-141, University of Chicago.

Müller P. (1991) *A Bayesian Vector ARCH Model for Exchange Rate Data*, WWZ discussion paper #9109, University of Basel.

Polasek W. (1994) Bayesian *VAR models with tightness priors*, WWZ discussion paper, University of Basel.

Polasek W. (1994) *Bayesian augmented ARCH and VARCH models*, WWZ discussion paper, University of Basel.

Polasek W. and Jin S. (1994) *Variable selection in regression models*, ISO University of Basel.

Engle R.F. (1982) *Autoregressive conditional heteroskelasticity with estimates of the variance of UK inflation*, Econometrica 50, 987-1008.

Smith A.F.M. and A. Gelfand (1992) *Bayesian statistics without tears: a sampling- resampling perspective*, American Statistician 46, 84-88.

Tierney L. (1991) *Markov chains for exploring posterior distributions*, Univ. of Minnesota, Tech. report No. 560, to appear in Annual of Statistics

A Class of Recursive Algorithms Using Non-parametric Methods with Constant Step Size and Window Width: A Numerical Study

GEORGE YIN and KEWEN YIN

Department of Mathematics, Wayne State University, Detroit, MI 48202
and
Department of Chemical Engineering, University of Minnesota, Duluth, MN 55812.

Abstract:

Motivated by a wide range of applications in chemical engineering problems, recursive algorithms using nonparametric methods combined with stochastic approximation procedures are developed in this work. Loosely speaking, the problems are to find roots of nonlinear functions provided that only noisy measurements are available. In addition, in the systems under consideration, not only the outputs but also the inputs are noise corrupted. Thus the conventional stochastic approximation algorithms become inapplicable. Algorithms using a kernel function with constant step size and constant window width are proposed and analyzed. After presenting the convergence result of the algorithms under general conditions, effort is directed to the numerical studies. Simulation results and numerical experiments are also given.

1 Introduction

Steady state estimation and system fault detection are important problems in chemical engineering. In order to operate and to control a process effectively, values of the system states (e.g., temperature, pressure, flowrate, concentration etc.) need to be known. Due to the presence of random disturbances, however, measurements of these quantities often involve certain errors. Therefore estimation is needed to provide more accurate/correct information of the system.

The problem under consideration can be related to the search for zeroes of a nonlinear

*Kitsos, C.P., and Müller, W.G., Eds., *Proceedings of MODA4*, Physica Verlag, Heidelberg, 1995

262

function $\bar{f}(x) = 0$. However either the form of the function is unknown or it is too complicated to compute. It is thus desirable to use stochastic approximation type of procedures to resolve the problem, i.e., construct a recursive algorithm based on the measurements $y_n = f(x_n, \xi_n)$, where x_n and ξ_n stand for the state and noise, respectively. Unfortunately, in the current situation, the sequence $\{x_n\}$ is also noise corrupted and is not at our disposal. As a result, some alternative methods have to be employed for solution. The problems resemble a system with partial observations in stochastic control. One possible approach is to use some kind of nonlinear filtering technique to pre-filter the measurements and to get an estimate of the state first. That method, however, is not desirable because of its high computational expense.

While a large number of powerful techniques are available for linear systems in chemical engineering, the theoretical development and practical application of detection, estimation and process control for nonlinear systems under random influence are still in a state of infancy. Recently several approaches have been developed and applied to various systems, e.g., Kalman filter approach, nonlinear optimization method, and analytical least squares solution among others. A review and comparison of some of the methods were provided in (4). It was concluded that the Kalman Filter approach performs satisfactorily in some processes. However, it is quite sensitive to several factors. The nonlinear programming approach, on the other hand, exhibits a superior performance in terms of robustness and speed of tracking, but is computationally more intensive. An estimation strategy based on the analytical solution to a least squares objective function via a two-level approach was proposed with a significant reduction in the computation time. A disadvantage of this approach, however, lies in its inability to handle variable bounds and inequality constraints (see (8) for more details). The development of an estimation algorithm which is suitable for real time computation in random nonlinear processes still remains a challenging task in chemical engineering and many other process industries.

In the late 80's, a method known as passive stochastic approximation was developed (3) (see also (7)). It is essentially a nonparametric methods with a combined usage of stochastic approximation and kernel estimation.

Enlightened by their results, we propose recursive algorithms which also use the combined approach of stochastic approximation and the nonparametric methods, but with constant step size and window width. The rest of the paper is arranged as follows. The formulation is given next together with the two classes of algorithms. We then present the main asymptotic results in Section 3, and carry out numerical experiments for a distillation column in Section 4. Finally, some further remarks are made in the last section.

2 Recursive Algorithms

Let $\{x_n\}$ be a sequence of $\mathbb{R}^r$-valued random variables which represents the measured states or inputs, $\{y_n\}$ a sequence of $\mathbb{R}^r$-valued random variables, that are measured outputs, with

$$y_n = f(x_n, \xi_n),$$

where $f(\cdot, \cdot)$: $\mathbb{R}^r \times \mathbb{R}^r \mapsto \mathbb{R}^r$, $\{\xi_n\}$ is a stationary sequence of $\mathbb{R}^r$-valued random disturbances.

Motivated by the idea of passive stochastic approximation, iterative algorithms with combined stochastic approximation and kernel estimation are suggested. The proposed class of algorithms is of the form:

$$z_{n+1} = z_n + \frac{\varepsilon}{\delta} K \left(\frac{x_n - z_n}{\delta} \right) y_n, \tag{1}$$

where $K(\cdot): \mathbb{R}^r \mapsto \mathbb{R}$ is a kernel function, $\varepsilon > 0$ is a small parameter representing the step size, and $0 < \delta = \delta(\varepsilon)$ is the window width satisfying $\varepsilon/\delta \xrightarrow{\varepsilon} 0$.

The algorithm above bears the same spirit as that of the passive stochastic approximation cases with decreasing step size. We compare the observable x_n with current estimate z_n, if they are far apart from each other, no weight will be put on the measurement y_n. If they are fairly close, the update is obtained by adding current measurement y_n nontrivially according to (1). The use of the constant step size and window width allows us to deal with situations that involve slight parameter and state variations and is more robust from computation point of view. In what follows, we shall demonstrate the convergence of the algorithms.

3 Main Results

To proceed, we make the following assumptions:

(A1) For each ξ, $f(\cdot, \xi)$ has continuous partial derivatives up to order 2 and $f_{zz}(\cdot, \xi)$ is bounded. For each z belonging to a bounded set, and each $T < \infty$, $\{f(z, \xi_j); j\varepsilon \leq T\}$ and $\{f_z(z, \xi_j); j\varepsilon \leq T\}$ are uniformly integrable.

(A2) Let E_n denote the conditioning up to time n, i.e., conditioning on the σ-algebra $\mathcal{F}_n = \sigma\{z_0, x_j, \xi_j; \ j < n\}$. For each z, there is a $\bar{f}(\cdot)$ such that

$$\frac{1}{N} \sum_{j=n}^{n+N} E_n f(z, \xi_j) \xrightarrow{N} \bar{f}(z) \text{ in probability.} \tag{2}$$

(A3) The kernel $K(\cdot)$ has a bounded support and satisfies

$$K(x) = K(-x), \quad \int K(x)dx = 1.$$

(A4) $\{x_n\}$ is a sequence of independent and identically distributed random variables, which is independent of $\{\xi_n\}$ such that $E|x_n| < \infty$. x_n has a density $\pi(\cdot)$, such that $\pi(\cdot)$ is continuous and $\pi(x) > 0$ for each x.

Remark: (A1) is an assumption on the function $f(\cdot)$ and the measurement noise $\{\xi_n\}$. It is a rather general condition and includes a wide variety of situations. A special case, commonly used in many applications is: $f(x, \xi) = \bar{f}(x) + \xi$, i.e., the measurement noise enters in an additive way. The uniform integrability condition is often easily verified, for example, in the additive noise case, if $\sup_n |\xi_n|^{1+\Delta} < \infty$ for some $\Delta > 0$, then the uniform integrability follows. The condition $f_{zz}(\cdot, \xi)$ being bounded can be weakened. In fact, we can assume $f(z, \xi) = (f_0(z) + \alpha) + f_1(z, \beta) + f_2(z)\gamma$ such that $f_i(\cdot)$, for $i = 0, 1, 2$, are twice continuously differentiable with respect to z. $\{\alpha_n\}$, $\{\beta_n\}$ and $\{\gamma_n\}$ are stationary and are independent of each other. $E\alpha_n = 0$, $E|\alpha_n|^{2+\Delta} < \infty$; $\{\gamma_n\}$ is a martingale difference sequence such that $E|\gamma_n|^{2+\Delta} < \infty$ for some $\Delta > 0$; $\{\beta_n\}$ is a sequence of bounded random variables with $Ef_1(z, \beta_n) = \bar{f}(z) - f_0(z)$ for each z and each n; $f_{0,zz}(\cdot)$ and $f_{2,zz}(\cdot)$ are bounded functions.

Condition (A2) is a main averaging assumption. Roughly, it is a law of large numbers type of condition. The condition is much weaker when the conditional expectation is inserted. To elaborate on it further, take for example the additive noise case when $\{\xi_n\}$

is a sequence of independent and identically distributed (i.i.d.) random variables with zero mean or a sequence of martingale difference. Then $E_n\xi_j = 0$ for $j \geq n$. Clearly the condition is satisfied. If $\{\xi_n\}$ is a sequence of stationary ϕ-mixing processes with zero mean, the condition is also verified by noting that $\{\xi_n\}$ is strongly ergodic.

Condition (A3) states that the kernel is symmetric and it integrates to unity. A typical example is as given in Section 4 in the sequel.

In fact, we have obtained the convergence under much weaker conditions. In particular, the sequence $\{x_n\}$ need not be i.i.d. However, to assume weaker condition on $\{x_n\}$, we need to utilize somewhat complex notation on the conditional expectations. For ease of presentation, we choose to use the simpler sitting here so as to make the main line of argument clear.

To proceed, define $z^\varepsilon(t) = z_n$ for $t \in [n\varepsilon, n\varepsilon + \varepsilon)$. This is a piecewise constant or 'stair case' function approximation.

Theorem 1. *For Algorithm (1), under (A1)-(A4) and the condition z_0^ε converges to z_0 weakly, $\{z^\varepsilon(\cdot)\}$ is tight in $D^r[0, \infty)$. Any weakly convergent subsequence satisfies*

$$\dot{z}(t) = \pi(z)\bar{f}(z), \quad z(0) = z_0 \tag{3}$$

provided the differential equation has a unique solution for each initial condition.

Theorem 1 above indicates that two averages are taking place. One of them is the average of the measurement noise, and the other is the average of the measured inputs. It gives a convergence result for t confined in a large but still bounded interval. It is of particular interest to know what happens when $\varepsilon \to 0$ and $t \to \infty$ in some appropriate sense.

Theorem 2. *Assume that the conditions of Theorem 1 are satisfied. Suppose that there is a point θ that is globally attracting for the ordinary differential equation (3) and is stable in the sense of Liapunov, and suppose*

$$\{z_n; n < \infty, \varepsilon > 0\} \text{ is bounded in probability.} \tag{4}$$

Let $\{t_\varepsilon\}$ be a sequence such that $t_\varepsilon \to \infty$ as $\varepsilon \to 0$. Then $z^\varepsilon(t_\varepsilon + \cdot)$ converges weakly to θ as $\varepsilon \to 0$.

The proofs of these theorems are contained in (14). For brevity, we shall omit them altogether.

4 Simulation and Numerical Experiments

A distillation column (Fig. 1) is used for separating two or more miscible liquids. Distillation is a highly nonlinear process with a large number of states involved, which are often strongly correlated. An algorithm applicable to this system can usually be expected to work well in other less complicated processes. A single feed stream with feed rate F and concentration x_f is introduced onto the feed tray. Heat is provided at the base of the tower. There are a certain number of trays installed in the column for heat and mass transfers between vapor and liquid. As the vapor rises along the tower, it is getting richer and richer in the more volatile component. While as the liquid goes down, it will contain more and more high boiling point component. Finally, the most volatile component comes off as vapor at the top of the tower and the least volatile one is drawn off at the base. The

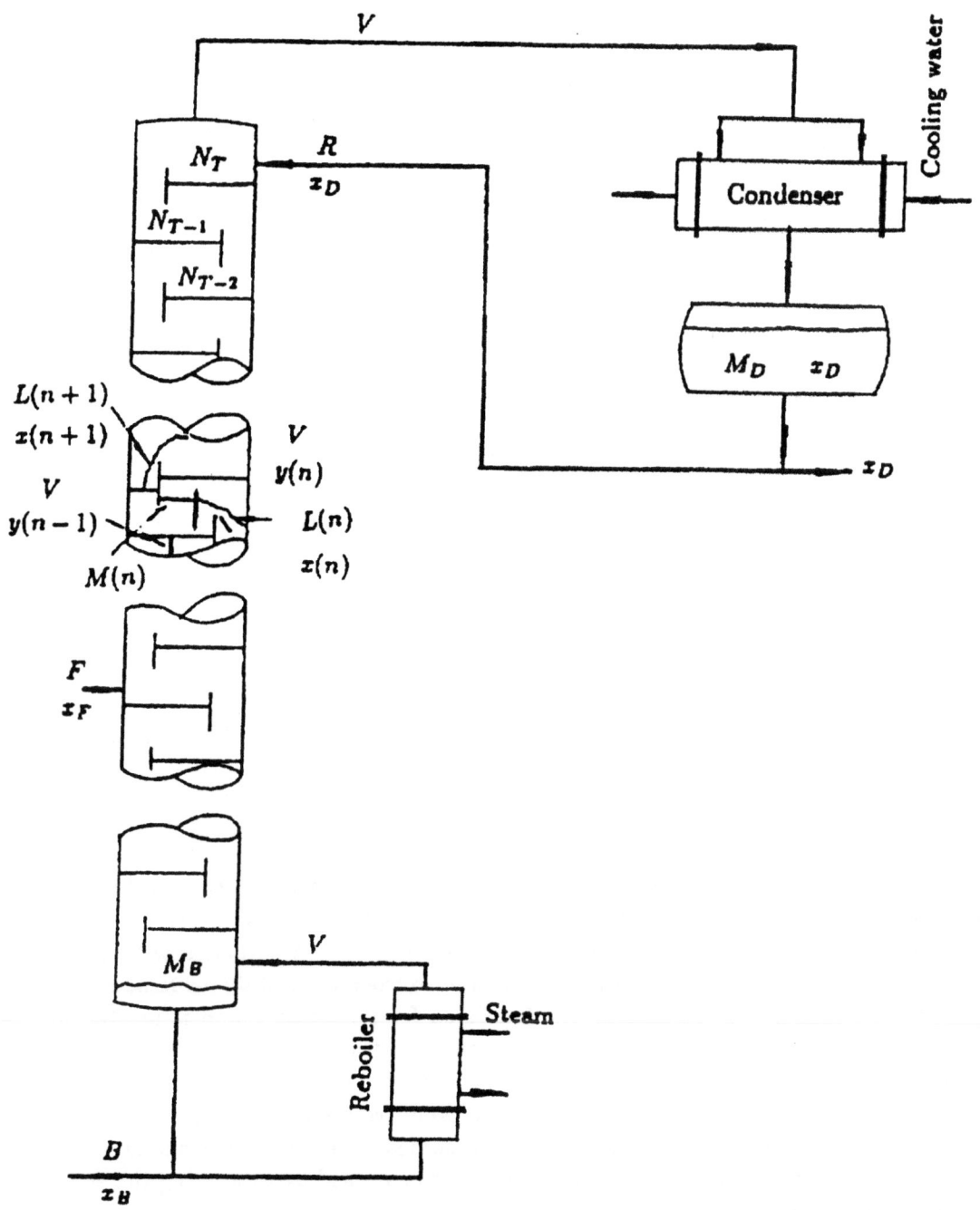

Figure 1   Distillation Column

overhead vapor is totally condensed in a condenser with composition x_d. At the base of the column, liquid bottom product is removed with composition x_b.

A moderate purity, 18 stage binary distillation column with a feed consisting of benzene/toluene has been chosen for study. Under a series of common assumptions, the system can be described by two ordinary differential equations per tray and two algebraic equations per tray. This mathematical model with total 72 equations will enables us to obtain information about the dynamic as well as steady state behavior of the column, which makes the further study on state estimation possible. Under nominal condition, feed was introduced into the column at the ninth plate with a molar concentration of benzene $x_f = 0.5$. It is desired to maintain the top and bottom concentrations simultaneously at constant values ($x_d = 0.98$, $x_b = 0.02$) despite disturbances in the feed. The discrete process data were generated by computer simulation. Feed concentration is the state of interest and to be estimated.

By using computer simulation of the distillation column, feed with its 'true' concentration x_f was used to calculate the top and bottom product concentrations x_d and x_b. Subsequently, a sequence of noise corrupted measurements $\{y_n\}$ was generated by adding $\{\xi_n\}$, a sequence of normal random variables,

$$y_n = x_{d,n} + \xi_{d,n} \text{ or } y_n = x_{b,n} + \xi_{b,n}.$$

A sequence of normal random variables $\{\eta_n\}$ was added to x_f to create $\{x_n\}$. It can be seen that y_n is some function of x_f. The function form is unknown in our case. A sequence of estimates $\{z_n\}$ was obtained from (2), where the Kernel function $K(\cdot)$ was taken as

$$K(x) = \begin{cases} 0.75(1 - |x|^2), & |x| \le 1; \\ 0, & |x| > 1. \end{cases} \quad (5)$$

When the estimate z_n is far away from x_n, i.e., $|x_n - z_n|/\delta$ is greater than 1, $K((x_n - z_n)/\delta)$ will be equal to zero, and nothing will be added to the estimate z_{n+1}, otherwise the observed y_n will be added non-trivially to the recursion in accordance with (6). The estimates z_n, as has been proven, should converge to the steady state value of x_f.

Since distillation is often used as the final step for purification and is located at the downstream stage of the whole process, any changes in the upstream stages might induce changes in feed composition to the distillation tower. This situation was simulated by introducing unwanted changes in the feed that would result in deviation of the product compositions from their desired values. It was assumed that correct measurements of the feed composition were available. The algorithm was examined by comparing the estimates with the true values.

In the first case, the real feed composition, which was higher than its set value due to the changes in the upstream, was measured and recorded. We used i.i.d. normal random variables $\mathcal{N}(0, \sigma^2)$ with $\sigma = 2.0 \times 10^{-3}$ to simulate disturbances in this measurements. Estimates could be obtained from either x_d or x_b, whose observations y_n were simulated by adding i.i.d. Gaussian noises with distribution $\mathcal{N}(0, \sigma^2)$. $\sigma_d = 10^{-3}$ for x_d and $\sigma_b = 10^{-4}$ for x_b were employed in accordance with the possible noise to signal ratio. Both sets of estimation results, computed from x_d and x_b and shown on the upper and lower parts of Fig. 2, are close to each other and close to the true values.

The same procedure was repeated for the case when the feed composition was becoming lower and lower due to some change in the upstream in the second case. The numerical experiments shown on Fig. 3, give satisfactory results. It should be pointed out that to

Figure 2 Estimation of the feed concentration

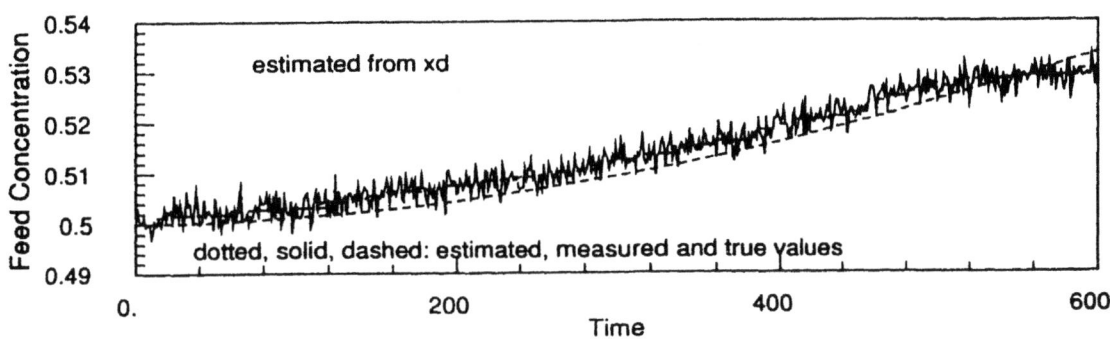

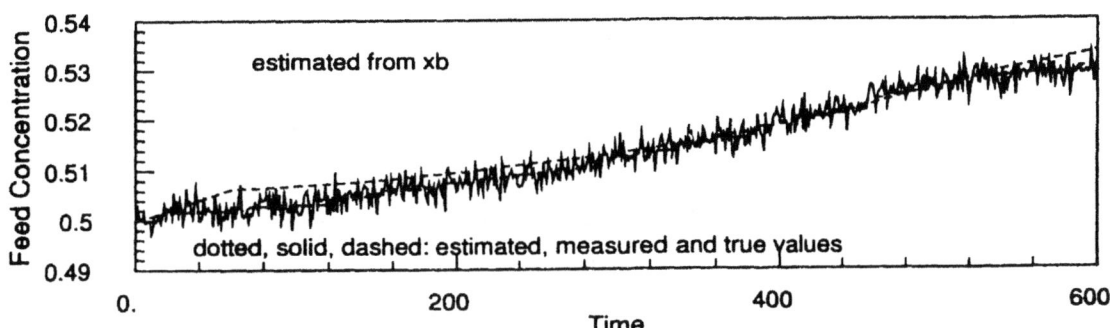

achieve better performance, different ε values had to be used, which can be attributed to the high nonlinearity of the system.

Then we proceeded to examine the situation where the state of interest, x_f, underwent changes in directions. The estimates exhibited sluggish response or lagged behind the real change at the turning point. Further study is called for to improve the performance of this algorithm in regard to this aspect.

This study has shown that the state — feed concentration x_f, can be estimated from observations of other related states in the system. This is quite useful in many chemical engineering systems where it is difficult to obtain accurate measurements for some of the states. Such situation is often encountered.

As discussed in the previous sections, the values of δ control the size of the window, and the ratio ε/δ determines the step size of the recursive estimation. Fig. 4 displays the impact of the different choices of δ on the estimation when $\varepsilon = .0004$ was used. For a fixed ε value, if δ is too small, $|x_n - z_n|/\delta$ will be greater than 1 all the time thus $K((x_n - z_n)/\delta)$ will always be zero. No improvement will be made on z_n (line of $\delta = .003$ in Fig. 4). On the contrary, if δ is too big compared to $|x_n - z_n|$, $|x_n - z_n|/\delta$ will be very small and the values of the kernel function will remain to be unchanged (0.75 in our case). The value of ε (in conjunction with δ) determines the rate of convergence. Thus if the ratio ε/δ is too small, the rate of convergence will be very low (line of $\delta = .3$ in Fig. 4). Fig. 5 shows the estimation results when $\delta = .03$ with different ε values. For a fixed δ, if ε is too small, the tracking rate will be very low (line of $\varepsilon = .00004$ in Fig. 5). On the other hand, a large ε value, which outbalances the kernel function greatly, will dominate the estimation therefore leads to incorrect results (line of $\varepsilon = .004$ in Fig. 5). This must be avoided.

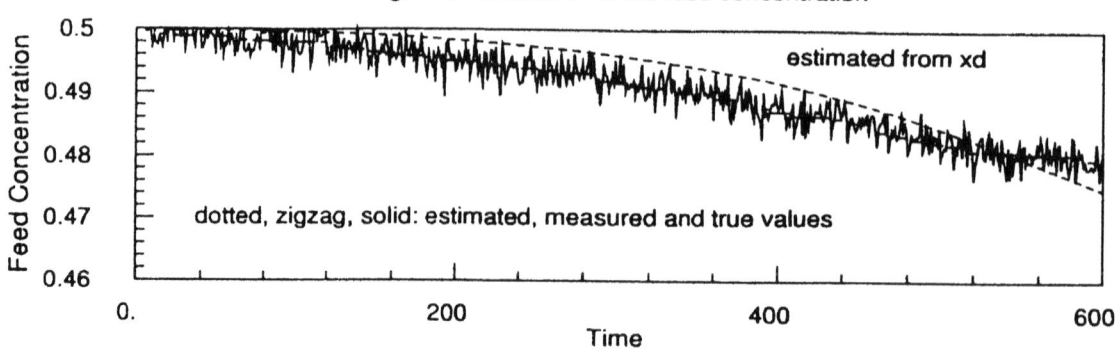

Figure 3 Estimation of the feed concentration

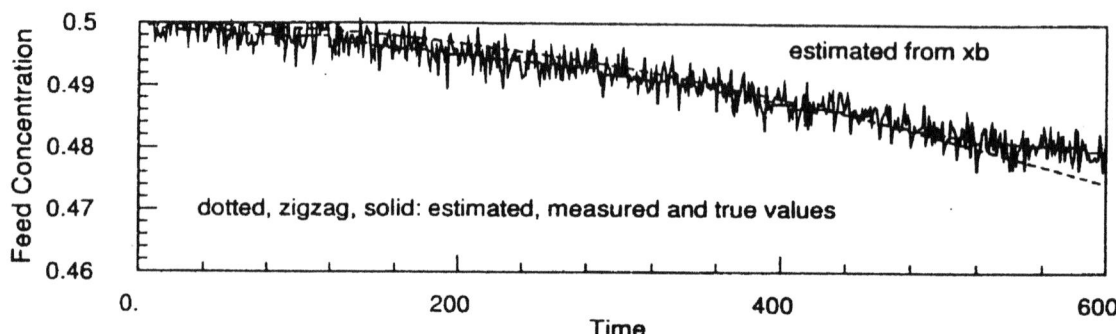

For a specific system, an appropriate combination of δ and ε (lines of $\delta = .03$ in Fig. 4 and $\varepsilon = .0004$ in Fig. 5) can be identified from a properly designed experiment which, we believe, is a necessary step in the estimation procedure.

5 Further Discussions

Recursive algorithms based on stochastic approximation and kernel estimation is introduced in the paper. Their application to chemical engineering processes — a distillation column, is examined. Simulation studies have shown very promising results. It can be seen that the algorithms developed herein are very useful for certain nonlinear systems in state estimation. It is able to not only differentiate the real and faulty measurements correctly, but also give good estimate of the steady state value promptly. Compared with many other methods, our approach appears to be much simpler in form, easier to implement and computationally less intensive. The algorithms are capable of estimating certain unmeasured states from measured input and measured output. This is particularly useful for those chemical engineering systems where not all variables are measured, and only limited information is available.

On the premise that in many chemical engineering applications, very often certain nominal values of the state are available and can be used as a reference for comparisons, the following algorithm can be used in lieu of (1)

$$z_{n+1} = z_n + \frac{\varepsilon}{\delta} K \left(\frac{x^* - z_n}{\delta} \right) y_n, \qquad (6)$$

where x^* is a nominal or reference value of the state or input. This algorithm takes

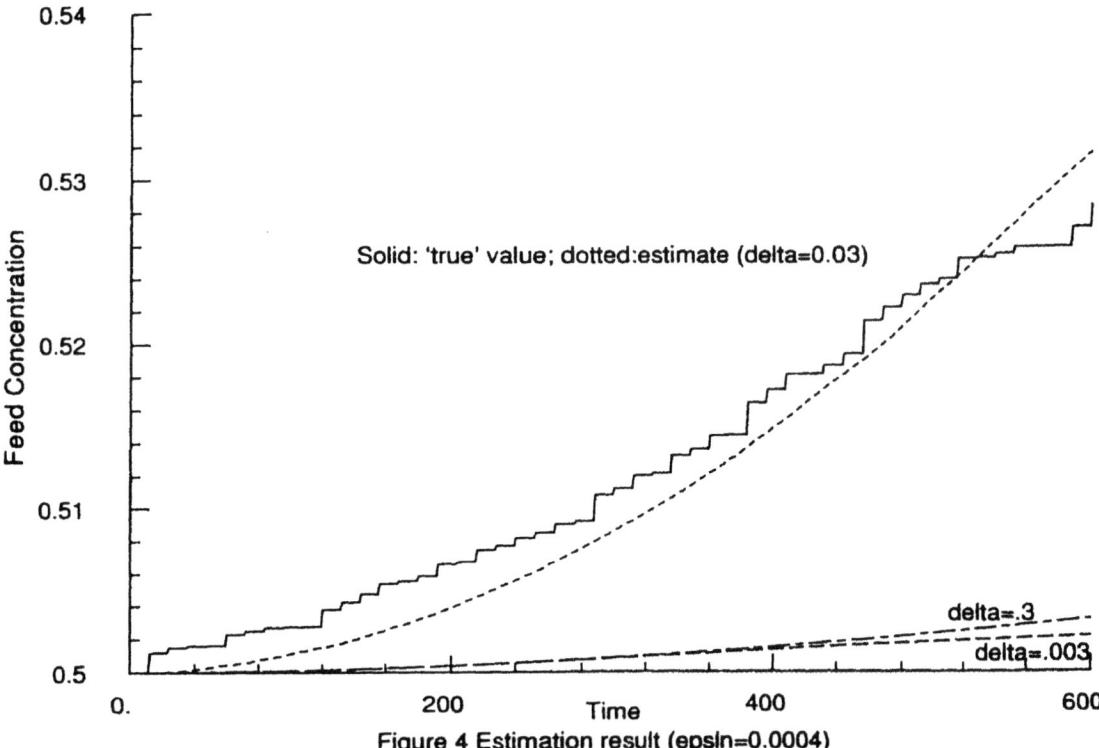

Figure 4 Estimation result (epsln=0.0004)

advantage of the known value x^*. In this case, the sequence $\{x_n\}$ is not needed in the actual computing, but it requires the system operate close to or does not deviate too much from its known nominal condition.

We have also carried out numerical experiments for Algorithm (6). The performance for both algorithms display similar behavior. Although seemingly less information is needed for Algorithm (6), stronger conditions are needed for the convergence analysis.

In the algorithm, y_n is some function of x_n. The function form is not necessarily known. Only observations are needed for implementation. To some extent, the algorithm considered is a stochastic system with multiple scaling, where the scaling factors are the step size of the recursive algorithm and the window width of the kernel function, which are determined by ε and δ in (1) and (6). The choices of ε and δ depend on each individual case. It was found that a number of different pairs of ε and δ yielded almost equally good result. Further study needs to be directed to the fine tuning of the step size ε and the window width δ, and to the development of adaptive tuning procedures.

In our approach $K(\cdot)$ is similar to a δ-function in feature but smoother. It should be noted that a number of different functions other than the one used in the simulation can be utilized. Will they result in significantly different results? What is the 'best' choice of the kernel function? All of these open up areas for further studies.

270

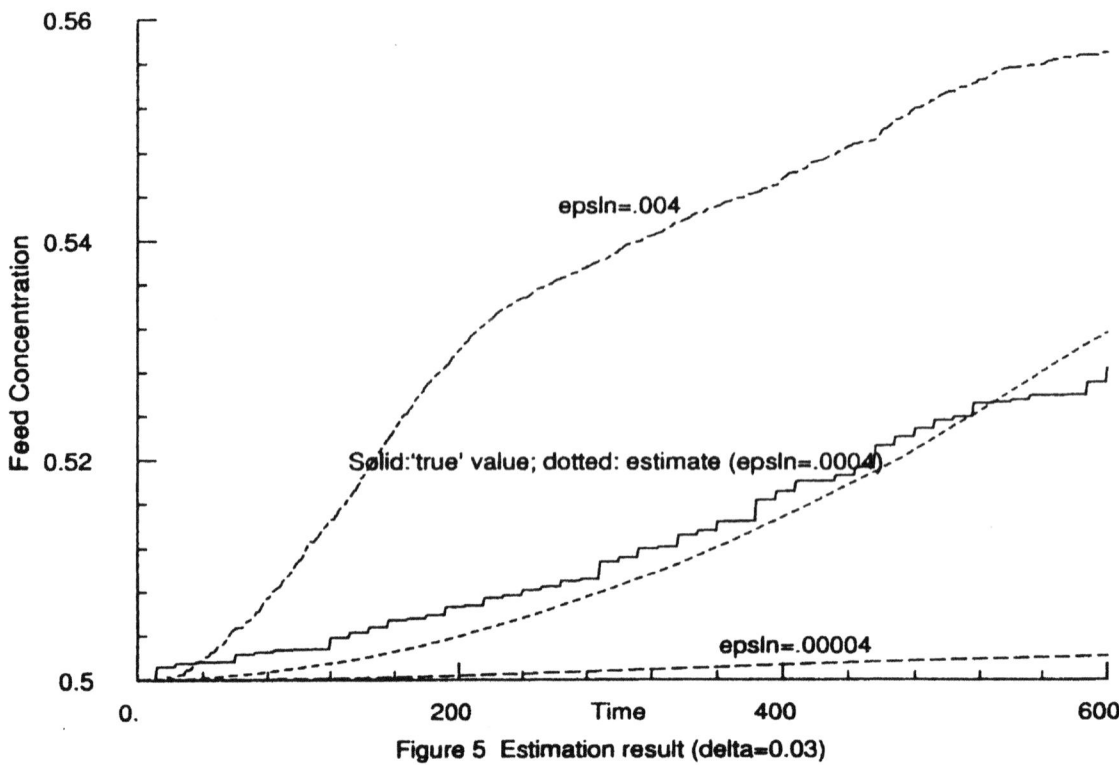

Figure 5 Estimation result (delta=0.03)

Acknowledgement

Research of the first author was supported in part by the National Science Foundation under grant DMS-9224372, and in part by Wayne State University. Research of the second author was supported in part by the National Science Foundation under Grant CTS-9403193, and in part by Minnesota Supercomputer Institute.

References

A. Benveniste, M. Métivier and P. Priouret, *Adaptive Algorithms and Stochastic Approximation*, Springer-Verlag, Berlin, 1990.

S.N. Ethier and T.G. Kurtz, *Markov Processes, Characterization and Convergence*, Wiley, New York, 1986.

W.K. Härdle and R. Nixdorf, Nonparametric sequential estimation of zeros and extrema of regression functions, *IEEE Trans. Inform. Theory* **IT-33** (1987), 367-372.

S. Jang, B. Joseph and H. Mukai, Comparison of two approaches to on-line parameter and state estimation of nonlinear systems, *Ind. Eng. Chem. Process Des. Dev.* **25** (1986), 809-814.

H.J. Kushner, *Approximation and Weak Convergence Methods for Random Processes, with applications to Stochastic Systems Theory*, MIT Press, Cambridge, MA, 1984.

L.C. Limqueco and J.C. Kantor, Nonlinear output feedback control of an exothermic reactor, *Comp. Chem. Eng.* **14** (1990), 427-437.

A.V. Nazin, B.T. Polyak and A.B. Tsybakov, Passive stochastic approximation, *Automat. Remote Control* **50** (1989), 1563-1569.

Y. Ramamurthi, P.B. Sistu and B.W. Bequette, Data reconciliation and fault detection in dynamic processes, *AIChE Annual Meeting*, Paper No 147d (1991).

W.H. Ray, New Approaches to the dynamics of nonlinear systems with implications for process system design, in *Chemical Process Control 2*, 246-267, D.E. Seborg and T.F. Edgar Eds.; United Engineering Trustees, New York, 1981.

H. Robbins and S. Monro, A stochastic approximation method, *Ann. Math. Statist.* **22** (1951), 400-407.

P.B. Sistu and B.W. Bequette, Nonlinear predictive control of uncertain processes: application to a CSTR, *AIChE J.* **37** (1991), 1711-1723.

A. Uppal, W.H. Ray and A.B. Poore, On the dynamic behavior of continuous stirred tank reactors, *Chem. Eng. Sci.* **29** (1974), 967-985.

G. Yin and K. Yin, Asymptotically optimal rate of convergence of smoothed stochastic recursive algorithms, *Stochastics and Stochastic Reports*, **47** (1994), 21-46.

G. Yin and K. Yin, Steady state estimation and detection via stochastic approximation and kernel estimation, Supercomputer Inst. Res. Report, UMSI 93/175, Univ. Minn. 1993.

Robust Design of Products Depending on Both Qualitative and Quantitative Factors

IVAN N. VUCHKOV and L. N. BOYADJIEVA

Laboratory of Organization and Automation of Experimental Research,
Higher Institute of Chemical Technology, 1156 Sofia, Bulgaria.

Abstract:

The paper considers a model-based approach to quality improvement of products or processes depending on both qualitative and quantitative factors. Models with dummy variables are used for this purpose. A procedure for model structure selection is proposed. Models of the mean value and var7iance of the performance characteristic in mass production are obtained. Some specific features of the optimization procedures for quality improvement for models with qualitative and quantitative variables are considered. An example concerning the quality of resistors is given.

1 Introduction

Performance characteristics of many products depend on both qualitative and quantitative factors. The levels of the quantitative factors can be numerically measured while the levels of the qualitative factors can only be named or numbered. Examples of qualitative factors are: the type of raw material, the method of treatment or measurement, the type of equipment used in an experiment, the operator, etc. The existence of both types of factors has to be taken into account in both design of experiments and data analysis. Some of the problems arising in the design of experiments are considered by Draper and John [1]. For example a standard design of experiments could not be possible to be used, because the number of the design levels might not correspond to the number of levels of qualitative factors. It is not also clear how to connect the levels of the qualitative factors with the levels of design for quantitative factors.

Problem also arises with the models. In the simplest case the intercept can not be one and the same for any choice of levels of the qualitative factors. In the most complex case the models are completely different . In an intermediate class of problems only a part of

---

*Kitsos, C.P., and Müller, W.G., Eds., *Proceedings of MODA4*, Physica Verlag, Heidelberg, 1995

274

model coefficients are different at each combination of the qualitative factors.

The problem with choosing a model is important because it is connected with the design of experiments and the number of runs. In the worst case the number of models is the same as the number of combinations of the qualitative factors. In this case the number of coefficients could be too large so that the number of the runs i s high.

Another important feature of robust design problem is that in mass production qualitative factors can be set without errors, while quantitative ones can be considered as random variables. This must be taken into account when modelling the mean value and the variance of the performance characteristic in mass production.

In this paper we suppose that both qualitative and quantitative factors can be set without errors during the experiment, while in mass production the quantitative factors can be considered as random variables. We also assume that external noise factors do not exist. There are some differences between the present results and those we obtained in [4]. They are due to the presence of qualitative factors.

Therefore the following problems are of interest concerning quality improvement of a product depending on both qualitative and quantitative factors:

-Definition of a model of performance characteristic depending on both qualitative and quantitative factors and model coefficient estimation;

-Model structure selection;

-Design of experiments;

-Derivation of models of the mean value and variance of the performance characteristic in mass production;

-Optimization procedures.

Many of these problems can be solved in a similar way as for the case with quantitative factors. We will refer to the results obtained for this case but the accent will be put on the differences coming from the inclusion of qualitative factors.

2 Models of Performance Characteristic Depending on Both Qualitative and Quantitative Factors

Consider a product (or process) with a performance characteristic depending on m quantitative factors $\mathbf{x} = (x_1, x_2, \ldots, x_m)^T$ and q qualitative factors $\mathbf{z} = (z_1, z_2, \ldots, z_q)^T$. Denote by $n_1, n_2, \ldots, n_q$ the numbers of levels of $z_1, z_2, \ldots, z_q$ respectively. The total number of combinations of the levels of the qualitative factors is $d = n_1 n_2 \ldots n_q$. The dummy (indicator) variables are used for presentation of the qualitative factors in a regression model of the performance characteristic. They must correspond to a specific combination of levels of the qualitative variables. This can be determined in many different ways. The qualitative factors are presented by r dummy variables, where $r = n_1 + n_2 + \ldots + n_q - q$ (Johnston [3]). In this case to present a qualitative factor z_i we use $n_i - 1$ dummy variables $w_1^{(i)}, w_2^{(i)}, \ldots, w_{n_i-1}^{(i)}$. For example if z_i is at its l-th level then $w_k^{(i)} = 1$ for $k = l$ and $w_k^{(i)} = 0$ for $k \neq l$, $k = 1, 2, \ldots, n_{i-1}, l = 1, 2, \ldots, n_i$.

In this paper, second order polynomial models which are often used in the engineering practice are considered. We discuss two important cases:

i) No interactions exist between qualitative and quantitative factors, although there might be interactions within each group of factors. In this case the regression model has the form:

$$y_1(\mathbf{x}, \mathbf{w}_j) = \beta_0 + \mathbf{w}_j^T \alpha + \mathbf{w}_j^T \mathcal{A} \mathbf{w}_j + \beta^T \mathbf{x} + \mathbf{x}^T \mathcal{B} \mathbf{x} + v \qquad (1)$$

where

$\mathbf{w}_j = (w_1^j, w_2^j, \ldots, w_r^j)^T$ is a vector of dummy variables corresponding to the j-th combination of levels of qualitative factors, $j = 1, 2, \ldots, d$;

v is output random noise with $E(v) = 0$ and constant variance σ_v^2; the values of v are independent between the observations;

β_0 is the intercept;

$\beta = (\beta_1, \beta_2, \ldots, \beta_m)^T$ is a vector of the coefficients in the linear terms of the quantitative factors, while the coefficients in the second order terms of the model are given by a $(m \times m)$-matrix $\mathcal{B}$ with elements:

$$[\mathcal{B}]_{st} = \left\{ \begin{array}{ll} \beta_{ss} & \text{for } s = t \\ \frac{1}{2}\beta_{st} & \text{for } s \neq t \end{array} \right. ; \qquad (2)$$

$\alpha = (\alpha_1, \alpha_2, \ldots, \alpha_r)^T$ is a vector of coefficients in the linear part of the regression with respect to qualitative factors and coefficients in the second order terms are shown in a $(r \times r)$- matrix $\mathcal{A}$ of the form

$$\mathcal{A} = \begin{pmatrix} \mathbf{0} & \mathcal{A}_{12} & \ldots & \mathcal{A}_{1q} \\ \mathcal{A}_{12}^T & \mathbf{0} & \ldots & \mathcal{A}_{2q} \\ \ldots & \ldots & \ldots & \ldots \\ \mathcal{A}_{1q}^T & \mathcal{A}_{2q}^T & \cdots & \mathbf{0} \end{pmatrix} \qquad (3)$$

with submatrices $\mathcal{A}_{ij}, i = 1, 2, \ldots, q, j = 1, 2, \ldots, q$ of dimensions $(n_i - 1) \times (n_j - 1)$ having elements of the form:

$$[\mathcal{A}_{ij}]_{st} = \frac{1}{2}\alpha_{st}^{(ij)}, s = 1, 2, \ldots, n_{i-1}, t = 1, 2, \ldots, n_{j-1}.$$

The changes of $\mathbf{w}_j$ do not influence the terms in (1) comprising quantitative factors $\mathbf{x}$. That is why model (1) can be rewritten in the form

$$y_1(\mathbf{x}, \mathbf{w}_j) = \beta_{oj} + \beta^T \mathbf{x} + \mathbf{x}^T \mathcal{B} \mathbf{x} + v \qquad (4)$$

where

$$\beta_{oj} = \beta_0 + \mathbf{w}_j^T \alpha + \mathbf{w}_j^T \mathcal{A} \mathbf{w}_j, \quad j = 1, 2, \ldots, d. \qquad (5)$$

Consequently without interactions between $\mathbf{x}$ and $\mathbf{w}$ the changes of qualitative factors influence only the intercept of the regression model.

ii) There are interactions between qualitative and quantitative factors. Denote by $\mathcal{C}$ a $(r \times m)$ - matrix with elements $[\mathcal{C}]_{st} = c_{st}$ which are the coefficients before the interaction terms $w_s x_t$, $s = 1, 2, \ldots, r, t = 1, 2, \ldots, m$ in the regression model. In this case we can write the following model, for the j-th combination of the levels of the qualitative factors:

$$y_2(\mathbf{x}, \mathbf{w}_j) = \beta_o + \mathbf{w}_j^T \alpha + \mathbf{w}_j^T \mathcal{A} \mathbf{w}_j + \mathbf{w}_j^T \mathcal{C} \mathbf{x} + \beta^T \mathbf{x} + \mathbf{x}^T \mathcal{B} \mathbf{x} + v$$

$$= \beta_{oj} + \beta_j^T \mathbf{x} + \mathbf{x}^T \mathcal{B} \mathbf{x} + v \qquad (6)$$

where β_{oj} can be computed by (5) and

$$\beta_j^T = \mathbf{w}_j^T C + \beta^T, \quad j = 1, 2, \ldots, d.$$

One can see that if there are interactions between the quantitative and qualitative factors then the intercept and the coefficients in the linear terms of (6) depend on the combination of the levels of qualitative factors. Further complications of the model occur if interactions of higher order exist. In this paper we consider only second order models. If the experiments are conducted without errors in the factor levels, then the model can be estimated using ordinary least squares method.

The engineer has to decide which one of the models (4) or (6) fits better the data. Model structure can be selected using partial F-criterion [2]. Rewrite the models (4) and (6) as follows:

$$y_1(\mathbf{p}, \Theta_1) = \mathbf{f}_1^T(\mathbf{p}) + v \tag{7}$$

and

$$y_2(\mathbf{p}, \Theta_2) = \mathbf{f}_1^T(\mathbf{p})\Theta_1 + \mathbf{g}^T(\mathbf{p})\delta + v = \mathbf{f}_2^T(\mathbf{p})\Theta_2 + v \tag{8}$$

where $\mathbf{p}^T = (\mathbf{x}^T, \mathbf{w}^T)$ is a vector comprising both quantitative and qualitative factors, $\Theta_1^T = [\beta_0, \alpha^T, (vec\mathcal{A})^T, \beta^T, (vecB)^T]$ and Θ it$_2^T = (\Theta_1^T, \delta^T)$ are vectors of coefficients of two models, while $\delta = vec\mathcal{C}$. We denote by $vec\mathcal{A}$, $vecB$ and $vec\mathcal{C}$ vectors consisting of the columns of the matrices $\mathcal{A}$, B and $\mathcal{C}$ correspondingly. The functions in $\mathbf{f}_1(\mathbf{p})$ and $\mathbf{f}_2^T(\mathbf{p}) = (\mathbf{f}_1^T, \mathbf{g}^T)$ are defined by the models (4) and (6).

We want to test the null hypothesis $H_0 : \delta = \mathbf{0}$ (the alternative hypothesis is $H_1 : \delta \neq \mathbf{0}$. For this purpose the F-statistics is computed

$$F = \frac{[RSS(y_1) - RSS(y_2)]/(\nu_1 - \nu_2)}{RSS(y_2)/\nu_2} \tag{9}$$

where $RSS(y_1)$ and $RSS(y_2)$ are the residual sums of squares for models (7) and (8) respectively, while ν_1 and ν_2 are their corresponding degrees of freedom. Let $F_T(1 - \alpha, \nu_1 - \nu_2, \nu_2)$ is the critical value of F-distribution with $\nu_1 - \nu_2$ and ν_2 degrees of freedom for level of significance α. One of the following conclusions can be made:

-If $F \leq F_T(1 - \alpha, \nu_1 - \nu_2, \nu_2)$ then H_0 is accepted. The conclusion is that there is not significant difference between (7) and (8) and simpler model (7) can be used.

-If $F > F_T(1 - \alpha, \nu_1 - \nu_2, \nu_2)$ then H_0 is rejected and the model (8) is accepted to be better than (7).

3 Models of Performance Characteristic's Mean Value and Variance in Mass Production

As we already noted a realistic assumption for the mass production is that the qualitative factors are set without errors on given levels, while the values of quantitative ones vary randomly within given tolerance intervals. We also assume that an experiment without errors in factors can be conducted and a regression model of performance characteristic obtained. Then using error propagation theorems we can obtain models of mean value and variance in mass production in the same way as in Vuchkov and Boyadjieva [5].

It was shown in Section 2 that the models of the performance characteristic (4) and (6) are functions only of quantitative factors $\mathbf{x}$, and their coefficients β_{0j} or β_j depend on qualitative factors. As the qualitative factors are set without errors in both preliminary experiment and mass production the coefficients can be estimated on the basis of data obtained without errors in factors. Consequently in this case the models of mean value and variance of the performance characteristic can be obtained in the same way as for quantitative factors.

Consider for example model (6). Note that model (4) is a special case of (6). In mass production the errors $\mathbf{e} = (e_1, e_2, \ldots, e_m)^T$ occur in quantitative factors so that actually the values $x_i + e_i$ are set instead of x_i, $i = 1, 2, \ldots, m$. Denote by $b_0, \mathbf{b}, \mathbf{a}, \mathbf{A}, \mathbf{B}$ and $\mathbf{C}$ the estimates of $\beta_0, \alpha, \beta, \mathcal{A}, \mathcal{B}$ and $\mathcal{C}$ in (6). The following assumptions for the noises are used:

$$E(\mathbf{v}) = \mathbf{0}, cov(\mathbf{v}) = \sigma_v^2 \mathbf{I}$$

where $\mathbf{v}$ is N- vector column of output noise and

$$E(\mathbf{e}) = \mathbf{0}, cov(\mathbf{e}) = \Sigma_e = diag(\sigma_1^2, \sigma_2^2, \ldots, \sigma_m^2).$$

The variances of errors in the parameters $\sigma_1^2, \sigma_2^2, \ldots, \sigma_m^2$ are assumed to be constant over the factor space. The noise vectors $\mathbf{v}$ and $\mathbf{e}$ are supposed to be independently distributed.

Under these conditions the mean values and the variance in mass production can be computed using the same models as in Vuchkov, Boyadjieva [5] taking into account that b_{0j} and $\mathbf{b}_j$ depend on qualitative factors:

$$\tilde{y}_j = E(\hat{y}_j) = b_{0j} + \mathbf{b}_j^T \mathbf{x} + \mathbf{x}^T \mathbf{B} \mathbf{x} + tr\mathbf{B}\Sigma_e, \tag{10}$$

and

$$\sigma_{yj}^{\tilde{2}} = (\mathbf{b}_j + 2\mathbf{B}\mathbf{x})^T \Sigma_e (\mathbf{b}_j + 2\mathbf{B}\mathbf{x}) + HM + \sigma_v^2 \tag{11}$$

The term HM takes into account the high order moments of error distribution and can be computed as follows:

$$HM = 2\sum_{i=1}^{m} b_{ii}\sigma_i^4 + \sum_{i=1}^{m-1}\sum_{j=i+1}^{m} b_{ij}\sigma_i^2\sigma_j^2 \tag{12}$$

where $b_{ij} = [\mathbf{B}]_{ij}$, $i, j = 1, 2, \ldots, m$. The coefficients b_{oj} and $\mathbf{b}_j$ in (10) and (11) can be calculated as follows:

$$b_{oj} = b_0 + \mathbf{w}_j^T \mathbf{a} + \mathbf{w}_j^T \mathbf{A} \mathbf{w}_j \tag{13}$$

and

$$\mathbf{b}_j^T = \mathbf{w}_j^T \mathbf{C} + \mathbf{b}^T, \quad j = 1, 2, \ldots, d. \tag{14}$$

4 Optimization Procedures

4.1 Problem Formulation

The quality improvement can be defined as follows: minimize the variance $\sigma_{yj}^{\tilde{2}}$ while keeping the mean value $\tilde{y}_j$ on a target τ_j. The values of $\sigma_{yj}^{\tilde{2}}$ and $\tilde{y}_j$ can be computed using (11) and (10) correspondingly. The target value could be different at each combination of levels of the qualitative factors or to be the same at each of them. The values of σ_v^2 and HM (see (12)) do not depend neither on qualitative nor on quantitative factors. Consequently only the following part of the variance (11) can be minimized through optimization procedures with respect to the parameters:

$$S_j^2 = (\mathbf{b}_j + 2\mathbf{B}\mathbf{x})^T \Sigma_e (\mathbf{b}_j + 2\mathbf{B}\mathbf{x}). \tag{15}$$

A problem which must be taken into account in the optimization procedures is that the qualitative factors have only discrete levels (usually their number is not very high), while the quantitative ones can take values in a continuous subspace. Further on we will use some results obtained by Vuchkov, Boyadjieva [4] for quantitative factors and will see which changes have to be done in the case when the product quality depends on qualitative factors as well. We consider two special cases that differ upon the definition of the target:

-the target coincides with the extremum of the performance characteristic.

-the target does not coincide with the extremum of the performance characteristic.

In both cases we do not put constraints on factors.

4.2 Optimization when the Target Coincides with the Extremum of the Performance Characteristic

To find the extremums of the mean value and variance of the performance characteristic in mass production we set the first derivatives of (10) and (15) with respect to $\mathbf{x}$ to be zero :

$$\frac{\partial \tilde{y}_j}{\partial \mathbf{x}} = \mathbf{b}_j + 2\mathbf{B}\mathbf{x} = \mathbf{0}, \tag{16}$$

$$\frac{\partial S_j^2}{\partial \mathbf{x}} = 4\mathbf{B}^T \Sigma_e (\mathbf{b}_j + 2\mathbf{B}\mathbf{x}) = \mathbf{0}. \tag{17}$$

The extremum of the variance is always in the minimum because Σ_e is nonnegative definite and the matrix $8\mathbf{B}^T \Sigma_e \mathbf{B}$ which is second order derivative of S_j^2 is always nonnegative definite. The stationary point for both models (10)and (15) coincides and its coordinates are:

$$\mathbf{x}_{sj} = -\frac{1}{2}\mathbf{B}^{-1}\mathbf{b}_j. \tag{18}$$

As $\mathbf{b}_j$ depends on the qualitative variables (see(14)) the stationary point is different at each combination of their levels. The intercept b_{0j} in (10) also depends on dummy variables. Therefore the following optimization procedure can be applied:

(i) compute $\mathbf{x}_{sj}$ using (18) for all possible combinations of levels of qualitative factors, $j = 1, 2, \ldots, d$.

(ii) Substituting $\mathbf{x}_{sj}$ for $\mathbf{x}$ in (10) find the values of extremums of $\tilde{y}_{sj}$ for corresponding combinations of qualitative factor levels, $j = 1, 2, \ldots, d$.

The problem can be solved easier if there are not interactions between the qualitative and quantitative factors. One can see from (14) that in this case $\mathbf{b}_j = \mathbf{b}$ i.e. the coefficients in the linear terms of the regression models do not depend on qualitative factors. Therefore the variance (11) does not depend on qualitative factors, so do the coordinates of the stationary point:

$$\mathbf{x}_s = -\frac{1}{2}\mathbf{B}^{-1}\mathbf{b}. \tag{19}$$

In this case extremum of the mean value of the performance characteristic can be found separately with respect to qualitative and quantitative factors. The optimization procedure is as follows:

(i) Compute the optimal coordinates $\mathbf{x}_s$ of the of the quantitative variables using (19).

(ii) Substituting $\mathbf{x}_s$ for $\mathbf{x}$ in (10) find the values of extremums of $\tilde{y}_{sj}$ for corresponding combinations of qualitative factor levels, $j = 1, 2, \ldots, d$.

4.3 Optimization when the Target does not Coincide with the Extremum of the Performance Characteristic

The optimization problem is now defined as follows: minimize S_j^2 while keeping $\tilde{y}_j$ equal to a target τ_j. A geometric solution of this problem for quantitative factors was proposed by Vuchkov, Boyadjieva [4]. This approach can be applied only for unconstrained optimization and is based on the fact that the contours of variance (15) are always ellipses or ellipsoids for $\sigma_i^2 \neq 0, i = 1, 2, \ldots, m$. The points where these contours tangent to the contour corresponding to the target value of the performance characteristic are solutions of the problem. There are two tangent points and consequently two solutions exist. This idea is illustrated in Fig.1, where the target value is $\tau = 4$ and the contour of minimal variance is $S^2 = 2.56.10^{-6}$.

For implementation of this idea the variance contours (15) are transformed into spheres which makes possible to find the tangent points with $\tilde{y} = \tau$. The proof of this procedure is given in Vuchkov, Boyadjieva [4]. We consider here only the final results.

Denote by μ_{max} the maximal eigenvalue of the matrix $\frac{1}{4}\Sigma_e^{-1}\mathbf{B}^{-1}$ and by $\mathbf{t}_{max}$ its corresponding eigenvector. Taking into consideration the presence of qualitative factors the coordinates of the optimum can be written as follows:

$$\mathbf{x}_{opt,j} = \frac{1}{2}\mathbf{B}^{-1}\mathbf{t}_{max} + \mathbf{x}_{sj} \tag{20}$$

where

$$\mathbf{x}_{sj} = \frac{1}{2}\mathbf{B}^{-1}\mathbf{b}_j \tag{21}$$

is the stationary point of both surfaces for $\tilde{y}_j$ and S_j^2 which are described by (10) and (15) correspondingly and

$$\tilde{y}_{sj} = b_{oj} - \frac{1}{4}\mathbf{b}_j^T\mathbf{B}^{-1}\mathbf{b}_j + tr\mathbf{B}\Sigma_e. \tag{22}$$

is the mean value of performance characteristic at stationary point $\mathbf{x}_{sj}$ for j-th qualitative factor level combination. The corresponding minimal value of the variance S_j^2 is

$$S^2(\mathbf{x}_{opt,j}) = \xi_{max,j}^2 \tag{23}$$

where

$$\xi_{max,j} = \pm\sqrt{(\tau_j - \tilde{y}_{sj})/\mu_{max}}. \tag{24}$$

The value under the square root must always be positive so that the signs of $\tau_j - \tilde{y}_{sj}$ and μ_{max} must always coincide . Consequently if the nominator of this ratio is negative, the minimal negative eigenvalue must be substituted for μ_{max} in (24) and its corresponding eigenvector for $\mathbf{t}_{max}$ in (20) (see example).

Following geometric interpretation of this solution is possible. The axes of the surfaces describing the mean value $\tilde{y}_j$ and variance S_j^2 depend on the matrices of the following quadratic forms: $\mathbf{x}^T\mathbf{B}\mathbf{x}$ and $\mathbf{x}^T\mathbf{B}^T\Sigma_e\mathbf{B}\mathbf{x}$. Since $\mathbf{B}$ and $\mathbf{B}^T\Sigma_e\mathbf{B}$ are independent on the qualitative factor levels the direction of the axes of these surfaces are one and the same for all their combinations. The coordinates of the stationary point (21), the value $\tilde{y}_{sj}$ of the performance characteristic (22) at the stationary point and the minimal variance (23) are what changes.

For the case with no interactions between qualitative and quantitative factors we can put $b_j = b$. Then the stationary point (21) is one and the same for all combinations of the qualitative factor levels. Despite of the fact that the surface of variance (15) is independent on qualitative factors, its tangent points with the surface describing the mean value depend on them because $\tilde{y}_{sj}$ is a function of $\mathbf{w}$ through b_{0j}. However ξ_{maxj} and $S^2(\mathbf{x}_{opt,j})$ depend on qualitative variables via b_{0j}, and the vector $\mathbf{x}_{opt,j}$ depends on $\mathbf{w}$ via sign of $\tau_j - \tilde{y}_{sj}$ and corresponding choice of μ_{max} and $\mathbf{t}_{max}$.

If the target is one and the same for all combinations of qualitative factor levels then τ_j can be replaced by τ in (24) and the best combination of levels is the one, which ensure minimum of $S^2(\mathbf{x}_{opt,j})$ computed by (23).

For constrained optimization this procedure can be used only if the extremum is within the region of interest. If this condition is not fulfilled then numerical optimization procedure can be used.

5 Example

Consider a vacuum thermal process in the production of resistors. The performance characteristic (resistance $y[\Omega]$) depends on two quantitative factors (current $x_1[A]$ and time of metal layer deposition $x_2[s]$) as well on one qualitative factor (type of ceramics z) with two levels. The qualitative factor is described by a dummy variable w with levels 1 and 0 which correspond to two type of ceramics I and II .

A second order design was sequentially generated so that to satisfy D- optimality criterion. This approach is convenient for factors having different number of levels. It also allows a flexible choice of number of runs. The design and the corresponding values of the performance characteristic obtained from the experiments are given in Table 1. The levels of quantitative factors are shown in Table 2.

Table 1. Results of experiments and calculations

No	x_1	x_2	w	y	$\bar{y}_j + tr\mathbf{B}\Sigma_e$	$S_j^2 + HM$
1	-1	1	1	4.08	3.987	0.03092
2	1	-1	1	3.64	3.546	0.11268
3	-1	-1	1	4.35	4.312	0.05777
4	1	1	1	3.30	3.221	0.10156
5	-1	-1	0	32.6	32.479	0.12835
6	1	1	0	28.00	28.434	0.19604
7	-1	1	0	30.50	31.074	0.07868
8	1	0	0	29.00	28.953	0.20551
9	0	-1	0	31.00	31.464	0.16990
10	0	0	1	3.53	3.889	0.04194
11	1	-1	0	30.00	29.839	0.22071
12	0.1	1	0	30.75	29.924	0.14224
13	-1	0	0	31.90	31.593	0.09996
14	-1	-1	1	4.35	4.312	0.05777

Table 2. Levels of quantitative factors

Code	-1	0	1
Factors			
$x_1[A]$	35	45	55
$x_2[s]$	30	40	50

Stepwise regression analysis was used and following models were obtained:

$$\widehat{y_1} = -0.8855x_1 - 0.4283x_2 - 0.34x_1^2 + 0.1285x_2^2 + \begin{cases} 3.8323, & \text{for } w = 1 \\ 30.6388, & \text{for } w = 0 \end{cases} \quad (25)$$

and

$$\widehat{y_2} = -0.3053x_1^2 + 0.1831x_2^2 + \begin{cases} -0.3827x_1 - 0.1627x_2 + 3.886, & \text{for } w = 1 \\ -1.3195x_1 - 0.7025x_2 + 30.5773, & \text{for } w = 0 \end{cases} \quad (26)$$

For selection of one of these models the partial F-criterion (9) was used:

$$F = \frac{(5.0889 - 1.7077)/2}{1.7077/6} = 5.9399.$$

The critical value of F-criterion is $F_T(1 - 0.05, 2, 6) = 5.143..$ The model (26) was chosen because $F > F_T(1 - 0.05, 2, 6)$. The residual sum of squares $\sigma_v^2 = 1.7077/6 = 0.2846$ can be considered as an estimate of error variance. The standard deviations of errors in the quantitative factors are $\sigma_1 = 0.1$ and $\sigma_2 = 0.1$ and their covariance matrix is $\Sigma_e = diag(0.01, 001)$.

The mean values of the performance characteristic shown in Table 1 are calculated by means of (10) and (26) taking into account that $tr B\Sigma_e = b_{11}\sigma_1^2 + b_{22}\sigma_2^2 = -0.001223$.

The next step was to find the optimal parameter values which ensure minimal variances under the condition that the mean values are equal to $\tau = 4[\Omega]$ and $\tau = 30[\Omega]$ for ceramics type I and II correspondingly. The method of Section 4.3 was used. Taking into account the coefficients in (26) and relationships (2), (3), (13) and (14) we can define b_{0j} and elements of b_j and B as follows:

$$b_{0j} = \begin{cases} 3.886 & \text{, for } w = 1 \\ 30.5773 & \text{, for } w = 0 \end{cases};$$

$$\mathbf{b_j} = \begin{cases} (-0.3827, -0.1627)^T, & \text{for } w = 1 \\ (-1.3195, -0.7025)^T, & \text{for } w = 0 \end{cases};$$

$$\mathbf{B} = \begin{pmatrix} -0.3053 & 0 \\ 0 & 0.1830 \end{pmatrix}.$$

The eigenvalues of matrix $\frac{1}{4}\Sigma_e^{-1}B^{-1}$ and their corresponding eigenvectors are equal to

$$\mu_1 = -81.882; \mathbf{t_1} = (10, 0)^T,$$

$$\mu_2 = 136.575; \mathbf{t}_2 = (0, 10,)^T.$$

The coordinates of the stationary point and the corresponding value of the performance characteristic were computed by use of (21) and (22). Following values were obtained:

$$\mathbf{x}_{sj} = \begin{cases} (-0.6267, 0.4444)^T, & for \ w = 1 \\ (-2.1609, 1.9189)^T, & for \ w = 0 \end{cases} ;$$

$$\tilde{y}_{sj} = \begin{cases} 3.9685 & , \text{for } w = 1 \\ 31.3278 & , \text{for } w = 0 \end{cases}$$

The optimal parameter vectors (Fig.1, Fig.2) and corresponding minimal values of $S_j^2 + HM$ were computed by use of (20) and (23)

$$\mathbf{x}_{opt,j} = \begin{cases} (-0.6267, 0.8590)^T, (-0.6267, 0.0299)^T, & for \ w = 1 \\ (-4.2464, 1.9189)^T, (-0.0755, 1.9189)^T, & for \ w = 0 \end{cases} ;$$

$$S_j^2 + HM = \begin{cases} 0.0002557 & , \text{for } w = 1 \\ 0.0162418 & , \text{for } w = 0 \end{cases}$$

where $HM = 0.000025345$ and was computed by means of (12).

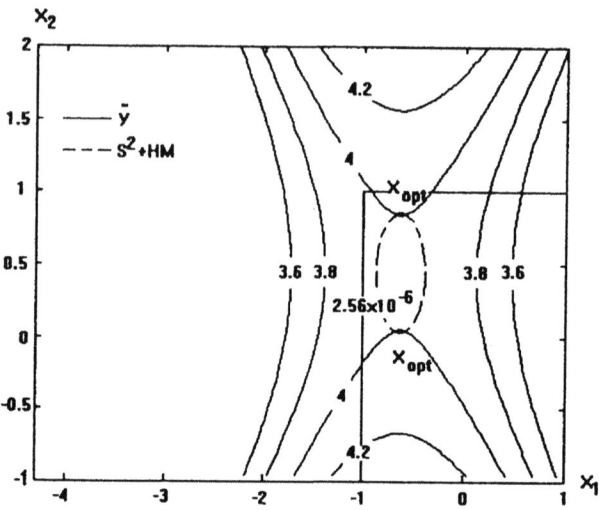

Fig.1 Optimal solutions for type of ceramics I (w=1).

One can see that for $w = 0$ the optimal solutions are outside of the region of interest defined by the inequalities $-1 \le x_i \le 1, i = 1, 2$. A graphical solution (Fig.2) was found in the point $x_{opt}^* = (0.043, 1)$ where the contour corresponding to $\tilde{y} = 30$ intersects the boundary of the region defined by $x_2 = 1$. In this point $S_j^2 + HM = 0.01926$.

The variances obtained in the optimal points for $w = 0$ and $w = 1$ are significantly smaller then the values $S_j^2 + HM$ given in the last columns of Table 1.

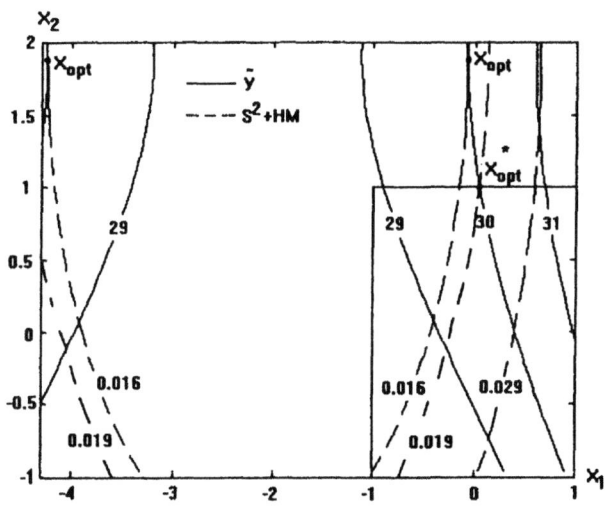

Fig.2 Optimal solutions for type of ceramics II (w=0).

Acknowledgement

This work was supported by the National Scientific Foundation and Ministry of Education and Culture of Bulgaria.

References

Draper N., J. John (1988). Response-surface designs for quantitative and qualitative variables. *Technometrics*, 30(4), p.423-428.

Draper N., H.Smith (1981). *Applied regression analysis*(2nd ed.), John Wiley.

Johnston J. (1972). *Econometric methods*(2nd ed), McGraw-Hill.

Vuchkov I., L.Boyadjieva (1988). The robustness against tolerances of performance characteristics described by second order polynomials. In *Optimal Design and Analysis of Experiments*, Y.Dodge, V.Fedorov and H.Wynn (eds), Amsterdam, North Holland, p.293-309.

Vuchkov I., L. Boyadjieva (1992). Quality improvement through design of experiments with both product parameters and external noise factors. In *Model Oriented Data Analysis*, W.G. Müller, V.V. Fedorov, I.N. Vuchkov (eds), Physica Verlag, Heidelberg, p.195-212.

Improving on Golden–Section Optimisation for Locally Symmetric Functions

Luc Pronzato, Henry P. Wynn and Anatoly A. Zhigljavsky

Laboratoire I3S, CNRS-URA 1376, Sophia Antipolis, 06560 Valbonne, France,
School of Mathematics, City University, Northampton Sq., London EC1V 0HB, UK,
Dept. of Mathematics, St.Petersburg University, Bibliotechnaya sq. 2, 198904, Russia.

Abstract:

We consider the minimisation of a uniextremal function $f(.)$ on $[0,1]$ using a "second-order" algorithm. At each iteration the current feasible region is rescaled to $[0,1]$, so that the optimizing value x^* in the initial $[0,1]$-interval varies from iteration to iteration, which defines a dynamic system. Many line-search algorithms exhibit chaotic behaviour when rescaling is applied. If $f(.)$ is symmetric around x^*, the associated dynamic system is time-homogeneous and often possesses an invariant density. In a first part, we show that the asymptotic behaviour of the classical Golden-Section algorithm is the same for locally symmetric functions as for pure symmetric functions. We believe that this property is also true for other line-search algorithms, with sometimes a better ergodic rate than the Golden-Section algorithm. In a second part, we consider the case where the number of iterations is fixed *a priori*, with a dynamic-programming approach, using a uniform prior density on $[0,1]$ for x^*.

1 Introduction

We consider the minimisation of a uniextremal function $f(.)$ on $[0,1]$ using a "second-order" algorithm, which can be defined as follows. Let $I_n = [a_n, b_n]$ be the current interval at iteration n. Two points x' are then placed between a_n $a_n \leq x' < x'' \leq b_n$. If $f(x') < f(x'')$, then I_n is no minimizer in $[a_n, x']$, $f(x') \geq f(x'')$ then I_n One can note that in both cases the next interval I_{n+1} contains either x' be evaluated once at each iteration. The way this evaluation is chosen defines the algorithm. For a more detailed introduction see for example (7).

---

*Kitsos, C.P., and Müller, W.G., Eds., *Proceedings of MODA4*, Physica Verlag, Heidelberg, 1995

The *Golden Section algorithm* selects x', x'' symmetrically placed between a_n

$$x' + x'' = a_n + b_n \,,$$

and such that

$$\frac{x'' - a_n}{b_n - a_n} = \lambda = \frac{\sqrt{5} - 1}{2} \simeq 0.61804 \,,$$

where λ is the positive root of the equation $\lambda^2 + \lambda - 1 = 0$ and is itself known as the Golden Section.

The performances of a line-search algorithm can be measured by its *reduction rate*. Denote $L_n = b_n - a_n$ the length of I_n, with $L_1 = 1$. The reduction rate of the n-th iteration is then defined as

$$r_n = \frac{a_{n+1} - b_{n+1}}{a_n - b_n} \,, \tag{1}$$

and the asymptotic rate as

$$r = \lim_{N \to \infty} (L_N)^{\frac{1}{N}} = \lim_{N \to \infty} (\prod_{n=1}^{N} r_n)^{\frac{1}{N}} \,, \tag{2}$$

if this limit exists for almost all x^* with respect to the Lebesgue measure.

The Golden Section algorithm has a constant rate λ: after n initial interval is reduced by a ratio λ^{n-1}.

We shall consider *renormalised processes*, obtained by renormalising I_n back to $[0, 1]$ after each iteration. Suppose that x^* is the unknown minimiser of $f(.)$, in $[0, 1]$ at the beginning, in I_n at iteration n, and denote x_n its renormalised location in $[0, 1]$ at iteration n. Instead of x^* being fixed and I_n moving, x_n moves in the fixed interval $[0, 1]$. We define for each n a function $g_n(.)$ mapping I_n back to $[0, 1]$:

$$g_n : I_n \longmapsto [0, 1] \,,$$

with $x_n = g_n(x^*)$. Second-order line-search algorithms, in the renormalised form, then compare function values at two points, first, e_n, from the previous iteration and second, e'_n, selected by the algorithm at the current iteration. The choice of the sequence $\{e'_n\}$ defines the algorithm, and will be called *strategy* in Section 3. Define

$$w_n = \min\{e_n, e'_n\}, \quad w'_n = \max\{e_n, e'_n\}, \quad c_n = \frac{w_n + w'_n}{2} \,,$$

then the deletion rule is:

$$\begin{cases} \text{(R)}: & \text{if } f_n(w_n) < f_n(w'_n) \text{ delete } (w'_n, 1] \\ \text{(L)}: & \text{if } f_n(w_n) \geq f_n(w'_n) \text{ delete } [0, w_n) \,, \end{cases} \tag{3}$$

with $f_n(.)$ the renormalised function on $[0, 1]$ defined by

$$f_n(x) = f(g_n^{-1}(x)) \,.$$

Here (R) and (L) stand for right and left deletion. The remaining interval is then renormalised to $[0, 1]$.

The *Golden Section algorithm* corresponds to

$$w'_n = 1 - w_n = \lambda \,,$$

which gives $x_{n+1} = h_n(x_n)$, with

$$h_n(x_n) = \begin{cases} x_n(1+\lambda) & \text{if } f_n(1-\lambda) < f_n(\lambda) \quad \text{(R)} \\ x_n(1+\lambda) - \lambda & \text{if } f_n(1-\lambda) \geq f_n(\lambda) \quad \text{(L)}. \end{cases} \qquad (4)$$

In the special case when $f(x)$ is symmetric around x^* the renormalised algorithm yields the time-homogeneous dynamic process

$$x_{n+1} = h(x_n) = \begin{cases} x_n(1+\lambda) & \text{if } x_n < \frac{1}{2} \\ x_n(1+\lambda) - \lambda & \text{if } x_n \geq \frac{1}{2}. \end{cases} \qquad (5)$$

This dynamic system exhibits chaotic behaviour, and has an invariant measure $\mu(.)$ such that

$$\lim_{k \to \infty} \frac{1}{k} \sum_{i=1}^{k} g(x_i) = \int_0^1 g(x)\mu(dx)$$

for any continuous function $g(.)$ on $[0,1]$ and μ-almost all starting points $x_0 \in [0,1]$, see (7). The measure $\mu(.)$ has the density (with respect to the Lebesgue measure on $[0,1]$) $p(x)$

$$p(x) = \begin{cases} 0 & \text{if } x \notin [\frac{1-\lambda}{2}, \frac{1+\lambda}{2}) \\ \frac{4+3\lambda}{5} & \text{if } \frac{1-\lambda}{2} \leq x < \frac{\lambda}{2}, \\ \frac{7+4\lambda}{5} & \text{if } \frac{\lambda}{2} \leq x < 1 - \frac{\lambda}{2} \\ \frac{4+3\lambda}{5} & \text{if } 1 - \frac{\lambda}{2} \leq x < \frac{1+\lambda}{2} \end{cases} \qquad (6)$$

Many other line-search algorithms exhibit chaotic behaviour when renormalisation is applied. If $f(.)$ is symmetric, the associated dynamic system is time-homogeneous and often possesses an invariant density, sometimes with a better ergodic rate than the rate λ of the Golden Section algorithm, see (6). An important question then concerns the definition of *optimal* algorithms in terms of the ergodic rate.

Consider the case where $f(.)$ is symmetric about x^*, and suppose $f(w_n) \geq f(w_n')$. Then symmetry implies that $x_n \geq c_n$. We can thus achieve a greater interval reduction than in the usual case by eliminating $[0, c_n)$ rather than only $[0, w_n)$. We shall refer to this as the *optimistic* rule. The recurrent relation for $\{x_n\}$ and the rates $\{r_n\}$ are given by the formulae

$$x_{n+1} = \begin{cases} \frac{x_n}{c_n} & \text{if } x_n < c_n \\ \frac{x_n - c_n}{1 - c_n} & \text{if } x_n \geq c_n \end{cases} \qquad r_n = \begin{cases} c_n & \text{if } x_n < c_n \\ 1 - c_n & \text{if } x_n \geq c_n \end{cases}$$

One can clearly achieve the ergodic rate $r = \frac{1}{2}$ if $r_n = \frac{1}{2}$ for all n. A way to do this is always to place the new observation point at one of the endpoints of the new interval obtained by the optimistic rule. An important result is then that we cannot achieve an ergodic rate less than $\frac{1}{2}$ for all x^* lying in a set of positive Lebesgue measure, see (3; 8). A similar result will be derived in Section 3 for the case when the number of iterations of the algorithms is fixed *a priori*.

Most results about the asymptotic behaviour of line search algorithms hold under the assumption that the function $f(.)$ is uniextremal and symmetric. In Section 2 the condition of symmetry will be relaxed, and we shall consider locally symmetric functions satisfying

$$f(x) = f(x^*) + C_1|x - x^*|^\gamma + O(|x - x^*|^{\beta+\gamma}) \quad \text{for some } \beta > 0, \qquad (7)$$

with $C_1 > 0, \gamma > 0$. Note that if $f(.)$ is smooth at x^*, then (7) holds with $\gamma = 2$. We then show that the asymptotic behaviour of the Golden Section algorithm is the same as for symmetric functions. We believe that the same property holds for some other line-search algorithms, for example those with a finite number of states considered in (4).

An algorithm that achieves the optimal ergodic rate of $\frac{1}{2}$ for functions satisfying (7) is detailed in (8). Following a similar line, ϵ-optimal algorithms with finite number of states are presented in (3; 4). The case when the number of iterations is fixed *a priori* is considered in Section 3, with a dynamic-programming approach.

2 Asymptotic Behaviour of the Golden Section Algorithm for Locally Symmetric Functions

In this section the function $f(.)$ will be assumed to be locally symmetric and uniextremal at x^*. It thus satisfies (7), which we rewrite in the following form:

$$C_1|x - x^*|^\gamma - C_2|x - x^*|^{\beta+\gamma} \leq |f(x) - f(x^*)| \leq C_1|x - x^*|^\gamma + C_2|x - x^*|^{\beta+\gamma} \quad (8)$$

for any x in $[0, 1]$, for some $\beta > 0$, $C_1 > 0$, $C_2 \geq 0$ and $\gamma > 0$. Throughout this section we shall still denote by $\{x_k\}$ the dynamic process (5) obtained for a symmetric function, but shall denote by $\{x'_k\}$ the process defined by (4), which corresponds to the actual function $f(.)$ satisfying (8). We thus define the sequence

$$x'_{n+1} = \begin{cases} x'_n(1 + \lambda) & \text{if } f_n(1 - \lambda) < f_n(\lambda) \quad (\text{R}') \\ x'_n(1 + \lambda) - \lambda & \text{if } f_n(1 - \lambda) \geq f_n(\lambda) \quad (\text{L}'). \end{cases} \quad (9)$$

This section aims at proving that the processes $\{x_n\}$ and $\{x'_n\}$ have the same asymptotic behaviour.

Before transforming (8) to the renormalised function $f_n(.)$, we first derive the expression of $f_n(.)$ for the Golden Section algorithm. Consider the deletion rule (3), and define at iteration n:

$$\alpha_n = \begin{cases} 0 & \text{if (R)}, \\ 1 & \text{if (L)}. \end{cases}$$

We then have $x_1 = x^* = \lambda x_2 + \alpha_1(1 - \lambda)$, and by induction

$$x^* = g_n^{-1}(x_n) = \sum_{i=1}^{n} \alpha_i(1 - \lambda)\lambda^{i-1} + \lambda^n x_n,$$

which defines $g_n^{-1}(z)$ and thus $f_n(.) = f(g_n^{-1}(.))$ for any z in $[0, 1]$.

Note that the same decision rules (R') or (L') are obtained if $f_n(.)$ at iteration n is replaced by

$$\bar{f}_n(.) = K_1(n)f_n(.) + K_2(n),$$

for some $K_1(n), K_2(n)$. $\bar{f}_n(.)$ will be our renormalised function, uniextremal at x_n, and we then have the following lemma.

Lemma 1. If $K_1(n) = (1 + \lambda)^{n\gamma}$ and $K_2(n) = -(1 + \lambda)^{n\gamma}f(x^*)$ then $\bar{f}_n(.)$ satisfies

$$m_n(x_n, z) = C_1|z - x_n|^\gamma - C_2\lambda^{n\beta} \leq \bar{f}_n(z) \leq C_1|z - x_n|^\gamma + C_2\lambda^{n\beta} = M_n(x_n, z). \quad (10)$$

Proof.

The substitution of $g_n^{-1}(z)$ for x in (8) and the multiplication by $K_1(n)$ gives

$$K_2(n) \quad + \quad (1 + \lambda)^{n\gamma}f(x^*) + C_1|z - x_n|^\gamma - C_2|z - x_n|^{\gamma+\beta}\lambda^{n\beta} \leq \bar{f}_n(z) \leq$$
$$K_2(n) \quad + \quad (1 + \lambda)^{n\gamma}f(x^*) + C_1|z - x_n|^\gamma + C_2|z - x_n|^{\gamma+\beta}\lambda^{n\beta}.$$

Since $|z - x_n| \leq 1$, and using the definition of $K_2(n)$, we get (10). □

We shall say that

(i) a *mistake* has been made at iteration n of the process $\{x'_k\}$ if $x_n = x'_n$ and $x_{n+1} \neq x'_{n+1}$;

(ii) the mistake at iteration n has been corrected in m iterations if $x_{n+l} \neq x'_{n+l}$ for $1 < l < m$ and $x_m = x'_m$;

(iii) we pay a penalty $m - 1$ for any mistake corrected in m iterations.

The penalty s_N paid up to iteration N counts the number of pairs (x_n, x'_n), $n \leq N$, in the sequences $\{x_n\}$ and $\{x'_n\}$ such that $x_n \neq x'_n$, that is

$$s_N = \{\text{number of } n \leq N \text{ such that } x_n \neq x'_n\}. \tag{11}$$

From the definition of the Golden Section algorithm and the uni-extremality of $f(.)$ no mistake can happen outside the interval $[1 - \lambda, \lambda]$. The next statement shows that as n increases, the size of the interval where mistakes can occur decreases exponentially.

Lemma 2. The interval V_n where a mistake can occur at iteration n is included in

$$U_n = [\frac{1}{2} - C_3 \lambda^{n/\beta}, \frac{1}{2} + C_3 \lambda^{n/\beta}], \tag{12}$$

where

$$C_3 = \frac{C_2}{C_1}(3 + 2\lambda)^{\gamma-1} \max\{1, \frac{2^{\gamma-1}}{\gamma}\}. \tag{13}$$

Proof.
No mistake is possible at iteration n if

$$M_n(x_n, \lambda) < m_n(x_n, 1 - \lambda) \tag{14}$$

or

$$m_n(x_n, \lambda) > M_n(x_n, 1 - \lambda), \tag{15}$$

where $m_n(x_n, z)$ and $M_n(x_n, z)$ are the minorant and majorants of $\bar{f}_n(z)$ defined in (10). We consider the case $x_n \in (1/2, \lambda)$ (the case $x_n \in (1 - \lambda, 1/2)$ can be treated in the same way). The inequality (14) is equivalent to

$$C_1(\lambda - x_n)^\gamma + C_2 \lambda^{n\beta} < C_1(x_n - (1 - \lambda))^\gamma - C_2 \lambda^{n\beta}.$$

Denote $z = x_n - 1/2$, with $0 < z < \lambda - 1/2$, and rewrite the last inequality in the form

$$\frac{2C_2 \lambda^{n\beta}}{C_1} < ((\lambda - \frac{1}{2}) + z)^\gamma - ((\lambda - \frac{1}{2}) - z)^\gamma = \phi_\gamma(z). \tag{16}$$

For $1 \leq \gamma \leq 2$ the function $\phi_\gamma(.)$ is concave, and a sufficient condition for (16) to hold is

$$\frac{2C_2 \lambda^{n\beta}}{C_1} < (\phi_\gamma(\lambda - \frac{1}{2}) - \phi_\gamma(0))z = 2^\gamma(\lambda - \frac{1}{2})^\gamma z \leq \phi_\gamma(z).$$

For $0 < \gamma \leq 1$ and $\gamma \geq 2$ the function $\phi_\gamma(.)$ is convex, and a sufficient condition for (16) to hold is now

$$\frac{2C_2\lambda^{n\beta}}{C_1} < \phi_\gamma(0)) + \phi_\gamma'(0)z = 2\gamma(\lambda - \frac{1}{2})^{\gamma-1}z \leq \phi_\gamma(z).$$

The solution of the left-hand side inequalities with respect to z yields the upper end of U_n given by (12-13). □

Define the intervals

$$A_k = [\frac{1}{2} - l_k, \frac{1}{2} + l_k] \subset [0,1],$$

with

$$l_k = (\lambda - \frac{1}{2})(2\lambda - 1)^k = P_k + \lambda Q_k, \quad k = 1,2,\ldots$$

where the generating functions for the coefficients P_k and Q_k are

$$P(z) = -\frac{1-z}{2(1 - 4z - z^2)}, \quad Q(z) = -\frac{1}{1 - 4z - z^2}.$$

The next lemma states a so-called *self-correcting property* for the Golden Section algorithm: any mistake within the interval $A_1 = [4\lambda - 2, 3 - 4\lambda]$ will be necessarily corrected in three iterations. Moreover, it shows that for all $k \geq 1$, any point in the set $A_k \setminus A_{k+1}$ moves to $A_{k-1} \setminus A_k$ in three iterations of (5).

Lemma 3. *Self-correcting property for the Golden Section algorithm.* For any $n \geq 1$, $k \geq 1$ if $x_n = x_n' \in A_k \setminus A_{k+1}$ and $x_{n+1} \neq x_{n+1}'$ then $x_{n+3} = x_{n+3}' \in A_{k-1} \setminus A_k$.

Proof.
Let

$$x_n = x_n' \in (\frac{1}{2} + l_{k+1}, \frac{1}{2} + l_k],$$ (17)

that is $x_n = x_n' \in A_k \setminus A_{k+1} \cap [\frac{1}{2}, 1]$. Applying three iterations of (5) we obtain

$$x_{n+1} = (1+\lambda)x_n - \lambda \in \left((\frac{1}{2} + l_{k+1})(1+\lambda) - \lambda, (\frac{1}{2} + l_k)(1+\lambda) - \lambda\right],$$

$$x_{n+2} = (1+\lambda)x_{n+1} = (2+\lambda)x_n - 1 \in \left((\frac{1}{2} + l_{k+1})(2+\lambda) - 1, (\frac{1}{2} + l_k)(2+\lambda) - 1\right],$$

$$x_{n+3} = (1+\lambda)x_{n+2} = (3+2\lambda)x_n - (1+\lambda)$$
$$\in \left((\frac{1}{2} + l_{k+1})(3+2\lambda) - (1+\lambda), (\frac{1}{2} + l_k)(3+2\lambda) - (1+\lambda)\right]$$
$$= (\frac{1}{2} + l_k, \frac{1}{2} + l_{k-1}],$$

where the last inequality follows from the definition of the A_k's. Assume (17) again and $x_{n+1}' \neq x_{n+1}$, that is a mistake is made at iteration n. Then we get

$$x_{n+1}' = (1+\lambda)x_n \in \left((\frac{1}{2} + l_{k+1})(1+\lambda), (\frac{1}{2} + l_k)(1+\lambda)\right]$$

by application of (R') in (9). Since $(\frac{1}{2} + l_{k+1})(1+\lambda) > \lambda$, the (L') rule is necessarily used at iteration $n+1$ of (9), and therefore

$$x_{n+2}' = (1+\lambda)x_{n+1}' - \lambda = (2+\lambda)x_n - \lambda \in \left((\frac{1}{2} + l_{k+1})(2+\lambda) - \lambda, (\frac{1}{2} + l_k)(2+\lambda) - \lambda\right].$$

Since $(\frac{1}{2} + l_{k+1})(2 + \lambda) - \lambda > \lambda$, (L') is necessarily used again at iteration $n + 2$ of (9), and

$$x'_{n+3} = (1 + \lambda)x'_{n+2} - \lambda = (3 + 2\lambda)x_n - (1 + \lambda) = x_{n+3} \in \left(\frac{1}{2} + l_k, \frac{1}{2} + l_{k-1}\right].$$

The case $x_n < \frac{1}{2}$ can be treated in the same way. $\qquad\square$

Theorem. Let $f(.)$ satisfy the condition (8). Then the asymptotic behaviours of the dynamic systems $\{x_n\}$ and $\{x'_n\}$ defined respectively by (5) and (9) coincide in the sense that for almost all $x^* \in [0, 1]$ and any Borel set $A \subset [0, 1]$

$$\lim_{N\to\infty} \frac{1}{N} \sum_{n=1}^{N} I_A(x'_n) = \mu(A),\qquad(18)$$

where $I_A(.)$ is the indicator function of A and μ is the invariant measure for $\{x_n\}$ with the density (6).

Proof.

Assume that the unknown target $x^* = x_1 = x'_1$ is such that for any Borel set $A \subset [0, 1]$ the limit

$$\lim_{N\to\infty} \frac{1}{N} \sum_{n=1}^{N} I_A(x_n)\qquad(19)$$

for the dynamic system (5) exists and equals $\mu(A)$. According to (7) and classical results in ergodic theory, the Lebesgue measure of the set $S^* \subset [0, 1]$ of such x^* equals 1.

Consider the intervals U_n of Lemma 2. In a similar way to the proof of Lemma 3, one can demonstrate that for any $x_* \in S^*$ all mistakes are corrected in a finite number of iterations. Let n_0 be such that $U_{n_0} \subset A_1$ and the mistakes of all previous iterations have been corrected.

Let $\delta > 0$ be a fixed number and $M = M(\delta)$ be the smallest integer such that

$$\mu(U_M) = 2C_3 \lambda^{\frac{M}{\beta}} \frac{7 + 4\lambda}{5} \le \frac{\delta}{2},$$

with the measure $\mu(.)$ obtained from (6) and C_3 given by (13). This gives

$$M = \max\left\{1, 1 + \lfloor \frac{\beta}{\log(1 + \lambda)} \left(\log\frac{1}{\delta} + \log C_3 + \log\frac{4(7 + 4\lambda)}{5}\right)\rfloor\right\}$$

where $\lfloor a \rfloor$ is the integer part of a. Set $N_* = \max\{n_0, M\}$, we shall give an upper bound on the number of mistakes in the process (9) by assuming that we make a mistake every time $x_n \in U_n$. According to Lemma 3, we pay a penalty 2 for each mistake, and we therefore obtain for the penalty (11) up to iteration $N > N_*$:

$$s_N \le n_0 + 2\{\text{number of } n \le N \text{ such that } x_n \in U_n\}$$
$$= n_0 + 2\sum_{n=n_0}^{N} I_{U_n}(x_n) \le n_0 + 2(N_* - n_0) + 2\sum_{n=N_*}^{N} I_{U_{N_*}}(x_n)$$

Using the inequality above and the fact that (19) equals $\mu(A)$ for any Borel set $A \subset [0, 1]$, we get

$$\limsup_{N\to\infty} \frac{s_N}{N} \le \limsup_{N\to\infty} \left[\frac{n_0 + 2(N_* - n_0)}{N} + \frac{N - N_*}{N} \frac{2}{N - N_*} \sum_{n=N_*}^{N} I_{U_{N_*}}(x_n)\right] \le 2\mu(U_{N_*}) \le \delta.$$

Since δ can be chosen arbitrarily small, we get

$$\lim_{N \to \infty} \frac{s_N}{N} = 0.$$

Finally, for any $x^* \in S^*$ and any Borel set $A \subset [0,1]$, we have

$$\left| \frac{1}{N} \sum_{n=1}^{N} I_A(x'_n) - \mu(A) \right| \leq \left| \frac{1}{N} \sum_{n=1}^{N} I_A(x_n) - \mu(A) \right| + \left| \frac{1}{N} \sum_{n=1}^{N} (I_A(x'_n) - I_A(x_n)) \right|$$

$$\leq \left| \frac{1}{N} \sum_{n=1}^{N} I_A(x_n) - \mu(A) \right| + \frac{s_N}{N} \to 0 \text{ as } N \to \infty,$$

which gives (18). $\qquad\qquad\qquad\qquad\qquad\qquad\qquad\qquad\qquad\qquad\qquad\qquad$ $\square$

3   The Dynamic-Programming Point of View

Here we shall adopt a Bayesian view of the problem, in a way similar to (5) and Section 3 of (7), and assume that the number N of function evaluations to be performed is fixed *a priori*. We also assume x^* to have a prior distribution uniform on $I_0 = [0,1]$. We restrict our attention to second-order algorithms, which are based on the comparison between two values of $f(.)$ evaluated in the current uncertainty interval. Throughout the section, $f(.)$ is assumed to be symmetric around x^*.

In previous algorithms, the rule for deletion is based on the uni-extremality assumption, the symmetry of $f(.)$ only being used to update the support of the distribution of x^*. Deleting parts of the interval where the probability density of x^* is zero leads to other procedures. Starting from a uniform distribution on $[0,1]$, the distribution of x^* remains uniform on all intervals I_k. Optimal procedures for fixed N are constructed by consideration of the backward-in-time problem. The last interval I_{N-1} is divided into equal parts when the last two evaluations are performed symmetrically, and this step is average optimal, that is the expected length of the last interval is then minimal. The same argument applies for previous iterations, which yields an average-optimal procedure. It can be initialised for instance by two evaluations in $I_1 = [0,1]$ at $(1/3, 2/3)$, which produces a cycle with two states. The final length L_N is equal to $L_1/2^{N-1}$, and the rate $(\prod_{i=1}^{N} r_i)^{1/N}$ is equal to $(1/2)^{(N-1)/N}$. Since the reduction rate r_i is constant, this approach is also optimal in the minimax sense.

Consider again a deletion rule based only on the uni-extremality assumption. This always yields larger intervals than the ones obtained previously, so that $(1/2)^{(N-1)/N}$ is a lower bound for the rate of any second-order algorithm. Procedures that reach this bound asymptotically have been considered in (8) for locally symmetric functions, and ϵ-optimal algorithms with a finite number of states are presented in (3; 4).

In what follows, we wish to minimize $E_{x^*}\{L_N(S_N)\}$ with respect to the strategy S_N, with N fixed. The restriction to the class of quasi-symmetrical strategies (i.e. to procedures for which all observations are taken symmetrically except the last two ones) has already been considered in (5). Replacing average optimality by minimax optimality leads to the well-known Fibonacci method (1; 2). Minimax-optimal and quasi-symmetrical average-optimal strategies tend asymptotically (when $N \to \infty$) to the Golden-Section algorithm.

The distribution of x^* in each interval I_k, conditional on the information available in the interval, is uniform on the optimistic interval $[u_k, v_k]$ defined in Section 1. Note the difference with (5), where for simplicity the distribution was taken uniform over I_k, that is the assumption of symmetry for $f(.)$ was not taken into account when propagating the conditional density of x^*.

Define e_k, e_k' as in Section 1. The minimization of $E_{x^*}\{L_N(S_N)\}$ corresponds to the solution of the following dynamic-programming problem

$$\min_{e_1, e_1'} \quad E_{x^*|u_1, v_1}\{r_2(e_1, e_1') \times \min_{e_2'} E_{x^*|u_2, v_2}\{r_3(e_2, e_2')\ldots$$

$$\times \min_{e_{N-2}'} E_{x^*|u_{N-2}, v_{N-2}}\{r_{N-1}(e_{N-2}, e_{N-2}') \times \min_{e_N} E_{x^*|u_{N-1}, v_{N-1}}\{r_N(e_{N-1}, e_N)\}\}\ldots\}\}.$$

Define $l_k(e_k, u_k, v_k)$ as

$$l_k(e_k, u_k, v_k) = E_{x^*|u_k, v_k}\{r_k(e_k, e_k') \times E_{x^*|u_{k+1}, v_{k+1}}\{r_{k+1}(e_{k+1}, e_{k+1}')\ldots$$

$$\times E_{x^*|u_{N-1}, v_{N-1}}\{r_N(e_{N-1}, e_N)\}\ldots\}\},$$

where we omit the dependence of l_k over the e_m''s, $k \le m < N$. Since the distribution of x^* is uniform on $[u_k, v_k]$, we get

$$l_k(e_k, u_k, v_k) = \begin{cases} \frac{v_k - c_k}{v_k - u_k}(1 - e_k)l_{k+1}(\frac{e_k' - e_k}{1 - e_k}, \frac{c_k - e_k}{1 - e_k}, \frac{v_k - e_k}{1 - e_k}) \\ + \frac{c_k - u_k}{v_k - u_k}e_k'l_{k+1}(\frac{e_k}{e_k'}, \frac{u_k}{e_k'}, \frac{c_k}{e_k'}) \quad \text{if } e_k' > e_k, \\ \\ \frac{v_k - c_k}{v_k - u_k}(1 - e_k')l_{k+1}(\frac{e_k - e_k'}{1 - e_k'}, \frac{c_k - e_k'}{1 - e_k'}, \frac{v_k - e_k'}{1 - e_k'}) \\ + \frac{c_k - u_k}{v_k - u_k}e_k l_{k+1}(\frac{e_k'}{e_k}, \frac{u_k}{e_k}, \frac{c_k}{e_k}) \quad \text{otherwise}, \end{cases}$$

with $c_k = \frac{e_k + e_k'}{2}$. The expected length $E_{x^*}\{L_N\} = l_1(e_1, 0, 1)$ can thus be calculated, at least in principle, for any choice of the e_k''s.

We follow again a backward-in-time approach, and consider the interval I_{N-1} in order to express $e_N = e_{N-1}'$ as a function of e_{N-1}, u_{N-1} and v_{N-1}. Assuming first that $e_N \le e_{N-1}$, e_N must be chosen so as to minimize

$$l_N^0(e_N) = (1 - e_N)(v_{N-1} - \frac{e_N + e_{N-1}}{2}) + e_{N-1}(\frac{e_N + e_{N-1}}{2} - u_{N-1}),$$

with the constraints

$$u_{N-1} \le \frac{e_N + e_{N-1}}{2} \le v_{N-1} \text{ and } 0 \le e_N,$$

or equivalently

$$e_N \in J_{N-1}^0 = [\max(0, 2u_{N-1} - e_{N-1}), \min(e_{N-1}, 2v_{N-1} - e_{N-1})].$$

The optimum e_N^0 is then either $1/2 + v_{N-1} - e_{N-1}$ if this point lies in J_{N-1}^0, or one of the bounds of J_{N-1}^0. Assuming now that $e_N \ge e_{N-1}$, e_N must be chosen so as to minimize

$$l_N^1(e_N) = (1 - e_{N-1})(v_{N-1} - \frac{e_N + e_{N-1}}{2}) + e_N(\frac{e_N + e_{N-1}}{2} - u_{N-1}),$$

with the admissible interval

$$J_{N-1}^1 = [\max(e_{N-1}, 2u_{N-1} - e_{N-1}), \min(2v_{N-1} - e_{N-1}, 1)].$$

The optimum e_N^1 is then either $1/2 + u_{N-1} - e_{N-1}$ if this point lies in J_{N-1}^1, or one of the bounds of J_{N-1}^1. Finally, the optimal choice for e_N is obtained by comparing the criteria

$l_N^0(e_N^0)$ and $l_N^1(e_N^1)$. This gives:

$$l_{N-1}(e_{N-1}, u_{N-1}, v_{N-1}) = \begin{cases} e_{N-1} - \dfrac{(2v_{N-1}-1)^2}{8(v_{N-1}-u_{N-1})} & \text{if } e_N = e_N^0 \\[2ex] 1 - e_{N-1} - \dfrac{(2u_{N-1}-1)^2}{8(v_{N-1}-u_{N-1})} & \text{if } e_N = e_N^1 \\[2ex] \dfrac{2e_{N-1}^2 - e_{N-1}(1+u_{N-1}+v_{N-1})+v_{N-1}}{v_{N-1}-u_{N-1}} & \text{if } e_N = e_{N-1} \\[2ex] e_{N-1} & \text{if } e_N = 2v_{N-1} - e_{N-1} \\[1ex] 1 - e_{N-1} & \text{if } e_N = 2u_{N-1} - e_{N-1} \end{cases}$$

Although the same approach could theoretically be used to express the optimal choice for e'_{N-2} as a function of e_{N-2}, u_{N-2} and v_{N-2}, it is clearly untractable. In what follows we shall therefore restrict our attention to fixed control policies, that is to cases where e'_k is a fixed function of e_k, u_k and v_k,

$$e'_k = g(e_k, u_k, v_k).$$

The only degree of freedom then lies in the choice of the initial point e_1, and replacing e'_k by its expression in $l_k(u_k, u_k, v_k)$ we can compute $l_1(e_1, 0, 1)$. Symmetry within the optimistic interval $[u_k, v_k]$ is a desirable property for the reduction rate to asymptotically converge to $1/2$, so that we consider the control function defined by

$$e'_k = u_k + v_k - e_k.$$

Figure 1 presents $l_1(e_1, 0, 1)$ as a function of e_1 for various values of N when e_N is chosen optimally as a function of e_{N-1}, u_{N-1} and v_{N-1}. Contrarily to the case considered in (5), no analytical expression is obtained for $l_1(e_1, 0, 1)$ around the optimum.

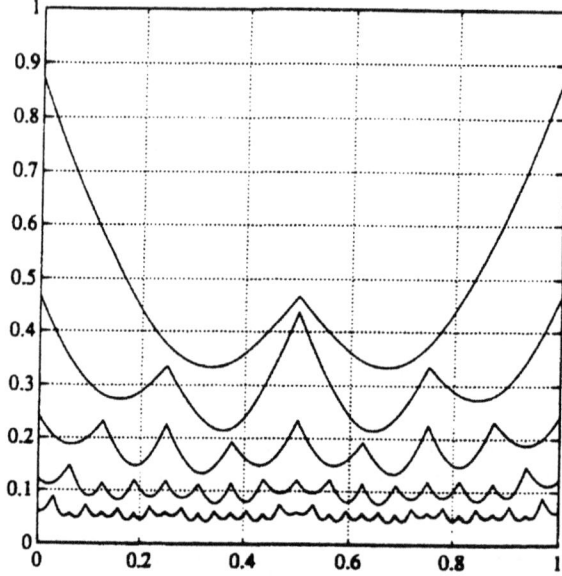

Figure 1: $l_1(e_1, 0, 1)$ as a function of e_1 when e_N is chosen optimally

Simplifying the choice of the last point according to the sub-optimal policy $e_N = e_{N-1}$, the following expressions can be derived

$$l_1(e_1, 0, 1) = \begin{cases} 2e_1^2 - 2e_1 + 1 & \text{if } N = 2 \\ 6e_1^2 - 8e_1 + 3 & \text{if } N = 3 \\ 14e_1^2 - 18e_1 + 6 & \text{if } N = 4 \\ 26e_1^2 - 71/2 e_1 + 49/4 & \text{if } N = 5 \end{cases}$$

which are valid around their associated optimum starting point (in $[1/2, 1]$). These optimum starting points are thus

$$e_1 = \begin{cases} 1/2 & \text{if } N = 2 \\ 2/3 & \text{if } N = 3 \\ 9/14 & \text{if } N = 4 \\ 71/104 & \text{if } N = 5 \end{cases}$$

The evolution of $l_1(e_1, 0, 1)$ as a function of e_1 for various values of N is given on Figure 2, to be compared with Figure 1. The value obtained when $e_1 = 0$ is easily calculated,

$$l_1(0, 0, 1) = \frac{1}{2^{N-3}}.$$

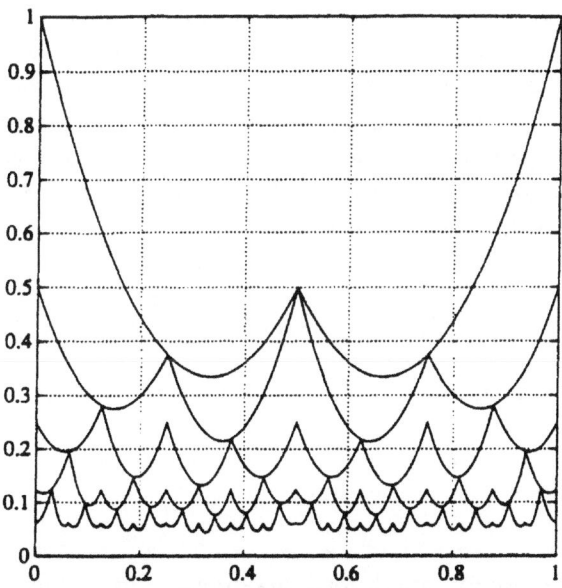

Figure 2: $l_1(e_1, 0, 1)$ as a function of e_1 when $e_N = e_{N-1}$

Finally, we consider the case where the last point e_N is also chosen according to a symmetrical policy, i.e. $e_N = u_{N-1} + v_{N-1} - e_{N-1}$. The evolution of $l_1(e_1, 0, 1)$ as a function of e_1 is presented on Figure 3.

The optimal starting point can then be obtained analytically,

$$e_1 = \frac{2}{3} \left(1 - \left(-\frac{1}{2} \right)^N \right),$$

296

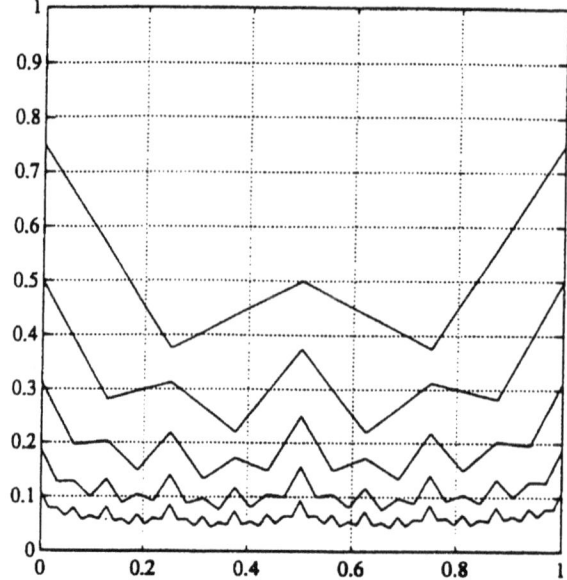

Figure 3: $l_1(e_1, 0, 1)$ as a function of e_1 when $e_N = u_{N-1} + v_{N-1} - e_{N-1}$

which tends to $2/3$ when N tends to infinity. This policy can be compared with the Fibonacci method $(1; 2)$, where at each step e'_k is taken equal to $1 - e_k$ (i.e. symmetrical with respect to the center $(a_k + b_k)/2$ of the unnormalised interval $[a_k, b_k]$). For both policies, when starting with the optimal value for e_1, each e_k is an optimal starting point for the sequence with $N + 1 - k$ evaluations allowed. When symmetry with respect to $(a_k + b_k)/2$ is considered, the procedure converges to the Golden Section method, which is the unique symmetrical algorithm with two states. When symmetry is taken with respect to $(u_k + v_k)/2$, the procedure asymptotically satisfies

$$e_k = \frac{u_k + v_k}{2} \pm \frac{v_k - u_k}{6},$$

which corresponds to the unique algorithm with two states in the optimistic interval $[u_k, v_k]$.

References

J. Kiefer. Sequential minimax search for a maximum. *Proc. Am. Math. Soc.*, 4:502–506, 1953.

J. Kiefer. Optimum sequential search and approximation methods under minimum regularity assumptions. *J. Soc. Indust. Appl. Math.*, 5(3):105–136, 1957.

L. Pronzato, H.P. Wynn, and A.A. Zhigljavsky. Dynamic systems in search and optimization. Technical report, Laboratoire I3S, CNRS-URA 1376, Sophia Antipolis, 06560 Valbonne, France, 1993.

L. Pronzato, H.P. Wynn, and A.A. Zhigljavsky. Section invariant numbers and ergodically fast line search. Technical Report 93-50, Laboratoire I3S, CNRS-URA 1376, Sophia Antipolis, 06560 Valbonne, France, 1993.

L. Pronzato and A.A. Zhigljavsky. On average-optimal quasi-symmetrical univariate optimization algorithms. In W.G. Müller, H. P. Wynn, and A.A. Zhigljavsky, editors, *Model-Oriented Data Analysis III, Proceedings MODA3, St Petersburg, May 1992*, pages 269–278. Physica Verlag, Heidelberg, 1993.

H.P. Wynn and A.A. Zhigljavsky. Chaotic behaviour of search algorithms. *Acta Applicandae Mathematicae*, 32:123–156, 1993.

H.P. Wynn and A.A. Zhigljavsky. Chaotic behaviour of search algorithms: introduction. In W.G. Müller, H. P. Wynn, and A.A. Zhigljavsky, editors, *Model-Oriented Data Analysis III, Proceedings MODA3, St Petersburg, May 1992*, pages 199–211. Physica Verlag, Heidelberg, 1993.

H.P. Wynn and A.A. Zhigljavsky. Achieving the ergodically optimal convergence rate for a one-dimensional minimization problem. Technical report, City University, Northampton Sq., London EC1V 0HB, UK, 1994.